PRINCIPLES OF NUCLEAR RADIATION DETECTION

by

GEOFFREY G. EICHHOLZ

JOHN W. POSTON

School of Nuclear Engineering
Georgia Institute of Technology
Atlanta, Georgia

ANN ARBOR SCIENCE
PUBLISHERS INC
P.O. BOX 1425 • ANN ARBOR, MICH. 48106

P.O. Box 1425, Ann Arbor, Michigan 48106

Library of Congress Catalog Card No. 79-88897
ISBN 0-250-40263-7

Manufactured in the United States of America

PREFACE

It has been about 15 years since the last general textbooks on nuclear radiation detection were written. In that time some fundamental changes have occurred, both with respect to the complexity and sophistication of counting equipment and to the more widespread use of radiation technology in medicine and industry. Many of the changes reflect the replacement of vacuum tubes by semiconductor devices and the consequent increase in speed and variety of data handling capability; many others arise from the development and maturing of detector systems that were only in their infancy then. For these reasons it seems appropriate to bring out a new book that represents more clearly the present status of the field.

This book is intended for senior undergraduate and beginning graduate students in physics, nuclear engineering, health physics and nuclear medicine, and for specialized training courses for radiation protection personnel and environmental safety engineers. To keep the size of the book manageable, material has been selected to stress those detectors that are in widespread use. Attempts have also been made to emphasize alternatives available in approaching various measurement problems and to present the criteria by which a choice among these alternatives may be made. Limited bibliographies are provided to encourage further reading and numerical problems to develop a feeling for the orders of magnitude of various system parameters. The SI system of units has been adopted as far as practical, though the curie and the roentgen have been retained where it seemed appropriate to reflect current usage.

The contents represent the outcome of courses presented at the Georgia Institute of Technology over many years, and we are grateful to out colleagues, B. Kahn, M. T. Ryan and M. D. Matheny, for their helpful comments.

A companion volume is provided to serve as a laboratory manual for radiation detection courses. The manual is cross-referenced to the present book, but either can be used separately as desired. Additional questions and problem sets will be found in the laboratory manual.

Geoffrey G. Eichholz is Regents' Professor of Nuclear Engineering at the Georgia Institute of Technology. He obtained both his BSc and PhD degrees at the University of Leeds, England, specializing in physics and has since continued in various areas of applied physics.

Dr. Eichholz headed the Physics and Radiotracer Subdivision of the Canadian Bureau of Mines for twelve years, where he focused on problems of uranium ore analysis, industrial instrumentation, radiation protection and industrial uses of radioisotope tracers. In 1963 he joined the faculty of Georgia Tech where his research and teaching focus on radiation detection problems, radiation effects on materials, environmental applications of radiotracers and radiation safety. Professor Eichholz is author of *Environmental Aspects of Nuclear Power,* published by Ann Arbor Science, and editor of the book *Radioisotope Engineering.*

John W. Poston is Associate Professor of Nuclear Engineering at Georgia Institute of Technology. He received his BS degree in mathematics from Lynchburg College and MS and PhD degrees from Georgia Tech specializing in radiation protection. He also attended the Oak Ridge School of Reactor Technology.

Dr. Poston was employed as an Experimental Reactor Physicist by the Babcock and Wilcox Company for six years and was a staff member in the Health Physics Division at the Oak Ridge National Laboratory for thirteen years. During this latter period he was involved in various aspects of external and internal dosimetry and was head of the Medical Physics and Internal Dosimetry Section. His research at ORNL concentrated on programs related to the activities of the Atomic Bomb Casualty Commission and the International Commission on Radiological Protection. He joined the faculty of Georgia Tech in 1977 where his teaching and research interests include radiation detection and applied radiation protection at nuclear facilities.

CREDITS

The authors are particularly indebted to the authors and publishers who so kindly gave their permission to reproduce copyrighted material from the sources listed below:

	Figure(s)
O. C. Allkofer, *Spark Chambers*, Verlag Karl Thiemig, Munich	7-9, 7-20
American Institute of Physics, *Physics Today*	4-17, 7-22
K. Becker, *Solid State Dosimetry,* CRC Press, West Palm Beach, FL	7-8, 7-11, 7-12
K. H. Beckurts and K. Wirtz, *Neutron Physics,* Springer-Verlag, New York	2-17, 9-3, 9-4a, 10-4
G. Bertolini and A. Coche, *Semiconductor Detectors,* North-Holland Publishing Company, Amsterdam	6-3, 6-4, 6-12
J. B. Birks, *The Theory and Practice of Scintillation Counting,* Macmillan Publishing Company, New York	5-10
E. D. Bransome, Ed., *The Current Status of Liquid Scintillation Counting,* Grune and Stratton, New York	5-1
J. R. Cameron, N. Suntharalinham and G. N. Kenney, *Thermoluminescence Dosimetry,* The University of Wisconsin Press, Madison, WI	6-13, 6-17, 6-20
H. Cember, *Introduction to Health Physics*, Pergamon Press, Ltd., Oxford	4-18, 4-19
R. L. Chase, *Nuclear Pulse Spectrometry,* McGraw-Hill Book Company, New York	5-14
G. Dearnaley and D. C. Northrop, *Semiconductor Counters for Nuclear Radiations,* 2nd edition, John Wiley & Sons, Inc., New York	6-1, 6-6
Health Physics Society, *Health Physics Journal,* Bethesda, MD	5-15c, 7-11, 9-7
W. R. Hendee, *Radiologic Physics, Equipment and Quality Control,* Yearbook Medical Publishers, Chicago	4-6, 4-8, 4-12, 4-15, 8-1

H. Kiefer and R. Maushart, *Radiation Protection Measurements,* Pergamon Press, Inc., Elmsford, NY	2-17a, 7-6, 7-7, 9-6, 9-8, Table 7-2
B. Leskovar, "Microchannel Plates," *Physics Today* (Nov. 1974):43	5-6b
K. Z. Morgan and J. E. Turner, *Principles of Radiation Protection*, John Wiley & Sons, Inc., New York	4-11
C. F. Powell and G. P. S. Occhialini, *Nuclear Physics in Photographs*, Clarendon Press, Oxford	
K. Siegbahn, Ed., α, β, γ *Spectroscopy,* North-Holland Publishing Company, Amsterdam	6-5, 6-9
R. L. Kathren, for the evaluation program for instruments, adapted on pages 333-338	

CONTENTS

CHAPTER 1

INTRODUCTION

Nuclear and atomic radiations find many applications in our society, from chest X-ray radiography to the control of the thickness of aluminum foil, from tracer measurements of stream flow or natural gas reservoirs to measurements of cardiac output from the heart, from determination of cosmic ray intensities on mountain tops to the monitoring of radiation levels around nuclear power reactors or supervoltage particle accelerators. In each case the measurement of radiation intensity or radiation exposure requires a thorough understanding of the characteristics of alternative detection systems available, an appreciation of the physical problem to be solved, and an informed selection of the detector type that is most suited to the particular application. Since nuclear radiations are not sensed by the common human senses, a detector is required to provide a conversion to a readily measurable phenomenon. The properties of these detectors and any refinements that can be supplied to improve their performance must form the core of any discussion of radiation detection systems. The performance of a detector system depends on the characteristics of all of its components, that is, the detector proper which responds to the incident radiation and converts it into a more tractable form, usually electric charges or voltages, the amplifier system, and the data display or evaluation system.

Nuclear radiations comprise several forms of energy transmission that are characterized by their ability to excite, ionize or otherwise interact with most atoms or molecules that they encounter. They derive their name from the fact that they are emitted as the result of the de-excitation processes that a nucleus undergoes following bombardment or capture reactions; the type and energy spectrum of the emitted radiation reflect the characteristic energies associated with these processes. The radiations of particular importance are called alpha (α), beta (β) and gamma (γ) radiations and neutrons, though a whole host of other charged and uncharged particles fit the same general description and may be encountered in specific research situations. X-Rays, whose characteristics are closely allied to gamma radiation, are usually included in considerations of nuclear radiation detectors as a matter of convenience, though strictly they are not of nuclear origin but arise from transitions between orbital electron energy levels.

Radiation detection may consist of different types of operations; these are variously referred to as *detection, monitoring, analysis or dosimetry* with each type requiring slightly different equipment and emphasizing a different aspect of the measurement process. In detection one tends to concentrate on verifying the presence or absence of a given kind of radiation above a certain level of intensity. In monitoring operations this process is typically related to a time-dependent observation or to established safety standards of exposure. Analysis of radiation requires further identification of various radiation sources, if present, and a determination of energy distributions or spectra, and, in some cases, discrimination in time sequence of events. Finally, dosimetry relates the type and energy distribution of the radiation field to the absorption of this energy in body tissue and to any related health effects that may arise from this exposure.

Consequently, one may divide detector systems into those that are intended and most suited for the measurement of radiation intensity, those dedicated to the measurement of radiation dose, and those capable of determining energy distributions or spectra. In general, each detector system will be capable of responding principally to only one type of radiation, though high levels of other radiation types may cause interference; hence, a wide variety of detector types and detector systems have been developed for different purposes.

All radiations represent a transfer of energy from the emitting source to the surrounding media where they may interact to an extent determined by the characteristics of the radiation and the nature and structure of the intervening medium. If the interaction between the radiation and the target medium is strong, energy is transferred at a rapid rate and the radiation is strongly attenuated. If radiation attenuation is weak, the radiation may have an appreciable range and may be detected at great distances from the source. Since all detection methods depend on finding an efficient interaction mechanism to transfer some of this energy so that it can be converted into readable form, these interaction processes are central to any discussion of radiation detection.

The energy of radiations emitted by nuclei is of the order of nuclear excitation energies or binding energies; these are of the order of kiloelectron-volts (keV) and megaelectron-volts (MeV), where 1 MeV = 0.16 picojoule (pJ).

The radiations will interact with materials, that is, they will lose kinetic energy to any solid, liquid or gas through which they pass by a variety of interaction mechanisms that depend on the energy of the incident radiation, the density and atomic number of the absorbing medium, and the relative probability for one or another process to take place. These probabilities are called cross sections and are usually expressed in barns (1 barn = 10^{-24} cm^2 = 10^{-28} m^2). The result of these interaction processes is a gradual slowing down of any incident particle until it is brought to rest or "stopped" at the end of its range. Although the terminology varies among the nuclear physicist, the health physicist, the radiobiologist and the radiation chemist, they all are interested in the rate of energy loss or transfer ("dE/dx, LET") as the incident particle is slowed down, the total energy lost or transferred by various secondary processes and the consequent effect per unit mass or unit volume on the target material.

Since charged particles lose energy at a fairly steady rate when transversing material substances, it is possible to relate their initial energy to the distance traveled in that medium. This is called the *range* of the particle and represents the average distance traversed in a given medium by a particle of specified energy.

The emission of radiation from nuclear processes occurs randomly in time; consequently all radiation detection processes deal with statistically random events and the data evaluation and interpretation of detector response must take into account the statistical nature of all events. This has considerable bearing on the sensitivity and accuracy of radiation detection systems and this aspect will be treated in some detail in Chapter 3.

RADIATION INTENSITY

The intensity of a radiation field can be described in several ways. The simplest is to describe it in terms of a particle flux or flux density, *i.e.*, the number of particles passing unit area per unit time, with all particle paths projected normally onto the reference plane. This is a convenient way of expressing field intensities for parallel beams of particles or for fields created by a point source. It is the customary way of describing neutron fields around a reactor where one also needs to distinguish between neutrons in different energy ranges. Flux density is usually expressed in number of particles/cm^2-sec.

For X-rays and gamma-rays, a more experimentally oriented approach is customary in which the field intensity is expressed in terms of the number of ions produced in a known mass or volume of detector. Such a measure of the ionization produced in air by X- or gamma radiation is called the *exposure*. The unit of exposure is the Roentgen (R). It has been defined as that quantity of X- or gamma radiation that would, through associated corpuscular radiation, produce one statcoulomb of charge of either sign in one cubic centimeter of dry air at STP.

$$1 \text{ R} = 2.58 \times 10^{-4} \text{ coulomb/kg of air}$$

The energy *absorbed* by that 1 cm^3 air volume represents an equivalent of 87.7 ergs per gram of air or 8.77 x 10^{-9} joule/kg; this quantity is useful in comparing exposure values and dose values, but it is important to maintain the distinction.

SOURCE STRENGTH

Radiation will originate from natural or artificial sources, which may be steadily decaying radioactive materials, nuclear reactors, astronomic sources or high-voltage accelerators of various types. The *source strength* is a measure of the rate of emission of radiation. More than one kind of radiation and more than one energy group of a given radiation type may be emitted; this fact needs to be considered in describing the source characteristics. Consequently there are several ways in which a radiation source can be described.

Emergent Flux

For machine sources and reactor cores it may often be easiest to describe source strength in terms of total number of particles or photons emitted per unit time into the total solid angle (4π). This procedure is not appropriate if the emission is not isotropic or if there is significant conversion of the radiation within the source volume. In reactors or machines any measurement of this quantity must allow for reflection or backscattering effects and, in the case of neutrons, for spectral shift. The result is usually a description of the *particle or photon flux* at a given external point near the source, which is assumed to be concentrated to a small volume.

In reactors the equivalent source strength of the core region may involve both prompt fission effects and delayed fission product activity; hence it is rarely practical to describe ambient radiation fields in terms of an equivalent source strength.

Radioactive Sources

For radioactive materials it is customary to describe the source strength in terms of the *source activity*, which is defined as the number of disintegrations per unit time occurring in a given quantity of this material. For any pure radioactive substance, the rate of decay is usually described by its half-life τ, *i.e.*, the time it takes for a specified source material to decay to half its initial activity. The activity A can be written as

$$A = \lambda N = \frac{\log_e 2}{\tau} N = \frac{0.693\ N}{\tau} \qquad (1\text{-}1)$$

where λ = decay constant (sec^{-1}) and N = total number of atoms of the radioactive species.

The SI unit of activity is the *becquerel* (Bq), equivalent to a source activity of one disintegration per second. More widely used is the traditional unit of activity, the *curie*, which is defined as:

$$1 \text{ curie (Ci)} = 3.70 \times 10^{10} \text{ disintegrations/sec} = 3.70 \times 10^{10} \text{ Bq}$$

Since activity is proportional to the number of atoms of the radioactive material, the quantity of any radioactive material is usually expressed in curies, regardless of its purity or concentration. The usual metric multiples apply to the curie. Since many nuclides have complex decay schemes, the activity does not in itself indicate the emergent flux from any source.

To calculate the radiation field for a given radioactive source it is necessary to know the decay scheme of the material, any branching ratios and the composition of the emitted radiation. Also note that the activity of a given source is a function of time since the number of radioactive atoms decreases exponentially:

$$N = N_o e^{-\lambda t} = N_o e^{\frac{-0.693t}{\tau}} \qquad (1\text{-}2)$$

where N_o = number of atoms at time t = 0. Hence,

$$A = A_o e^{\frac{-0.693t}{\tau}} \qquad (1\text{-}3)$$

In the case of radioactive materials contained in living tissue an additional allowance has to be made for the reduction in observed activity due to regular processes of elimination of the respective chemical or biochemical substance from the organism. This introduces a rate constant called the biological half-life, which is approximately the same for both stable and radioactive isotopes of a given element.

Under such conditions the time required for a radioactive element in any living organism to be halved as a result of the combined action of radioactive decay and biological elimination is the effective half-life:

$$\tau_{eff} = \frac{\tau_{biol} \times \tau_{RA}}{\tau_{biol} + \tau_{RA}} \qquad (1\text{-}4)$$

To identify a radioactive material, four types of measurement may have to be conducted:

1. determination of the mass number, or atomic number,
2. determination of the source strength or activity,
3. determination of the half life,and
4. determination of the energy or energy spectrum, types of radiation and decay schemes.

To establish the effect of the radiation field on a given target material, a different type of measurement is required:

1. measurement of the radiation intensity (or flux),
2. measurement of the energy loss,
3. measurement of absorbed energy (dose), and
4. measurement of the dose effect on a given target material.

ABSORBED DOSE

Since different types of radiation interact differently with any material through which they pass, any attempt to assess their effect on the human body or on plants and animals should take into account these differences. The quantity of interest is the absorbed dose, D, which is defined as the energy imparted by the incident radiation to unit mass of the target material or

$$D = \frac{\Delta E_D}{\Delta m} \tag{1-5}$$

The conventional unit of absorbed dose is the *rad*; 1 rad = 100 erg/gram = 0.01 J/kg. The SI unit of absorbed dose is the *gray*. 1 gray (Gy) = 1 J/kg. The rate of energy absorption is called the absorbed dose rate $\Delta D/\Delta t$ and is expressed in rad/hr, rad/min, Gy/sec, as appropriate.

The physical meaning of dose has concerned health physicists for many years and has led to many attempts at clarification. The problem arises from two sources: one, to measure internally the amount of energy actually transferred to an organic material and to correlate any observed effects with this energy deposition, and two, to account for and predict any secondary processes, such as collision effects or biologically triggered effects, that are an indirect consequence of the primary interaction event. Measuring instruments designed specifically to indicate absorbed dose are referred to as *dosimeters.*

CHARACTERISTICS OF NUCLEAR RADIATIONS

Although the term "radiation" is applied to all forms of energy emission from nuclear events, there are in fact significant and characteristic differences between the various types of radiation that are emitted or can be emitted under various circumstances of nuclear excitation. Basically one can distinguish between emission of electromagnetic radiation ("photons"), resulting from transitions between excited states of a given nuclide, and particulate radiations representing transformations or ejection of one or more nucleons from an atomic nucleus. The particles consist of charged and uncharged particles and are differentiated on the basis of mass, charge and spin, and among the more exotic particles by additional characteristics such as "charm," "strangeness" and "color." Most of the latter particles are produced only under very-high-energy excitation conditions and require particularly sophisticated methods of data evaluation for their detection. However, the methods available for their detection in principle are not significantly different from methods employed at lower energies and little specific reference will be made in this book to the high-energy region and its radiations.

Table 1.1 lists the characteristics of the radiations commonly encountered in nuclear engineering and in medical and industrial radiation applications. The names reflect a historical order of discovery for alpha, beta and gamma radiation. Because of their varied characteristics, methods of detection appropriate to each type have to be selected in every case. In general, no one detector method or system will be equally applicable to more than one radiation type or even to more than a limited energy range of that radiation. Thus, different detector systems may be called for in the case of "thermal" (low-energy) and "fast" (high-energy) neutrons or for "soft" or "hard" beta-rays or X-rays.

In general terms one distinguishes three radiation types for purposes of selecting interaction mechanisms and detector types:

1. Electromagnetic radiations, γ- or X-rays, that interact primarily through spin coupling with the orbital electrons of irradiated targets, causing ionization, excitation or Compton scattering events.
2. Charged particles, including heavy particles such as protons, deuterons, tritons and alpha particles as well as heavier charged ions, and the lighter particles such as electrons (β- and β+) and mesons, which have rest masses substantially below atomic masses. Charged particles can interact both with orbital electrons and the central nucleus.
3. Uncharged particles, principally neutrons, which have masses comparable to protons, and neutrinos, which have vanishingly small rest masses. Neutrons interact almost entirely with the central nucleus, since they are not screened off by the electrons, and may undergo capture or a scattering event, leading to transfer of some kinetic energy to the nucleus and deflection of the incident neutron.

DETECTOR SYSTEMS

The detector systems developed for each of these radiations attempt to utilize these interaction characteristics by interposing a medium of appropriate properties in which, by a process of excitation, ionization or momentum transfer, a more readily detectable signal is produced which can be counted, accumulated or sorted to provide the desired information about the radiation field being studied.

Table 1.1 Characteristics of Most Commonly Encountered Nuclear Radiations

Radiation	Rest Mass	Charge (e)	Spin	Typical Energy Range	Path Length (Order of Magnitude) Air	Path Length (Order of Magnitude) Solid	General Comments
α	4.00 amu	2+	0	4-10 MeV	5-10 cm	25-40 μm	Identical to ionized He nucleus
β (negatron)	5.48×10^{-4} amu 0.51 MeV	-	1/2	0-4 MeV	0-1 m	0-1 cm	Identical to electron
Positron (β positive)	5.48×10^{-4} amu 0.51 MeV	+	1/2	0-4 MeV	0-1 m	0-1 cm	Identical to electron except for charge
Proton	1.0073 amu 938.26 MeV	+	1/2	0-100 MeV	-	short	Identical to ^{1}H nucleus
Neutron	1.0086 amu 939.55 MeV	0	1/2	0-15 MeV	0-100 m	0-100 cm	Free half-life: 16 min
X (e.m. photon)	-	0	1	eV-100 keV	0.1-10 m[a]	0-1 m[a]	Photons from electron transitions
γ (e.m. photon)	-	0	1	10 keV-3 MeV	0.1-10 m[a]	1 mm-1 m	Photons from nuclear transitions
Neutrino	-	0	1/2	-	$\gg$1 km		Very little interaction with matter

Note: amu = atomic mass unit [12 amu = mass of carbon -12 atom].
[a]Exponential attenuation in the case of electromagnetic radiation.

Since the "ranges" of the more strongly interacting radiations are very short, problems often arise in finding an appropriate configuration that will permit the incident radiation to enter a sensitive volume without a change in characteristics. Often care must be taken that the introduction of the detector system itself does not modify the radiation field being studied. This may be of particular importance in medical dosimetry and in neutron measurements.

The design of a detector system accordingly involves the following steps:

1. identification of the type and energy range of the radiation to be detected;
2. estimation of the intensity to be expected and, therefore, the anticipated event rate;
3. choice of most appropriate interaction medium and detector type;
4. determination of geometry, *i.e.*, source-detector configurations, to define detector size and shape;
5. recognition of other radiations to which the preferred detector may be sensitive and estimation of the extent to which interfering events can be discriminated against;
6. selection of operational mode as a counter or as a ratemeter; *i.e.*, a mode of operation in which each interacting event is recorded separately or a mode in which the average rate of occurrence or charge production is determined;
7. assembly of the desired data indication or evaluation system, resulting in an indication of event counts, an energy spectrum display, an average count rate reading, or a computerized evaluation in terms of previously recorded information, such as peak identification or background subtraction; and
8. calibration procedures using an absolute or relative method of standardization under reproducible conditions of source-detector geometry and overall system configuration.

Since the dimensions of any radiation detector are finite in extent and a radiation field also is extended in space, the uniformity of the radiation field and its spatial relationship with the detector volume are of importance. The latter factor is usually referred to as the *geometry* of the source-detector configuration and is usually expressed in terms of the fraction of the solid-angle of the source emission that is intercepted by the detector. Since few sources are true point emitters and few detectors can be considered as points in space, usually the geometry may be rather complex and in practice one may have to determine it in relative terms. Figure 1.1 shows some representative geometries to illustrate this point.

Another factor of significance is the intrinsic efficiency of the detector. This factor, which depends on the particular mechanism of interaction between radiation and detector and on the energy of the incident radiation is a characteristic property of the detector material and independent of the geometry. The intrinsic efficiency ϵ is defined as

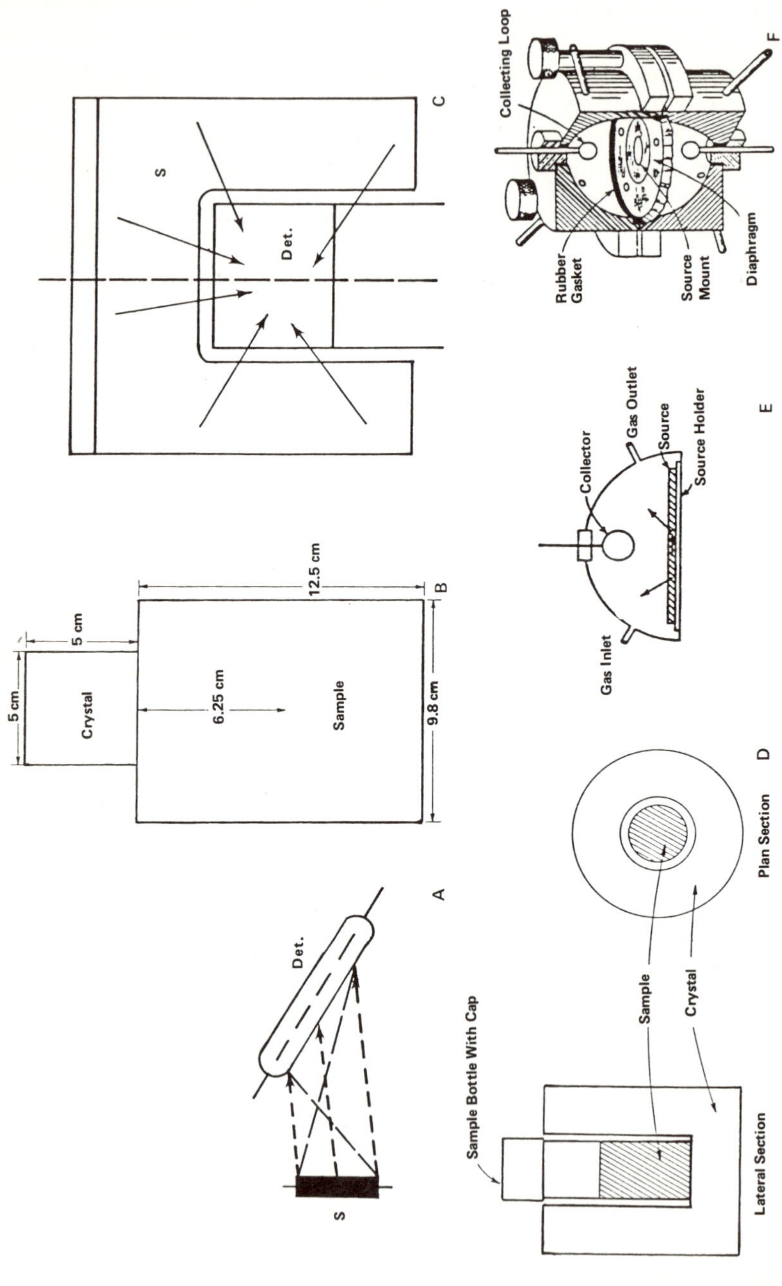

Figure 1.1 Examples of various source (S)-detector geometries: (a) poor geometry; (b) small solid angle; (c) Marinelli beaker surrounding detector; (d) well-counter (nearly 4π); (e) windowless flow counter with 2π geometry; and (f) 4π geometry counter for absolute beta determinations.

$$\epsilon = \frac{\text{number of events counted or indicated}}{\text{number of detectable particles (or that type) passing through the detector volume}} \tag{1-6}$$

The efficiency may vary with energy of the incident radiation but, in general, will be independent of intensity.

The number of events detected or recorded by the detector system, then, can be expressed as the *count rate* C.

This is given by

$$C = \phi \cdot G \cdot \epsilon \cdot \Sigma f_d \cdot f_i \tag{1-7}$$

where ϕ = flux of incident radiation (number of particles/cm^2sec)
f_d = correction factors for detector performance
f_i = correction factors for electronic instrument performance
G = geometry factor.

The flux ϕ can be related to source strength for simple source configurations; in the case of neutron fields, description of the radiation field in terms of flux in conventional. The correction factors f_d and f_i arise from count-rate limitations imposed by dead-time effects, pulse rise-time limitations and circuit nonlinearities, as well as certain secondary effects, all of which will be discussed in later chapters.

The product $\eta = G \cdot \epsilon$ is sometimes referred to as the *operational efficiency* of the detector. It is dependent on detector dimensions and source-detector configuration and, hence, is not an inherent property of any given detection material. This is also sometimes called the sensitivity of the detector; however, it is preferable to define the sensitivity as

$$S = \frac{\Delta C}{\Delta \phi} \tag{1-8}$$

In practice, the count rate obtained will include a certain number of events that may not be due to the particular radiation field being measured, but are due to other radiation sources or to electronic fluctuations inside the detector system. Such extraneous but unavoidable events are called background counts C_b.

$$C_b = C_{ext} + C_{int} \tag{1-9}$$

The external background may be due to cosmic radiation, natural radiation from ^{40}K, Th or U in structural materials or other components, or contamination and may be controlled to some extent by careful siting and shielding of the detector. The internal background may be due to voltage fluctuations, inadequate grounding and thermal noise in resistors and vacuum tubes, and may be controlled in part by voltage stabilization, cooling of the system and anticoincidence systems.

As a result, the total count rate obtained by the detector system C_g

$$C_g = C_s + C_b \tag{1-10}$$

where C_g is often called the gross count rate, and C_s is the net count rate and constitutes that part of the recorded count rate that is related to the source of interest.

Hence,

$$C_s = C_g - C_b \tag{1-11}$$

and the effective sensitivity, $S = \Delta(C_g - C_b)/\Delta\phi$, is significantly affected by the magnitude of the background.

Since both C_s and C_b are statistical in nature, the error in C_s is given by

$$\sigma_s^2 = \sigma_g^2 + \sigma_b^2 \tag{1-12}$$

and since the standard deviation σ^2 is proportional to the total number of events it is important to reduce background as much as possible (see Chapter 3). Thus, we find that the minimum sensitivity, that is, the lowest radiation intensity that can be meaningfully detected, is also dependent on both background level and geometry.

THE DETECTOR SIGNAL

The radiation detector is a device that converts the incidence of an energetic photon or particle into a more "visible" or readable phenomenon, usually referred to as a count. The nature of this outcome of the interaction event must be fully understood in order to appreciate any limitations imposed on the detection process and any requirements for further treatment or analysis of the information contained in each count signal.

In gas-filled detectors the interaction process usually results in the formation of positive and negative ions that can be collected by applying a voltage between two electrodes. The electric charge thus collected from all the ions produced directly or indirectly as a consequence of the passage of a single ionizing particle represents the count event or signal associated with that particle. In some detectors the voltage amplitude of the pulse of each event is a measure of the energy lost by the particle within the detector volume; the rate of collection or rise time of the pulse provides some information on the mobility of the ions in the detector gas. Since this mobility is slow, especially for positive ions, the number of events that can be counted in quick succession is limited by the drift time of the ions because the gas must return to neutral conditions before the next event can be detected. This recovery time is called the dead time of the detector and may range from hundreds of milliseconds in gas-filled detectors to nanoseconds in some solid-state detectors.

In some detectors, such as scintillation detectors, the initial interaction event gives rise to a light flash which is then converted to an electrical charge which appears as a brief voltage pulse for each detected event. Again the amplitude of the pulse contains information on the energy of the incident radiation while the count rate or number of events per unit time depends on the intensity of the radiation.

The rise time and the decay time may be different for different radiations, and by suitable electronic circuits one may use these parameters to differentiate between radiations. Similarly, simultaneous arrival, or arrival within a specified delay interval, may assist in selecting nuclear events that are genetically related or events belonging to a specific energy or velocity group within a wider range of particles. All of these functions must be performed by specially designed circuits that amplify, count, average or discriminate pulses of predetermined characteristics.

The simplest and most common form of data collection is *counting.* Counting implies the determination of the number of all events in the output of a detector that arise from radiation interaction. The events usually are transformed in an amplifier and a pulse-shaping circuit into pulses of specified shape and at least some limited amplitude. Since many detectors can detect thousands of events per second whereas a mechanical register can record only a few tens per second, it is customary to employ a *scaler* which, for convenience, reduces the number of counts registered by some power of ten. Many detector systems consist simply of the detector and a unit containing a high-voltage supply for the detector, an amplifier, one or more scaler units, a count indicator or register, and a timer unit. Such a system then supplies information on the number of counts per unit time ("preset time") or the time it takes for a fixed number of counts to be collected ("preset counts"). The latter system results in a constant statistical accuracy, but may result in rather long counting times, particularly with low-intensity sources.

For low-level counting or in dosimetry where relatively large fluctuations in count rate may occur, it is often more practical to average the count rate by "integrating" the charge collected. This is done by allowing the charge to build up on a capacitor which is discharged periodically at a fixed frequency, *e.g.,* in a vibrating-reed electrometer, or is allowed to leak away through a large parallel resistance (Figure 1.2). In both cases the voltage appearing across the capacitance is a measure of the average charge collected during the charge-discharge period and can be used as a measure of radiation intensity or radiation exposure. Since the voltage is proportional to the rate of arrival of charge pulses during the time of collection, assuming pulses of uniform amplitude, such a system is called a "ratemeter."

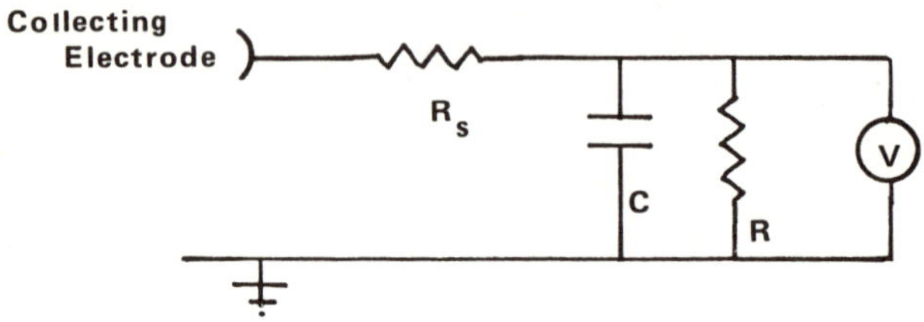

Figure 1.2 Ratemeter circuit.

If the charge q per pulse is constant, the equilibrium value of the voltage V is

$$V = rqR \tag{1-13}$$

where r is the average count rate (pulses per second). The time for the ratemeter to reach a steady value within one standard deviation for a single reading is given by

$$t_o = RC[\tfrac{1}{2} \log_e 2r\,RC + 0.394] \tag{1-14}$$

If the range of count rates to be covered is rather large, the ratemeter may have to have a logarithmic response. This can be accomplished by employing an amplifier with a logarithmic gain characteristic ahead of the integrating electrometer circuit.

It is evident that electronic circuits fulfill a number of crucial functions in translating the detector signal into a form that can be displayed, read, recorded or analyzed. In recent years, online computers have been incorporated extensively to perform many of these operations. As a result the overall performance of a detector system depends on the proper matching of the performance characteristics of both the detector and the associated electronic circuits. The most sensitive detector or the fastest circuit response may not necessarily represent the most appropriate detector system for a given application.

Development of compact electronic systems and of more rapid data-handling capabilities in recent years has promoted the use of increasingly complex and more automated detection systems. However, intelligent use of such systems still requires a thorough understanding of the underlying principles involved; this will be stressed in the following chapters and in the laboratory experiments outlined in the accompanying laboratory manual.

PROBLEMS

1.1 Identify the decay products of the following nuclides given their mode of decay

$^{90}_{38}Sr(\beta)$, $^{238}_{92}U(\alpha)$, $^{147}_{61}Pm(\beta)$, $^{78}_{35}Br(\beta^+)$, $^{79}_{36}Kr(E.C.)$, $^{222}_{86}Rn(\alpha)$, $^{99m}_{43}Tc(\gamma)$,

$^{252}_{98}Cf(S.F.)$

1.2 A given accelerator produces neutrons by the (d,n) reaction on beryllium. If the beam current is 50 milliamps and the conversion efficiency is one percent, calculate the neutron production rate.

1.3 A certain plutonium-238 source has an activity of 50 millicuries. If all the alpha particles are self-absorbed in the source material, calculate the heat production [alpha energy 5.5 MeV].

1.4 In the previous question, estimate the amount of helium generated in the source in one year. If the source capsule has internal dimensions of 5 mm diameter by 2 mm thick and 5% void space, calculate the helium pressure after one year.

1.5 A 10 MBq cobalt-60 source consists of the thin wire 2 mm in length. Calculate the gamma-ray intensity at 30 cm and 2 m distance from that source.

1.6 Calculate the mass, carrier-free, 10 mCi of the following nuclides (look up the half-lives!).

^{226}Ra, ^{131}I, ^{65}Zn, ^{24}Na

1.7 Calculate the amount of ^{99m}Tc (half-life 6.0 hr) that is present in a 50 mCi ^{99}Mo source (half-life 67 hr) (a) 6 hr, (b) 12 hr, (c) 30 hr, (d) 67 hr, (e) 150 hr after the molybdenum source was freshly prepared.

1.8 A 100-mg silver foil is irradiated in a neutron flux of 10^{13} neutrons/ cm^2 sec. ^{109}Ag has an abundance of 48.17% and cross sections of 4 b and 88b, respectively, to produce ^{110}Ag(253 d) and ^{110m}Ag(24 sec). Calculate the activity of this foil after 5-min exposure and after 3-min and 3-hr cooling times.

1.9 A certain X-ray machine provides an output of 60 R/min at 1 m. Estimate the exposure at 75 cm from the target and the approximate dose in rads and grays.

1.10 Iodine-131 has a radioactive half-life of 8 days and a liver excretion half-time of 55 min. Plot the effective decay rate for 1 mCi of labeled material in the liver.

1.11 Copper-64 decays by positron, internal conversion and negatron emission. Identify the decay products and the amounts accumulating over the life ($7\tau_{1/2}$) of the Cu source (half-life 12.8 hr). Branching ratios are β 38%, β^{+}19%, E.C. 43%.

1.12 Calculate the solid angle subtended by a 5 cm i.d. cylindrical detector facing a point source 10 cm away.

1.13 A Geiger counter has a background count of 60 cpm. It is placed in a uniform gamma-radiation field. If the detection efficiency is 3%, calculate the minimum intensity required for a net count rate of five times background.

1.13 A small liquid sample vial can be placed on top of a 2 in. (5 cm) i.d. detector or inside a 1.5-cm well, 3.5 cm deep in the same size detector. If the detection sensitivity is porportional to the amount of detector material, compare the geometries involved in both cases and the operational efficiencies.

1.14 A thin beta emitter is in the form of a 2.5 cm i.d. plaque, 2 mm thick and has a specific activity of 50 μCi per cubic centimeter. The half-value layer for that beta energy is equivalent to 6 mm of the source material. Calculate the approximate count rate if the source is mounted in a proportional counter in a 4-π geometry. Assume a 30% detection efficiency and that the activity is concentrated (a) in the center line of the source; (b) uniformly throughout it.

1.14 Determine the relative intensities of the (a) 1.29 and 1.098 MeV γ-ray emissions in the decay of Fe-59, and (b) the 0.177 and 0.284 MeV lines of I-131.

1.15 Natural antimony consists of Sb-121 (abund. 57.25%, $\sigma_{n\gamma}$ 6 b) and Sb-123 (abund. 42.75%, $\sigma_{n\gamma}$4b). Neutron activation produces Sb-122 ($\tau_{1/2}$ 2.8 d) and Sb-124 ($\tau_{1/2}$ 60 d). Calculate the activities produced if a 5-g sample of antimony is exposed in a neutron flux of 10^{13} n/cm^2-sec for (a) 2 hr, (b) 2 days, (c) 1 year and determine how long the source would have to cool down in each case for the Sb-122 activity to constitute less than 10% of the total activity.

CHAPTER 2

INTERACTION OF RADIATION WITH MATTER

EXCITATION PROCESSES

Since nuclear radiations cannot be seen or felt by the human senses, their detection involves the use of a medium that is capable of absorbing some of their energy and converting it into a form, most often that of an electric charge, that can be amplified and transformed into a signal that can be counted, averaged or displayed. Much of the continued development of radiation detection methods has centered on the discovery of such conversion mechanisms and on their optimum deployment as increasingly higher sensitivity and more accuracy was desired.

In order to understand the various detection processes one must understand the ways in which any medium, made of atoms or molecules, can absorb energy and how any subsequent events arise. It must also be remembered that many of these processes are likely to be different for essentially free atoms or molecules, as may be found in a gas at low pressures, or for atoms in a solid where they are interrelated in a three-dimensional crystal structure.

An atom is usually pictured as consisting of a central, positively charged nucleus surrounded by a number of "orbital", negatively charged electrons whose total number balances the positive charge of the nucleus, +Ze. The nucleus is made up of Z protons, determining the nuclear charge and the atomic number, and an equal or greater number of neutrons bound together by nuclear binding forces. These binding forces are on the order of several million electron-volts (MeV) and, in general, energy of that order of magnitude is required to split off a nuclear particle (*nucleon*) or to form a slightly heavier nuclide. Complete splitting of a nucleus, such as in uranium fission, liberates binding energy and kinetic energy of product particles totaling about 200 MeV.

The electrons orbiting the nucleus are not randomly arranged but occupy certain energy levels that meet quantum mechanical requirements. They are bound to the nucleus by electrostatic interactions balanced by their orbital motional energy. Each electron has a "spin" of $\pm 1/2\,\hbar$, where $\hbar$ = Planck's constant divided by 2π, and other quantum numbers associated with it, all of which must meet Pauli's exclusion rule that no two electrons can occupy the same quantum state. The two spin states lead to the pairing of electrons. Orbital electrons are found in "shells" of 2, 8, 8, . . . electrons, each subsequent position in a shell representing an additional electron and each closed shell corresponding to a row in the Periodic Table of Elements. The shells are conventionally designated as K, L, M, N, O shells, with K being the closest and containing the most tightly bound electrons. Electrons occupying an unfilled shell are somewhat more easily detached and are capable of interacting with electron levels of neighboring atoms; these electrons are called *valence electrons*, and their number and accessibility governs, to a large extent, the chemical properties of the element.

It must be remembered that the orbital picture of the electronic structure of the atom should not be taken too literally. The *state* of each electron merely represents a description of the energy condition in which it finds itself, that is, its binding energy with respect to its nucleus, its relation to other energy states within that atom, that may or may not be "filled" and represent potential alternative energy states for that electron, and its affinity and possible transfer to adjoining atoms that may offer identical or energetically compatible energy states. The process of increasing the energy of a given electron, especially a valence electron, involves raising it to a higher *state* or *energy level*. This is called "excitation" of that electron and may occur whenever an external process imparts enough energy to the atom to supply the requisite energy for the electron to reach a quantum-mechanically possible higher state. In general, the energy required to excite an electron in this fashion is of the order of a few electron volts (eV); the precise value and the spacing of such higher, but unoccupied, levels varies from nuclide to nuclide. The excitation process may be the result of mechanical bombardment of the atom by a moving particle, electrostatic interaction and energy transfer by the movement of a nearby charged particle, heat absorption, *i.e.*, transfer of vibrational energy among adjoining atoms, in a bulk material, or absorption of electromagnetic radiation by means of any of the available interaction processes.

In general, electrons will occupy the lowest available energy levels; however, only at absolute zero will that condition be completely satisfied. At all other temperatures, electrons will fill some higher levels in accordance with Fermi statistics, but even so at moderate temperatures only a few electrons will normally be found in the higher energy states associated with excited conditions. This will be clear when it is realized that the thermal energy, kT, associated with free movement at 300°K, is of the order of 0.02 eV only, where k = Boltzmann's constant. (This also is the energy associated with thermal neutrons.) The excited atom may retain its electron in that elevated energy state for some time or it may return to its normal lower condition right away. In either case the difference in energy, to return to the lower state, is available to be radiated away as a light quantum, or to be shared among adjoining atoms as mechanical energy, that is, heat. If the energy is radiated off, the photon energy, $h\nu$, must equal the energy difference, giving rise to light emission of characteristic frequencies, ν, or wave lengths, $\lambda = c/\nu$, which reflect the energy-level structure of the emitting atoms and can be used to identify them; this is the basis of optical spectrometry. Depending on the atoms or molecules involved and the particular excitation level, this light emission may occur in the infrared, visible or ultraviolet portions of the electromagnetic spectrum.

If de-excitation and light emission follow the excitation process almost immediately, the process is referred to as *fluorescence*. If the excited state can exist for a prolonged period, ranging from nanoseconds to minutes or hours, before a return to a lower level, this process is called *phosphorescence*. Both processes are utilized in certain types of radiation detectors. It is also possible for an excited atom to pass all or some of this excess energy to the electrons of a neighboring atom or molecule ("exciton transfer"). The newly excited atom or molecule can either pass on the exciton to another neighbor or undergo de-excitation to one of its own lower energy states accompanied by emission of light, now characteristic of its own level structure. In this way, wavelengths can be "shifted", that is, by means of exciton transfer from the original target atom or molecule to a minor impurity atom or molecule of more acceptable spectral emission characteristics, phosphorescent light of a selected wavelength can be obtained after one or more exciton transfers, hopefully without excessive energy losses in the process.

If the atom absorbs enough energy to raise an electron to an energy where its kinetic energy exceeds the electrostatic attraction to the nucleus, it becomes free to move away and may become an independent particle or may move subject to any external electric field such as those existing within the lattice of a metallic conductor. An atom that has lost an electron in this fashion is said to have been "ionized". The loss of a negatively charged electron leaves the atom as a whole positively charged; if it is an isolated atom, it is referred to as a "positive ion". A proton is a positively charged hydrogen ion, H^+; an alpha particle is a doubly-charged helium ion, He^{++}. The energies required to ionize an atom or molecule again depend on the particular energy-level structure. To separate the positive ion and the negative ion, *i.e.*, the electron, sufficient energy must be supplied to move them out of each other's reach, at least temporarily. Table 2.1 lists the ionization energy W, that is, the energy that must be absorbed to form one ion pair, in various gases.

Table 2.1 Energy Absorption per Ion Pair in Gases

Gas	W (eV)	Gas	W (eV)
Hydrogen	36.8	Carbon Dioxide	32.7
Helium	41.3	Air	34.2
Argon	26.4	Methane	28.1
Krypton	24.4	Boron Trifluoride	35.3

If there is a sufficient density of positive and negative ions, recombination occurs and the atoms return to a normal uncharged state with the emission of the appropriate photon energy, in this case usually in the ultraviolet region. Recombination can be delayed if positive and negative charges are separated deliberately under the influence of an electric field.

Another process involving energy transfer within the electronic structure is that associated with X-ray excitation. In this case the atom absorbs enough energy, often resonantly, to remove an electron from one of the inner, more tightly bound electron states. The now-vacant state can then be filled by the transfer of an electron from a higher electron shell, accompanied by emission of the energy difference in the form of electromagnetic radiation. In this case, this energy difference is of the order of kiloelectron-volts (keV) and the photons are in the X-ray portion of the spectrum and considerably more penetrating. The precise X-ray energy emitted again is characteristic of the atom involved; since it involves a lower, tightly bound energy level, it is unaffected by the chemical state of the atom. Different X-ray energies can be emitted corresponding to different identifiable transfer possibilities between the various shells and their sublevels involved and the final destination level. Accordingly, one distinguishes $K_{\alpha,\beta,\gamma}$, $L_{\alpha,\beta,\gamma}$, $M_{\alpha,\beta,\gamma}$, etc., X-rays for each atom. In some cases the major energy jump giving rise to emission may be preceded by some energy re-arrangement within the higher level, giving rise to two-stage transfers or some competitive transfer routes with several X-rays of varying intensity. This is called an Auger process. If the emission of characteristic X-rays of predetermined energy is the result of the absorption of higher-energy X-rays of perhaps less well-defined energy, then this is referred to as *X-ray fluorescence*. When a heavy charged particle is stopped in a medium by losing its energy to the target atoms by electron excitation, the prompt X-ray emission that accompanies the return of the excited electrons to the normal levels is called *bremsstrahlung* (German for "braking radiation").

In the case of bulk insulators and semiconductors, the scheme of potential energy levels for the electrons is not continuous from the occupied levels to the highest excited states. Instead there is a region, referred to as the "forbidden gap," where no stable electron states are normally found. The region where electrons normally occupy energy levels is called the *valence band*. Above the forbidden gap, electrons are only loosely bound to their parent atom and relatively free to drift through the crystal lattice. They are then said to be in the *conduction band* (Figure 2.1). The excitation process in this case implies supplying enough energy to an electron to lift it past the forbidden gap into the conduction band. The forbidden gap is of the order of only a few electron volts and this excitation energy can be supplied by electric forces, intense light absorption, heating, or bombardment with nuclear particles. The excited electron will return to the valence band, again with emission of the energy difference in the form of light. This is the process underlying semiconductor light sources and laser emission. De-excitation may be prompt or the electron may stay in the conduction band for a characteristic lifetime, τ_n, which depends on the gap width, the crystal structure and the number of positive charges left behind in the process. This positive charge, viewed as the potential location left behind in the valence band when the electron was excited, is called a *hole*. Such holes, possessing effective positive charge, also have a certain mobility through the lattice associated with them and a characteristic lifetime, τ_p, until they get annihilated by the de-excitation of an electron. The movement of a hole-electron pair through

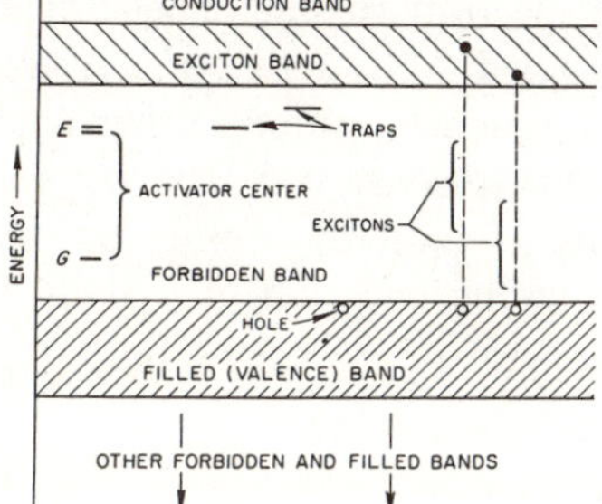

Figure 2.1 Energy band diagram for a semiconductor.

a material represents exciton transfer, and the relative population densities of holes and electrons and their differential lifetimes form the basis for semiconductor electronics.

The lifetime of both holes and electrons can be modified by the introduction of atomic impurities, called *dopants*. The effect of these impurities is to introduce additional energy levels into the forbidden gap that may facilitate production of conduction electrons or selectively trap holes or electrons in intermediate energy levels from which they may return only slowly or with considerable delay (Figure 2.1). One effect of these intermediate levels among others is to change the energy of transition associated with de-excitation and, hence, the frequency or color emitted in that process. For this reason, such trapping levels are also called color centers and they result in characteristic emission in a given matrix element. This process underlies the use of fluorescent materials in fluorescent lamps and TV tubes.

Some materials may remain ionized or in a trapped state for a long time, either because there is a dearth of oppositely charged ions or because, energetically, return to the ground state results in only a minor reduction in total energy. For instance, in the case of photographic emulsions, which consist of silver bromide grains dispersed in gelatin, the grains are made developable by being ionized by incident electromagnetic radiation (light, X-rays, etc.). Since gelatin is a good insulator it supplies only very few electrons and the ionized grains remain developable over long periods of time. However, as a result of thermal effects, stray electrons in the gelatin, and cosmic radiation, some grains slowly lose some of their charge and the latent image loses some of its initial sharpness; the process is called *fading* of the image.

Similarly, in some semiconductors it is possible to load up some of the traps with electrons and they remain there unless additional energy is supplied to raise them back into the conduction band or the material is heated which has the effect of narrowing the forbidden gap and facilitating the return of the electron to the valence band. This emptying of the traps by heating is called "annealing" and is accompanied by emission of light of appropriate frequency. Among other applications, it forms the basis of *thermoluminescent dosimetry*. If the traps are emptied by being hit by another "stimulating" radiation pulse, the material is said to be "lasing," again resulting in the emission of a flash of light.

ENERGY TRANSFER MECHANISMS

As stated in the Introduction, all detection methods depend on a means of observing the rate of deposition and the fraction of energy deposited in a target medium by the radiation field to be detected and the conversion of this energy into some form of heat, light or electrical charge that can be sensed and analyzed by appropriate means. The various mechanisms of interaction with the target medium, therefore, are central to any understanding and evaluation of the various detector systems. It is also evident that it is appropriate to talk of *detector systems* rather than just detectors since the whole procedure, to be useful, must match and consider not only the primary interacting medium but also the subsequent sensing, amplifying and indicating devices.

There are two ways of looking at the mechanism of energy transfer. One is to observe the loss of energy or intensity of a well-defined, possibly collimated, beam of radiation by measuring the incident intensity and the emerging intensity or the intensity at some specified distance along the path. For beta and gamma radiation for a collimated parallel beam the reduction in intensity with distance x in a finite medium is proportional to the intensity, *i.e.*, the number of particles or photons crossing unit area per unit time, or

$$-\frac{dI}{dx} = \mu I \qquad (2\text{-}1)$$

which leads to an exponential attenuation law of the form

$$I = I_o \, e^{-\mu x} \qquad (2\text{-}2)$$

where I_o is the incident intensity at x = 0, I is the intensity at any depth x into the medium, and the proportionality constant, μ, is called the *linear attenuation coefficient*. This "constant" μ does not specify the mechanism of intensity loss and in general varies with the energy of the incident radiation.

For an isotropic point source of radiation, radiating particles or photons equally into all directions (4π), the intensity in air at any point at a distance r from the source follows the well-known inverse-square law since the flux density or intensity is reduced as the effective surface area traversed expands, *i.e.*,

$$I_{(r)} = \frac{S_o}{4\pi r^2} \qquad (2\text{-}3)$$

where S_o is the source emission rate or source activity.

At two points P_1 and P_2, at distances r_1 and r_2 from the source, the intensity ratios then would be

$$\frac{I_1}{I_2} = \frac{\dfrac{S_o}{4\pi\, r_1^{\,2}}}{\dfrac{S_o}{4\pi\, r_2^{\,2}}} = \frac{r_2^{\,2}}{r_1^{\,2}} \tag{2-4}$$

For two points at the incident and emerging surfaces of a material of thickness x in such a divergent field

$$I_2 = I_1\, e^{-\mu x}\, \frac{r_1^{\,2}}{r_2^{\,2}} = I_1\, e^{-\mu x} \left(\frac{(r_2 - x)^2}{r_2^{\,2}} \right) \tag{2-5}$$

In a uniform attenuating medium encompassing the source the intensity at any distance r is

$$I_{(r)} = S_o\, \frac{e^{-\mu r}}{r^2} \tag{2-6}$$

The loss of *energy* in a collimated beam is much more complicated and depends on the various mechanisms by which the energy transfer can occur. For charged particles this transfer can occur by electromagnetic interaction with the orbital electrons, leading to their excitation or ionization, or by repulsive, electrostatic (Coulomb) interaction resulting in a deflection of the incident particle and some transfer of momentum to the target atom. This process is known as Rutherford scattering and is depicted in Figure 2.2.

The angle between the asymptotes not containing the hyperbola defines the deflection of the particle θ;

$$\frac{Ze^2}{2Eb} = \tan\frac{\theta}{2} \tag{2-7}$$

where $b = L/(2mE)^{1/2}$ is the impact parameter, defined as the distance between the center of force and the limiting line of flight of the particle, for large values of r.

The number of protons dn having an impact parameter between b and b + db is $2\pi b\, db = dn$, where, from Equation 2-7,

$$db = -\frac{Ze^2}{4E}\,\frac{d\theta}{\sin^2(\theta/2)} \tag{2-8}$$

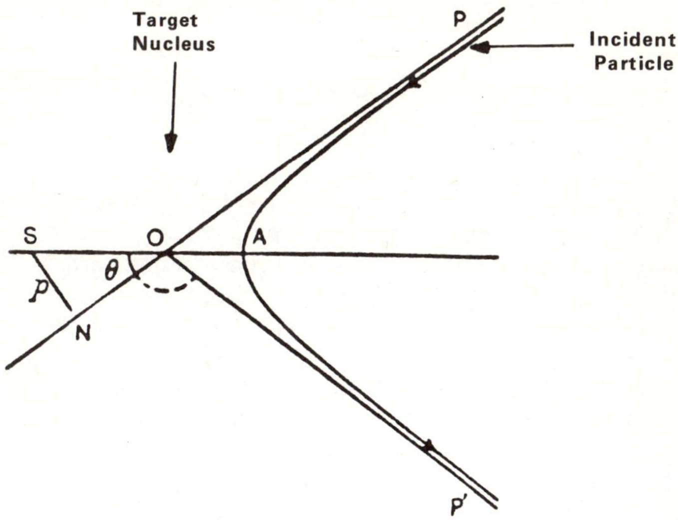

Figure 2.2 Rutherford scattering.

Hence

$$|dn| = \pi \left(\frac{Ze^2}{2E}\right)^2 \frac{\cos(\theta/2)}{\sin^3(\theta/2)} |d\theta| \tag{2-9}$$

This is the number of particles deflected through an angle between θ and $\theta + d\theta$. They pass with uniform density between two cones of aperture θ and $\theta + d\theta$. The solid angle included between these cones is

$$d\omega = 2\pi \sin\theta \, d\theta \tag{2-10}$$

and

$$\frac{dn}{d\omega} = \frac{1}{4}\left(\frac{Ze^2}{2E}\right)^2 \frac{1}{\sin^4(\theta/2)} \tag{2-11}$$

The quantity $dn/d\omega$ is dimensionally an area, and one can show that the differential-scattering cross section is

$$\frac{d\sigma}{d\omega} = \frac{1}{4}\left(\frac{e^2 Z}{mv^2}\right)^2 \frac{1}{\sin^4(\theta/2)} \tag{2-12}$$

This is the famous Rutherford scattering formula. Put in convenient numerical form it is

$$\frac{d\sigma}{d\theta} = 2\pi \sin\theta \frac{d\sigma}{d\omega} = \frac{0.8139Z^2}{E^2\ (\text{MeV})} \frac{\sin\theta}{\sin^4(\theta/2)} \times 10^{-26}\ \text{cm}^2 \text{ per nucleus} \qquad (2\text{-}13)$$

Electromagnetic radiations can interact with the orbital electrons as well as with the nucleus, losing energy in the process and transferring energy and momentum. Uncharged particles, like neutrons, do not interact significantly with the orbital electrons but lose energy by *elastic scattering*, where total energy and momentum are conserved, *inelastic scattering,* where some energy and momentum are lost to some excitation states of the target nucleus, and *capture processes,* where the incident neutron is swallowed up, giving rise to more or less stable nuclear excited states that may lead to immediate de-excitation, indicated by the emission of a prompt "capture gamma" photon, as well as delayed emission processes when one or more nuclear particle may be emitted [(n,n′), (n,2n), (n,p), (n,α), (n,fission) etc. reactions in the usual notation]. As the incident particle loses energy, it slows down and its probability of interaction, called the *cross section,* may increase until there is no more kinetic energy left to transfer and, finally it comes to rest. The distance a particle of given energy travels in an attenuating medium is called its range; this is a useful quantity to be used with beta rays and heavier charged particles.

For relatively thin layers of absorbers compared with the particle range, the energy may be considered almost constant and, hence, the attenuation coefficient μ is nearly constant. Under those conditions the linear rate of energy loss dE/dx is nearly constant and a measure of the energy E. The attenuation properties of the medium can be expressed in terms of a half-value layer, that is the thickness of the medium by which the incident intensity is halved:

$$\frac{I_2}{I_1} = \frac{1}{2} = e^{-\mu x_{1/2}} \qquad (2\text{-}14)$$

or

$$\mu x_{1/2} = -\log 2 = \mathbf{0.693}$$

$$\text{HVL} = x_{1/2} = \frac{0.693}{\mu} \qquad (2\text{-}15)$$

in appropriate units.

In health physics, the rate of energy loss, dE/dx, is often expressed in terms of the linear energy transfer $(\text{LET})_L$, which has been defined as the average energy dE_L, locally imparted to a medium by a (charged) particle of specified energy in traversing a distance $d\ell$:

$$L = \frac{dE_L}{d\ell} \tag{2-16}$$

Radiation is often described as having high or low LET.

All of these energy loss descriptions are merely concerned with the rate and distribution of energy transfer to the atoms and electrons of the attenuating medium. They do not deal with the consequences of this energy transfer; yet those consequences are the distinguishing marks between different target media and various detection processes. These consequences may be manifold, and can be classified into four categories:

1. *Excitation processes.*

As we have seen, excited electrons may get de-excited with the emission of light, the emission of X-rays and the transfer of excitons from atom to atom leading to delayed and often modified emissions.

2. *Ionization.*

Ionization processes in gases, in particular, lead to the production of mobile positive- and negative-charged ions and, hence, transfer of electric charge. In solids ionization may change conduction properties, and in many media ionization may result in the formation or degradation of chemical compounds.

3. *Collision processes.*

Where the interaction results primarily in the transfer of collisional energy (recoil) the hit atom may move through the medium on its own, triggering a cascade of secondary and tertiary interaction events, themselves causing excitation, ionization or recoils until all the energy is dissipated into heat or lattice energy. Such collisional events give rise first to scattering events, *i.e.,* deflection of particles from their original course, then perhaps to chemical or mechanical property changes, generation of color centers, and crystal defects. Some of these defects may be irreversible, such as the phenomena underlying the track-etch detection method. For fast neutrons, collisions with hydrogen atoms are particularly important since the near-equal mass of the hydrogen atom can lead to substantial recoil energy transfers from the incident neutron.

Another important collision event occurs near the surface of a solid. If such a surface is bombarded by charged particles, energetic electrons or even light quanta, enough energy may be transferred to electrons in the solid to give rise to emission of secondary electrons. For this to occur, the incident particle must possess enough energy to overcome the binding energy or work function of the surface material. If more than one secondary electron is emitted, then charge amplification has occurred and such a surface may be of use in electron multipliers.

Another collisional process gives rise to gas amplification. In this process, often observed in ionized gases, ions accelerated by electric fields may collide with, and ionize, a number of gas atoms before they themselves come to rest or are neutralized. As a result one energetic gas ion may cause an avalanche of secondary ions, greatly facilitating the detection of the initial ionizing event.

4. *Capture processes.*

If the incident particle is absorbed by the target nucleus, a large variety of subsequent events may occur. For charged particles, capture reactions require that the incident particle have enough energy to overcome the Coulomb repulsion of the nucleus, *i.e.,* there is a minimum "threshold" energy below which capture is improbable. For electromagnetic radiations, so-called photonuclear events can occur, but with relatively low cross sections.

The most important capture events are those associated with neutrons. Thermal neutrons, of energy kT $\sim$ 0.024eV, have a high cross section for simple capture, with de-excitation leading to the emission of a capture γ-ray [(n,γ) reaction]. The newly formed nuclide is now one atomic mass unit heavier and may not be stable. In that case, it will decay with its own characteristic half-life, and the process is referred to as *activation*; typically the decay will occur with delayed emission of gamma-rays and beta particles.

For heavy actinide elements, notably isotopes of U and Pu, neutron capture may lead to fission resulting in the disintegration of the compound nucleus into two or more heavy nuclides, each having enough energy to cause further intense interaction in any surrounding media.

Capture may also occur at higher energies leading to subsequent ejection of heavy charged particles, such as protons, deuterons or tritons, provided the incident energy exceeds the threshold value needed to overcome the binding energy in the nucleus for these particles. Again the emitted particles will give rise to their own interaction events. The energy transfer processes are usually classified in terms of the incident radiation being "detected": heavy charged particles (such as fission products, protons, deuterons, alpha particles and other atomic ions), light charged particles (leptons: electrons, pions, muons, etc.), neutrons, and electromagnetic radiations (gamma- and X-rays).

HEAVY CHARGED PARTICLES

Alpha particles were the first heavy charged particles to be studied in detail. They are still of interest since isotopic sources of alpha emitters, such as U, Th, Pu, Am and Po sources, are commercially available. The energy of alpha particles emitted is in the general range of 4.1-7.7 MeV, so their characteristics are relatively uniform. Other heavy charged particles, notably protons and deuterons, can be produced with a wide range of energies in various types of accelerators and their detection, though similar, must take into account the particular energy range. For this reason, alpha particle interactions will be described first.

All heavy charged particles, alpha particles included, interact principally with the orbital electrons of the target atoms, leaving a dense trail of excited and ionized atoms in their path. Some energy loss occurs as the positively charged incident particle raises the energy, electrostatically, of orbital electrons giving up to the target atom an amount of energy related to the closeness of approach and the energy needed to create an ion. If there are N target atoms per unit volume, the rate of energy loss per unit path length x is given by the Bethe-Livingston equation:

$$-\frac{dE}{dx} = \frac{4\pi e^4 z^2}{m_e v^2} NZ \left[\log_e \frac{2m v^2}{I} - \log_e (1 - \beta^2) - \beta^2 \right] \qquad (2\text{-}17)$$

where
- e = electronic charge
- m_e = electronic mass
- z = atomic number of incident particle
- Z = atomic number of target atoms
- v = velocity of incident particle
- β = v/c = fraction of velocity of light
- I = average ionization energy of target atoms

If $v \ll c$, the last two terms in the brackets are negligible. For velocities $\beta \ll 0.03\ z$, the particle interacts essentially with the neutral atoms with almost all the transferred energy going to the excited electron. In contrast, at higher energies the incident particle may penetrate the electron screen and interact with the nucleus directly, either causing nuclear reactions, such as (d, p), (p, α), (α, n) reactions, or undergoing nuclear or Rutherford scattering. In the latter case, the incident particle is deflected by Coulomb charge repulsion from the nucleus and the change in momentum imparts some recoil energy to the scattering nucleus. If one then observes the energy transfer by counting the fraction of collimated incident particles emerging after passing through a given thickness of target material, one finds that their number has dropped either by losing enough energy along the way so that they cannot continue, or by being scattered out of the initial beam direction into other directions. Any particle not emerging is said to have been "stopped". The target thickness just sufficient to stop most of the particles represents their range in that medium. For non-relativistic particles Equation 2-17 can be expressed as

$$-\frac{dE}{dx} \approx \frac{M z^2}{E} \qquad (2\text{-}18)$$

since the logarithmic term varies only slowly with E. Hence the energy loss and, therefore, the range vary as 1/E and with the square of the charge on the incident particle. For alpha particles, the loss in energy per collision is small so that (dE/dx) remains nearly constant for most of its range until the very end when E becomes small and the specific ionization increases rapidly before the particle is finally brought to rest. This is shown in Figure 2.3 which plots specific ionization against residual

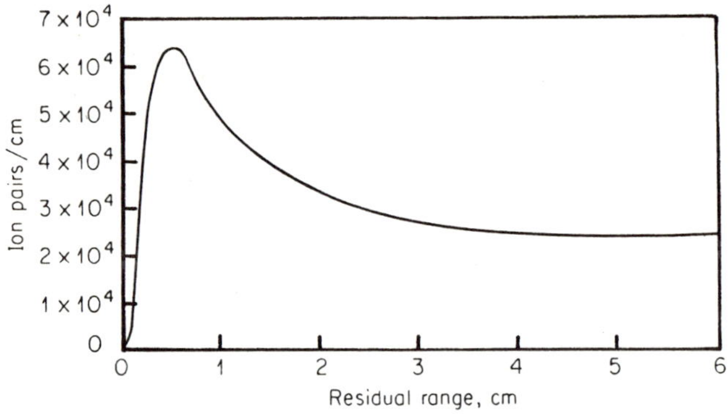

Figure 2.3 Bragg Curve for α particles.

(remaining) range. This is usually referred to as the Bragg Curve. Figure 2.4 shows how the range of monoenergetic α-particles can be determined by plotting the number of particles transmitted against absorber thickness or density. The range so obtained is not sharp because the sources used

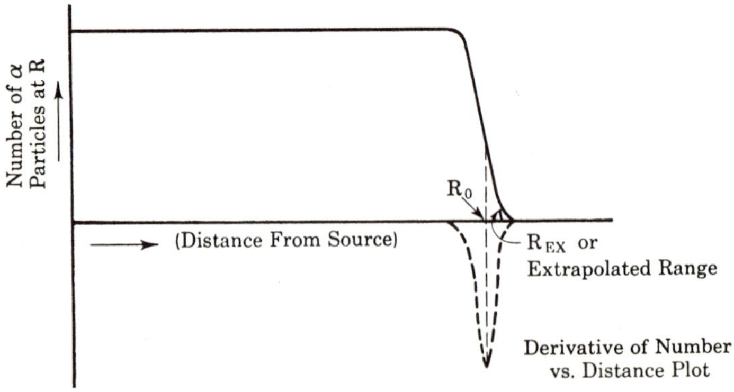

Figure 2.4 Range distribution for monoenergetic α particles.

rarely produce monoenergetic particles (due to self absorption and window effects) and because of the contribution of small-angle scattering events which leads to a phenomenon known as *straggling.*

If one takes the differential distribution curve to be Gaussian, the integral distribution curve

$$P(R) = \int_R^{\infty} P(R)\ dR = \frac{1}{2}\left[1 - \mathrm{Erf}\left(\frac{R - R_0}{2\ \alpha}\right)\right] \qquad (2\text{-}19)$$

shows an inflection point at a specific R value that is called the *mean range* R_0; α^2 is the dispersion of the curve. The intersection of the tangent drawn for this inflection point with the abscissa axis yields point R_1, lying to the right of R_0 at a distance

$$S = \sqrt{\frac{\pi}{a}}\ \alpha \qquad (2\text{-}20)$$

The abscissa value of R_1 is called the *extrapolated range.* S, denoting straggling, is another characteristic of the fluctuations and of the difference between the two "range" values R_0 and R_1 (Figure 2.5).

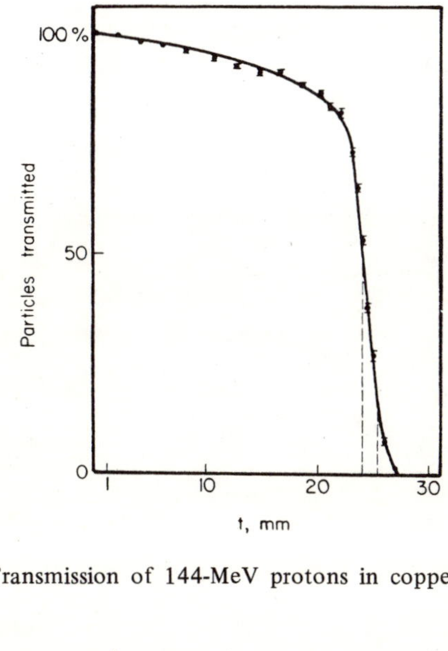

Figure 2.5 Transmission of 144-MeV protons in copper (from Koschmieder).

Multiple scattering becomes more and more important at higher energies, as seen in Figure 2.5 which shows the observed transmission curve for 144 MeV protons in copper. Since the range varies as 1/E such an absorption measurement can be used to determine the energy of the particles using calibrated foils. Conversely, a known source of charged particles with a suitable detector can be used to determine the density or composition of thin material layers or gases. The absorption of alpha particles in gases has been studied in considerable detail and in mixtures it has been found that each component contributes to the stopping of charged particles in strict proportion to its concentration (Bragg-Kleeman rule). The specific ionization I is usually expressed in terms of the average energy W that has to be absorbed to form an ion pair. Then, considering only ionizing events,

$$-\frac{dE}{dx} = n\,W \qquad (2\text{-}21)$$

where n = specific ionization. Table 2.2 lists values for W and I for various gases of interest.

Table 2.2 Specific Ionization and W Values for Some Gases

Material	Z	I(eV)	$\log_e$ I	W(eV)
Air				33.73 (β, γ) 35.0 (α)
H_2	1	18	2.8904	36.6 (β, γ) 36.2 (α)
He	2	42	3.7377	41.5 (β, γ) 46.0 (α)
N_2	7	78	4.3567	36.0
O_2	8	100	4.6052	32.5
Ar	18	184	5.2149	26.2
CO_2				32.9 (β, γ) 34.1 (α)
CH_4				27.3 (β, γ) 29.1 (α)
BF_3				35.6 (α)

Because so little energy is transferred in each interaction most heavy particles with $Z > 1$ move essentially along straight-line paths; this has been observed in photographic emulsions and cloud chambers. For alpha particles of energies 4-7 MeV, the range has been measured very carefully. In air it can be expressed as

$$R_{air} = 0.309\ E^{3/2}\ \text{cm} \tag{2-22}$$

and more generally

$$R_{\alpha} = 0.173\ E^{3/2}\ A^{1/3}\ \text{mg/cm}^2 \tag{2-23}$$

where A is the atomic weight of the absorbing material, subject to the Bragg-Kleeman rule. The relative stopping power of any material can then be expressed by comparison to dry air at normal temperature and pressure:

$$\text{Stopping power } S = \frac{(dE/dx)\ \text{mat.}}{(dE/dx)\ \text{air}} \tag{2-24}$$

In practical terms this means that the range of 5-MeV alpha particles is of the order of 4-6 cm in most gases and of the order of only 10-30 μm in most solids and liquids. Table 2.3 lists the ranges of some low-energy

Table 2.3 Ranges (in mg/cm^2) of Charged Particles in Aluminum

Particle	Energy 0.5	1	2	3	4	7	10 MeV
Proton	1.3	3.6	10.5	19.6	35.1	90.0	174.0
Deuteron	1.1	2.6	6.9	13.3	21.2	55.9	100.0
Alpha	0.5	0.8	1.6	2.0	3.7	8.8	15.6
Beta (e^-)	111	383	926	1470	2010	—	—

charged particles in aluminum (S = 1,680), density 1.6 mg/cm^2. Because of their strong interaction and relatively short range, alpha particles or fission products are easily stopped by thin foils or windows; consequently, most methods for their detection must employ internal counting, where the source material confronts the detecting medium without any intervening window or barrier. Failing this, careful correction must be made for self-absorption and cover-foil effects on both energy and range. The same applies when semiconductor detectors are employed which normally have a thin ohmic electrode layer and a dead layer that must be traversed before the sensitive volume is reached.

BETA RADIATION

Beta radiation is the term used to describe electrons of nuclear origin, both negatively or positively charged (positrons), though the same processes apply also to high-energy, non-nuclear electrons produced in accelerators such as synchrotrons. The principal difference between electrons and heavy charged particles arises from the small mass of the electron (1/1,836 amu), so that little momentum is transferred elastically should it collide with a nucleus. Consequently the main interaction effects observed are large-angle nuclear scattering, and excitation and ionization of orbital electrons. Although excitation is by far the most probable interaction event, most detection methods for (negative) electrons rely on ionization events; this leads to an apparent higher value of W than is actually the case if one excludes exciting events.

For nonrelativistic electrons, the Bethe formula (Equation 2-17) takes the forms

$$-\frac{dE}{dx} = \frac{4\pi e^4 z^2}{m v^2} NZ \left[\log_e \frac{1.66 \, m v^2}{2I}\right] \text{for } E < 0.5 \text{ MeV} \qquad (2\text{-}25)$$

and

$$-\frac{dE}{dx} = \frac{2\pi e^4}{m c^2} NZ \left[\log_e \frac{E}{2m c^2 I} + \frac{1}{8}\right] \text{for } E \gg mc^2 \approx 0.5 \text{ MeV} \qquad (2\text{-}26)$$

For low energies, again, the rate of energy loss decreases with rising energy, whereas above 0.5 MeV it increases slowly with E.

At a comparable energy, an electron moves much faster than a heavy ion; consequently it has a lower probability of interaction and, hence, a lower specific ionization. That means that the mean free path between interactions is long, their range is relatively long and, since they are easily deflected in scattering events, their path can be very tortuous. One can define a range and a stopping power, but now the range merely defines an *average* distance traveled in the initial forward direction, not the actual path traversed. It must also be remembered that beta particles are not monoenergetic (though accelerator-produced electrons can be), but represent an energy spectrum whose maximum energy, $E_{\beta max}$, is usually specified and whose mean energy, E_m, can be calculated. Because the energy transfer per collision is high, there is a continuous reduction in energy for each electron. Taking a typical beta-ray spectrum distribution it is found that the attenuation or stopping effect of a medium of thickness x can be described approximately by the attenuation law

$$I_x = I_o \, e^{-\mu x} \qquad (2\text{-}27)$$

where I_o is the incident intensity (number of electrons per unit area per unit time), I_x is the emergent intensity, and μ is a proportionality constant, called the linear attenuation coefficient. μ is a function of absorber composition and of $E_{\beta max}$. If the thickness of the absorber is expressed in mg/cm^2, it is preferable to use the *mass attenuation coefficient* μ' where $\mu' = \mu/\rho$ (ρ = density). This is convenient because electron density Z/A varies little across the periodic table of elements and for $E_{\beta max}$ between 0.15-3.5 MeV

$$\mu' = 0.017\ E_{\beta max}^{1.43} \tag{2-28}$$

Since the range of electrons depends on their energy and the attenuation on the absorber thickness, it is possible to determine $E_{\beta max}$ by measuring intensity variation with a series of foils of known composition and thickness. Such an absorption curve, shown in Figure 2.6 requires a well-collimated source and a set of carefully weighed absorber foils. Since many detectors are also sensitive to the gamma-rays and to bremsstrahlung emitted following absorption, the range is usually obtained by extrapolation from the logarithmic portion of the curve. From the extrapolated range, the energy $E_{\beta max}$ can be derived using one of the approximate range energy relations:

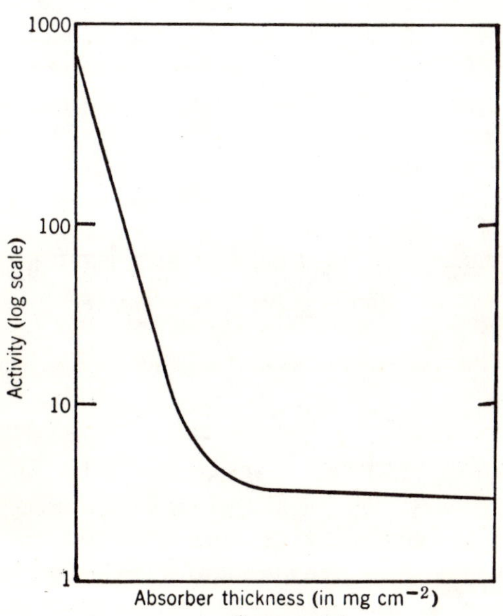

Figure 2.6 Beta-ray absorption curve.

$$E = 1.2 + 5\ R^{0.5} \qquad \text{for } R < 0.03\ \text{g/cm}^2$$
$$E = 6.67\ R + 0.0186 \qquad 0.02 > R > 0.002$$
$$E = 1.92\ R^{0.725} \qquad 0.3 > R > 0.02$$
$$E = 1.85\ R + 0.245 \qquad R > 0.3\ \text{g/cm}^2$$

Conversely the range in air can be estimated from the Glendenin expressions for the range-energy relationships which are as follows:

$$R = 0.542\ E_{max} - 0.133 \qquad \text{for } E_{max} \text{ greater than 0.8 MeV} \tag{2-29}$$

and

$$R = 0.407\ E_{max}^{1.38} \qquad \text{for } E_{max} \text{ between 0.15 and 0.8 MeV} \tag{2-30}$$

The units are the same as given above. The Glendenin relationships are plotted in Figure 2.7 and are satisfactory to quite low energies. It follows

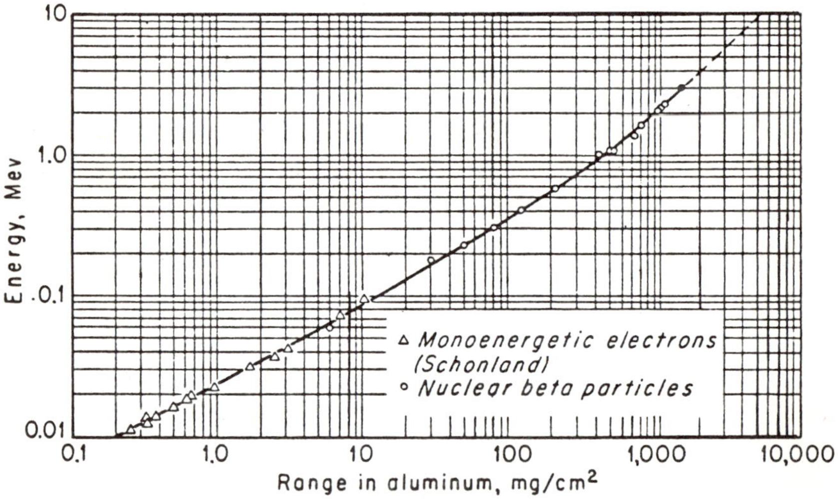

Figure 2.7 Range of electrons of various energies in aluminum (after Glendenin).

that electrons can be stopped by millimeter to centimeter thicknesses of most low-Z materials and by lesser thicknesses of lead or other heavy metals. Except for very high energies, electrons are usually not capable of dislodging atoms in solids, but their initial emission from a nucleus may be accompanied by a small recoil motion.

Free positive electrons (positrons) will behave much the same way in low-density media as negatrons. However, their behavior is overshadowed by their high probability of being annihilated when encountering a negative electron. The annihilation process frees an energy $E = 2\ m_e\ c^2 = 1.02$ MeV; therefore, the passage of a positron is signaled by the release of 1.02 MeV, usually in the form of two oppositely traveling photons of 0.51 MeV each.

X-RAYS AND GAMMA-RAYS

X-Rays originate from transitions in the electronic structure of the atom and, hence, have energies in the range of 5-90 keV, with corresponding wave lengths of the order of angstroms (Å) (1 Å = 10^{-10} m). Gamma-rays arise from nuclear reactions and have energies associated with nuclear excitation levels, typically of the order of 30 keV-3 MeV, though higher energies do occur such as the 7.11 MeV gamma-ray emitted by nitrogen-16. Since they represent specific transitions, each γ photon will be emitted with a well-defined energy $h\nu$ appropriate to that transition. However, a given source may emit gamma-rays through several alternative de-excitation pathways, giving rise to a gamma-ray spectrum of several lines with fixed relative intensities. This is illustrated in Figure 2.8 which shows some branching schemes and the corresponding relative emission intensities. To interpret gamma-ray spectra analytically, these transition probabilities must be known. Furthermore, since most detectors have sensitivities that vary for different energies, corrections for this must be applied when detecting gamma-rays of widely differing energies.

Since photons carry no charge, their main interaction is with the electromagnetic field of the atom or its electrons, though photonuclear reactions can occur, but with rather low probability. The primary interactions of X-rays and γ-rays with matter result in the production of energetic secondary charged particles, usually electrons, and it is these charged particles and their interaction with matter that account for the actual transfer of energy and the detectability of the initiating photon. The secondary particles produce most of the ionizing events, whereas the primary photon itself is only weakly ionizing. Consequently, all biological and radiochemical effects of X-rays and γ-rays must be ascribed to their secondary charged particles.

Energetic photons interact with matter by means of six principal processes:

- photoelectric effect
- Rayleigh scattering
- Compton effect
- pair production
- resonance elastic scattering
- photodisintegration

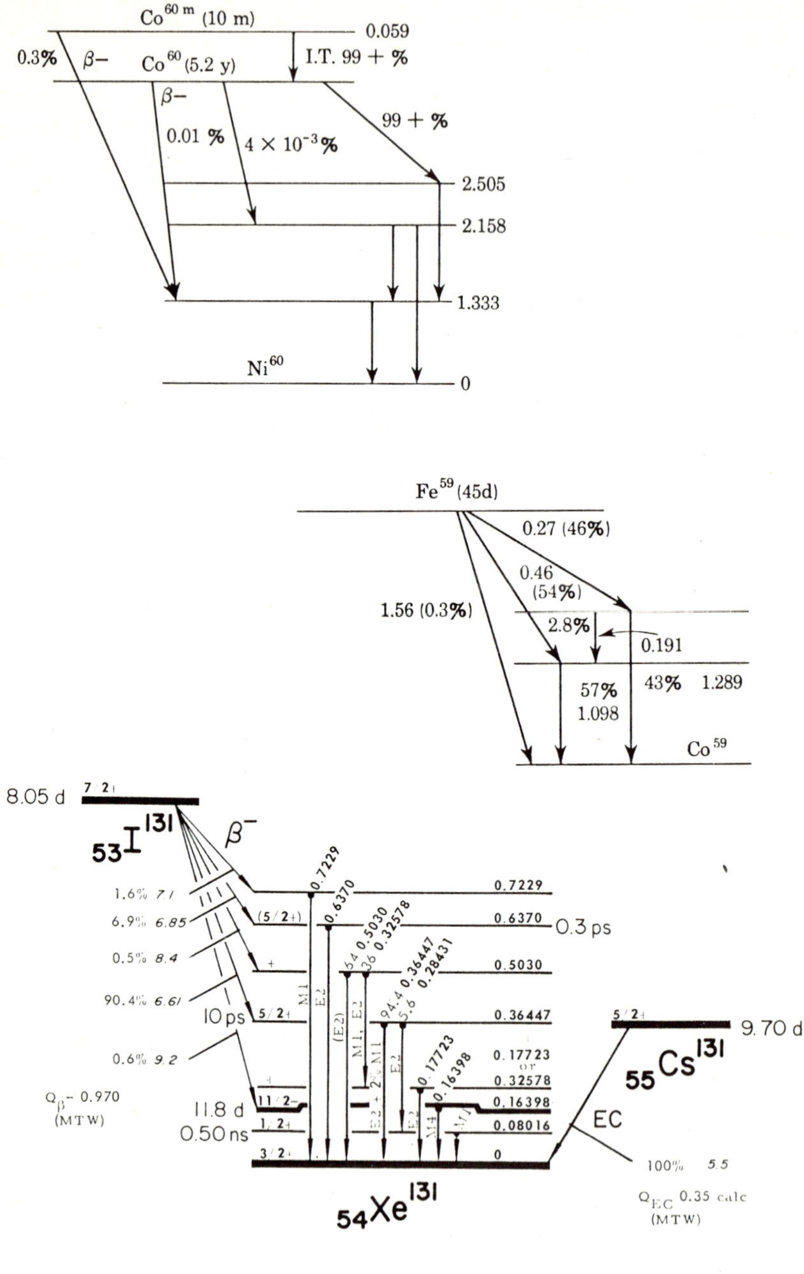

Figure 2.8 Typical nuclear decay diagrams showing excited levels and branching ratios for β^- and β^+ emitters.

The first two involve interaction with the whole atom, the next two with one electron in the atom, and the last with the nucleus. For detection purposes only the first, third and fourth are of major interest. The probability of any incident photon initiating any of these processes can be expressed as a *cross section* per atom; the probability for a photon interacting at all would be the *total cross section,* being the sum of the individual cross sections. If we multiply the total cross section per atom by the number of atoms per unit volume, we again obtain a linear attenuation coefficient, μ, which is a measure of the number of photons removed by one of the processes, from a parallel beam of photons per unit path length in the absorbing medium. Thus, for gamma-rays we also have an attenuation law of the form

$$I = I_o \, e^{-\mu x} \qquad (2\text{-}31)$$

where μ will vary with energy, but in a more complicated fashion than in the case of β particles since more different mechanisms are involved here. For convenience, one can again define a mass absorption coefficient $\mu' = \mu/\rho$. The thickness x that just halves the incident activity is called the *half-value layer* $x_{1/2}$, and is a convenient means of comparing γ-ray energies and relative shielding requirements:

$$\text{Half-value layer (HVL) } x_{1/2} = \frac{\log_e 2}{\mu} = \frac{0.693}{\mu} \qquad (2\text{-}32)$$

The removal of photons from the beam can proceed by elastic processes such as Rayleigh scattering or nuclear resonance scattering (Mössbauer effect) or by effects where all or part of the photon energy is absorbed, such as the photoelectric interaction or the Compton effect. Hence the attenuation coefficient μ is the sum of the scattering coefficient μ_s and the absorption coefficient μ_a:

$$\mu = \mu_s + \mu_a \qquad (2\text{-}33)$$

The Compton effect leads to a scattered photon as well and in many cases μ_s represents essentially the Compton scattering coefficient.

The Compton Effect

The Compton effect, named for A. H. Compton, was one of the first demonstrations of the wave-particle duality in physics. When an incident photon of energy $E_o = h\nu_o$ and momentum $p = (h\nu_o/c)$ interacts with an orbital electron, that electron can be considered a free particle if its binding energy is small compared with $h\nu_o$. (Since ionization energies are of the order of tens of electron-volts, whereas X- and γ-photons have energies above 10 keV, this is easily the case.) The interaction can then be described as an ordinary elastic collision, as shown in Figure 2.9. The electron recoils with an energy $E = 1/2\ mv^2$ and the scattered photon emitted at an angle θ now has an energy $E' = h\nu' = h\nu_o - E$. Consequently, there has been a change in photon energy and frequency, in contrast to the usual quantum absorption processes for photons. This frequency or wavelength shift can be derived from the conservation equations and is

$$\frac{1}{h\nu'} - \frac{1}{h\nu_o} = \frac{1}{m_o c^2}(1 - \cos\theta)$$

or

$$\lambda' - \lambda_o = \frac{h}{m_o c}(1 - \cos\theta); \qquad (2\text{-}34)$$

substituting $\alpha = \dfrac{h\nu_o}{m_o c^2}$, we get

$$\frac{h\nu'}{h\nu_o} = \frac{1}{1 + \alpha(1 - \cos\theta)} \qquad (2\text{-}35)$$

The recoil angle ϕ and the photon scatter angle θ are related by

$$\cot\phi = (1 + \alpha)\tan\frac{\theta}{2} \qquad (2\text{-}36)$$

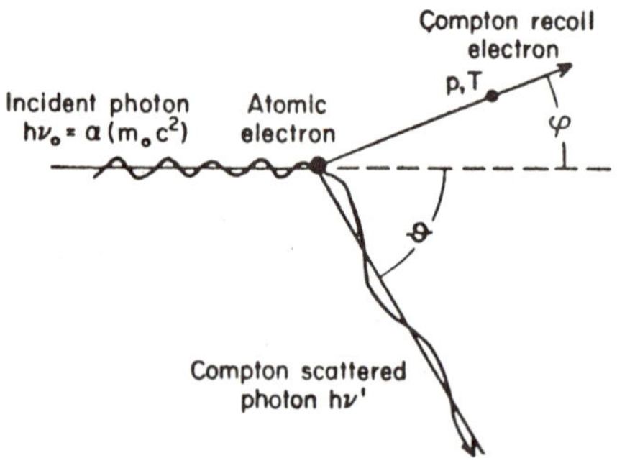

Figure 2.9 Trajectories in the scattering plane for the Compton Effect

and the recoil energy E

$$E = h(\nu_o - \nu') = h\nu_o \frac{\alpha(1 - \cos\theta)}{1 + \alpha(1 - \cos\theta)} \tag{2-37}$$

This reaches a maximum value

$$E_{max} = h\nu_o \left(\frac{2\alpha}{1 + 2\alpha}\right) \text{ for } \theta = 180^\circ \tag{2-38}$$

The fraction of photons that are scattered in a particular direction is given by the collision differential cross section per unit solid angle $d\sigma/d\Omega$ and is shown in Figure 2.10. From this can be derived the number versus angle distribution of scattered photons; which has a forward maximum that lies near $\theta = 20^\circ$ for $E_o = 3$ MeV, and moves to large angles for lower values of E_o. Similarly, one can obtain the angular distribution for the recoil electrons, which in practice are mainly responsible for any detectable ionization. The number distribution with angle has two maxima near

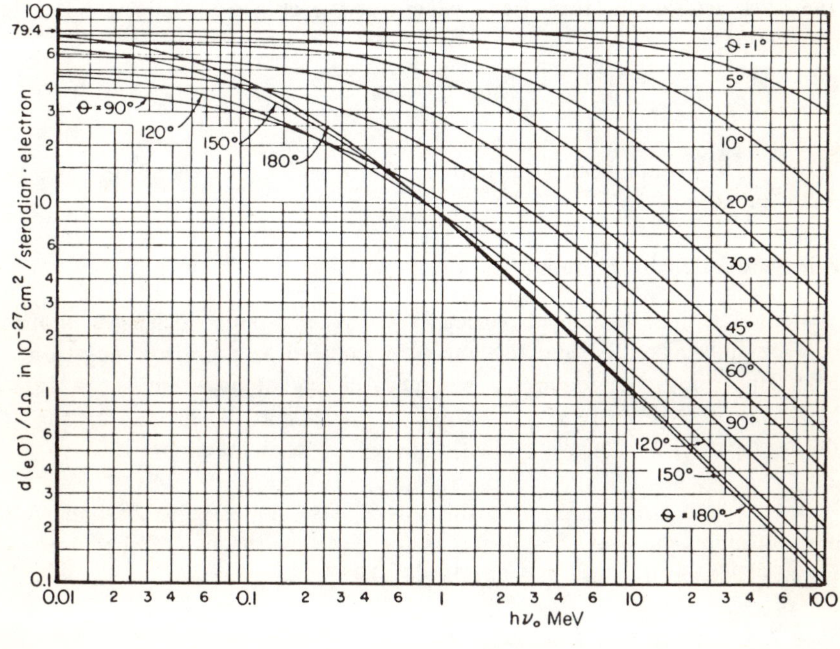

Figure 2.10 Collision differential cross section $d\sigma/d\Omega$ for the number of photons scattered per unit solid angle in direction θ (after Evans).

$\phi = 20°$ and $60°$ for low photon energies below 0.5 MeV; for high energies, a single low-angle maximum occurs at low values of ϕ. In terms of energy, the largest number of recoil electrons occurs at the maximum energy E_{max}(Equation 2-38); this sharp change in the distribution is called the Compton edge in γ-ray spectroscopy (see Chapter 5).

The Photoelectric Effect

In photoelectric collisions, the entire incident photon energy $h\nu$ is absorbed by the struck atom and one electron, usually from the K or L shell, is emitted immediately with a kinetic energy E,

$$E = h\nu - B_e \tag{2-39}$$

where B_e is the binding energy of the ejected electron. The vacancy in the inner shell is then filled from higher levels, resulting in the emission of characteristic X-rays or Auger electrons, while momentum is conserved by the backward recoil of the residual atoms. For γ-rays, $h\nu >> B_e$ and the emitted electron carries off energy nearly equal to the photon energy. The cross section for this effect, except at the absorption edges,

$$\sigma_{pe} = \text{Const.} \frac{Z^4}{(h\nu)^3} \tag{2-40}$$

Values of σ_{pe} have been tabulated by White and Grodstein (NBS Circ. 583) and others.

Pair Production by Photons

In the field of a charged particle, such as an atomic nucleus, a photon may be fully absorbed and the energy re-emitted as a positron-negatron pair. This conversion requires a minimum energy of $h\nu = 2m_o c^2 = 1.02$ MeV; any energy above this threshold is available to be divided among the two particles to provide kinetic energy up to a maximum of $h\nu - 2m_o c^2$.

The atomic cross section for this effect, σ_{pp}, increases as Z^2 and with photon energy $h\nu$. Tables and graphs for σ_{pp} have been published by Evans (1955) and White and Grodstein (1957).

Mass Attenuation Coefficients

The linear attenuation coefficient μ is the sum of all the various contributions,

$$\mu = \sigma_a + \sigma_s + \sigma_{pe} + \sigma_{pp} \tag{2-41}$$

where σ_a and σ_s are the Compton absorption and scattering cross sections, respectively. Since each of these cross sections varies differently with energy and absorber composition, different effects predominate in different regions of the photon spectrum. This is shown in Figure 2.11 for the main range of nuclear photon energies.

The situation is different if one looks at the Z-dependence of the various effects. It is usually more convenient to work with mass attenuation coefficients $\mu' = (\mu/\rho)$ which have the advantage of being independent of the actual density and physical state of the absorber. For instance, Compton mass absorption coefficients are nearly independent of Z because the number of electrons per gram is almost constant for $Z > 1$.

Figures 2.12-2.15 present mass attenuation coefficients for photons for water, aluminum, sodium iodide and lead (after Evans, 1955). Note the dominant role of the photoelectric effect ($\sigma_{pe} = \tau$) at low energies and the relatively high energy, >> 1.02 MeV, before pair production ($\sigma_{pp} = \kappa$) makes a significant contribution. Rayleigh scattering is not discussed here as its contribution invariably is only minor.

In summary, gamma-rays are usually detected by observing the electrons they produce. These have a fairly well-defined energy if produced by the

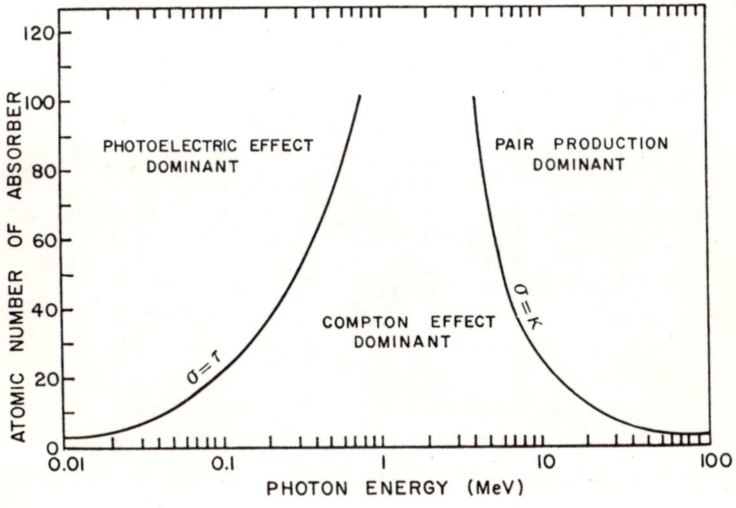

Figure 2.11 Relative importance of the three major types of γ-ray interaction as a function of photon energy and atomic number.

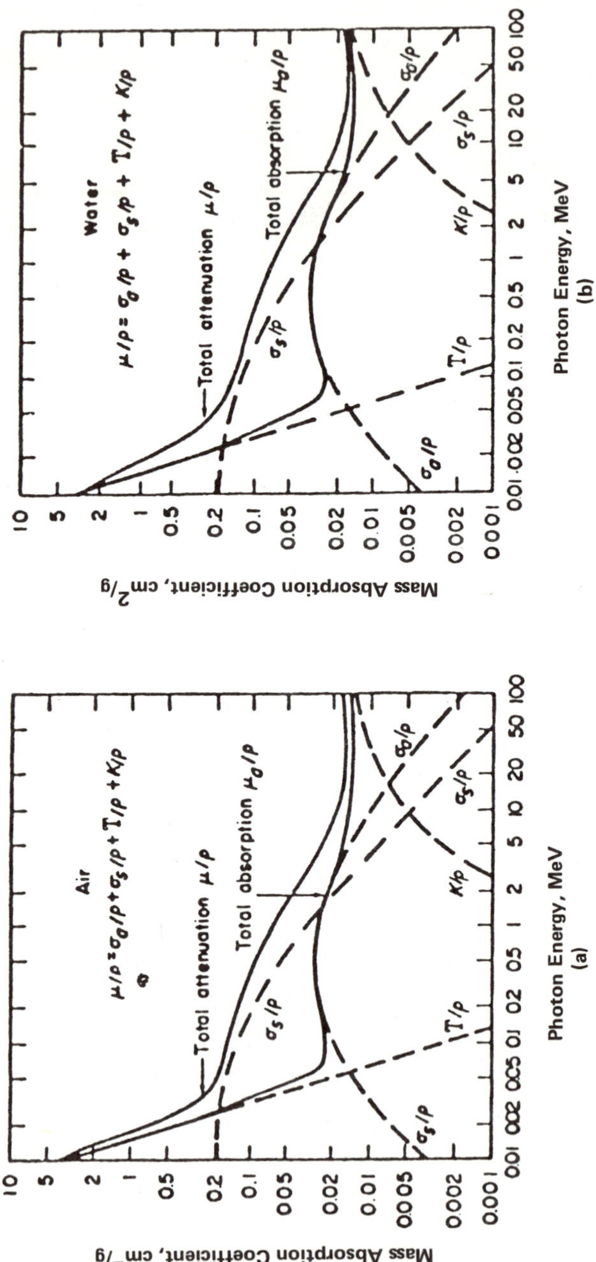

Figure 2.12 Mass attenuation coefficients of photons in air and water.

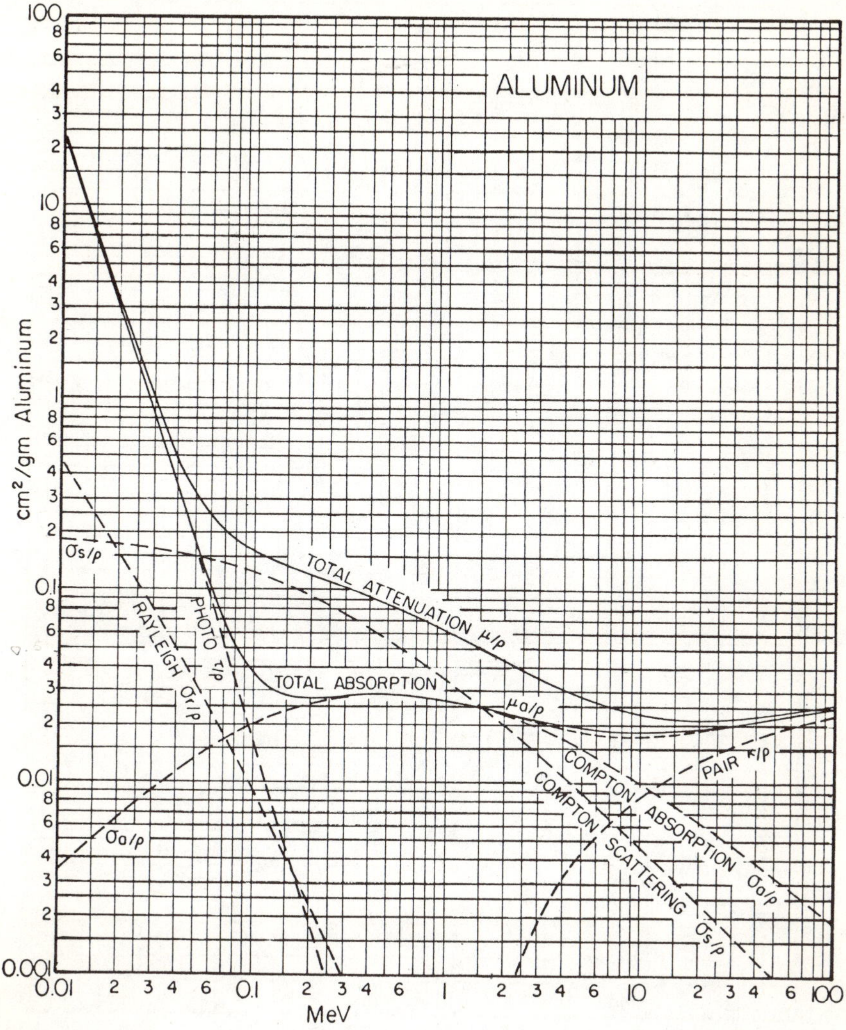

Figure 2.13 Mass attenuation coefficients of photons in aluminum.

Compton effect or pair production, though in the latter case the annihilation photon itself may again produce a photoelectron.

NEUTRONS

Neutrons, being uncharged particles of mass ~1 amu, interact only with the nuclei of target materials. Because of the small size of the nuclei

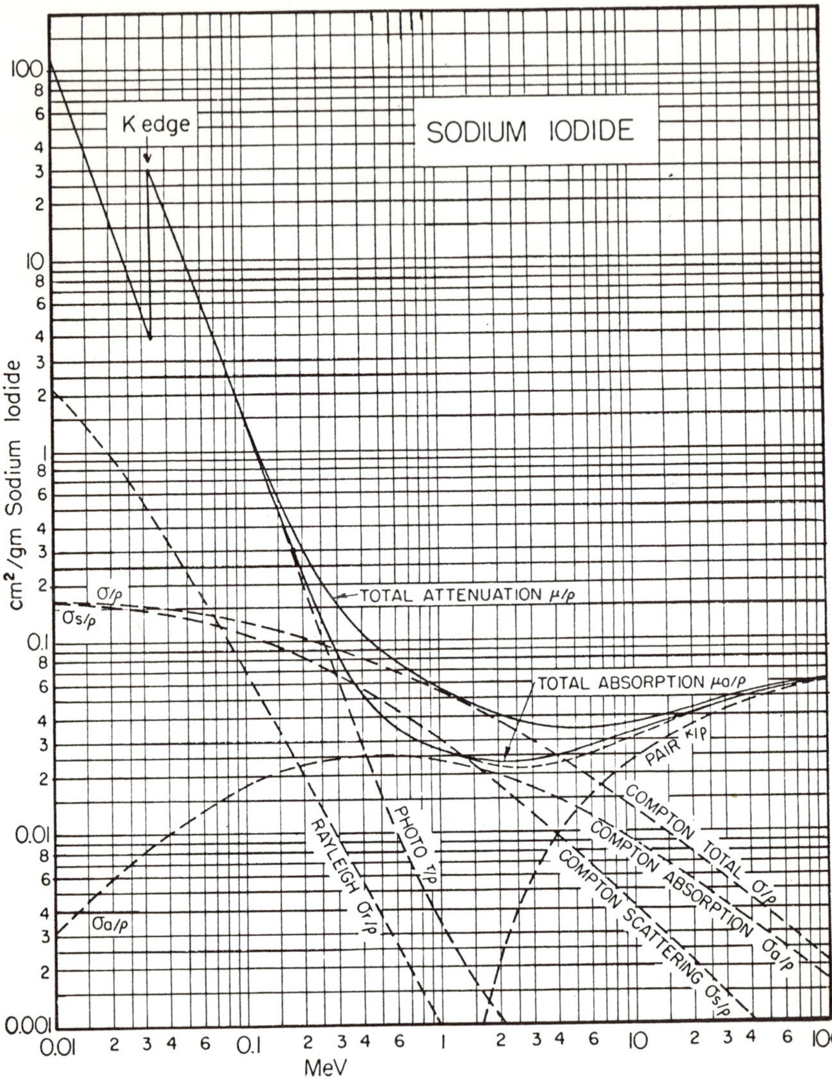

Figure 2.14 Mass attenuation coefficients of photons in sodium iodide.

compared with atomic dimensions, the probabilities for all types of neutron interactions are quite small. These cross sections are usually expressed in terms of a special unit, the barn; 1 barn = 10^{-24} cm^2 = 10^{-28} m^2. As in the case of photons, neutrons do not ionize atoms themselves and have to be detected by means of ionizing events caused by secondary particles.

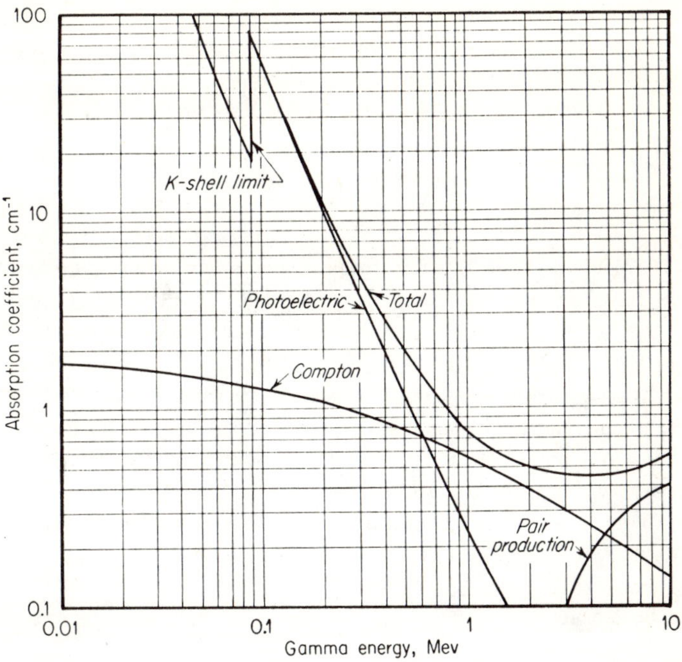

Figure 2.15 Mass attenuation coefficients of photons in lead.

Such secondary particles can be caused by elastic scattering collisions, inelastic scattering, capture events and, at very high energies, spallation.

In elastic scattering the neutron bounces off the struck atom in a billiard ball collision, transferring in the process an amount of energy

$$E_p = E_n \frac{4\ M_1\ M_n}{(M_p + M_n)^2} \cos^2\theta \tag{2-42}$$

where θ = the angle of recoil in the laboratory system ($\theta = (\pi - \xi)/2$, where ξ is the scattering angle in the center of mass system)
M_n = mass of neutron
M_p = mass of (primary) hit atom
E_n = neutron energy

If scattering is isotropic in the center-of-mass system, *i.e.*, all recoil directions are equally probable, the average energy transferred to the struck atom would be

$$\bar{E}_p = \frac{2\ E_n\ M_1\ M_n}{(M_p + M_n)^2} \tag{2-43}$$

For hydrogen whose mass very nearly equals that of the neutron

$$\bar{E}_p = \frac{E}{2} \tag{2-44}$$

i.e., on the average, half the neutron energy is transferred and would appear as recoil energy of the proton. Actually, for hydrogen the recoil angle in the laboratory system of coordinates will be nearly half the scattering angle in the center-of-mass system, *i.e.*, θ ranges from 0 to $\pi/2$. For heavier atoms $\theta = \xi$ and ranges from 0 to π. In condensed systems, the recoil energy is balanced by the crystalline binding energy of the struck atom and, hence, significant recoil effects from scatterers become observable only for epithermal and fast neutrons, $E_n > 300$ eV. In general, macroscopic scattering cross sections decrease slowly with energy as shown in Figure 2.16.

Inelastic Scattering

If a neutron is absorbed and re-emitted without conservation of energy and momentum, it is said to have undergone inelastic scattering, or an (n,n′) reaction. The excited nucleus will emit a high-energy gamma-ray within a short time interval. For most elements, (n,n′) reactions are threshold reactions, that is, they do not occur below certain minimum neutron energies, usually above several MeV. The cross sections for these reactions are low and they play only a minor part in detection processes.

Capture Processes

Because of the absence of a repulsive Coulomb potential, neutrons are readily absorbed by most nuclides, with a probability close to the geometrical cross section, of the order of a few barns. The simplest process retains the neutron, with the prompt emission of a capture gamma-ray, which represents the difference in energy state of the target atom before capture and the lowest energy state of the new compound nucleus, now one mass unit heavier. Such reactions are called (n,γ) reactions and for most nuclides they occur without threshold, with a cross section σ_c that varies as $1/v$, *i.e.*, $\sigma \propto E^{-1/2}$. In the epithermal range, capture cross sections may undergo abrupt fluctuations, as shown in Figure 2.16; that region is referred to as the resonance absorption region.

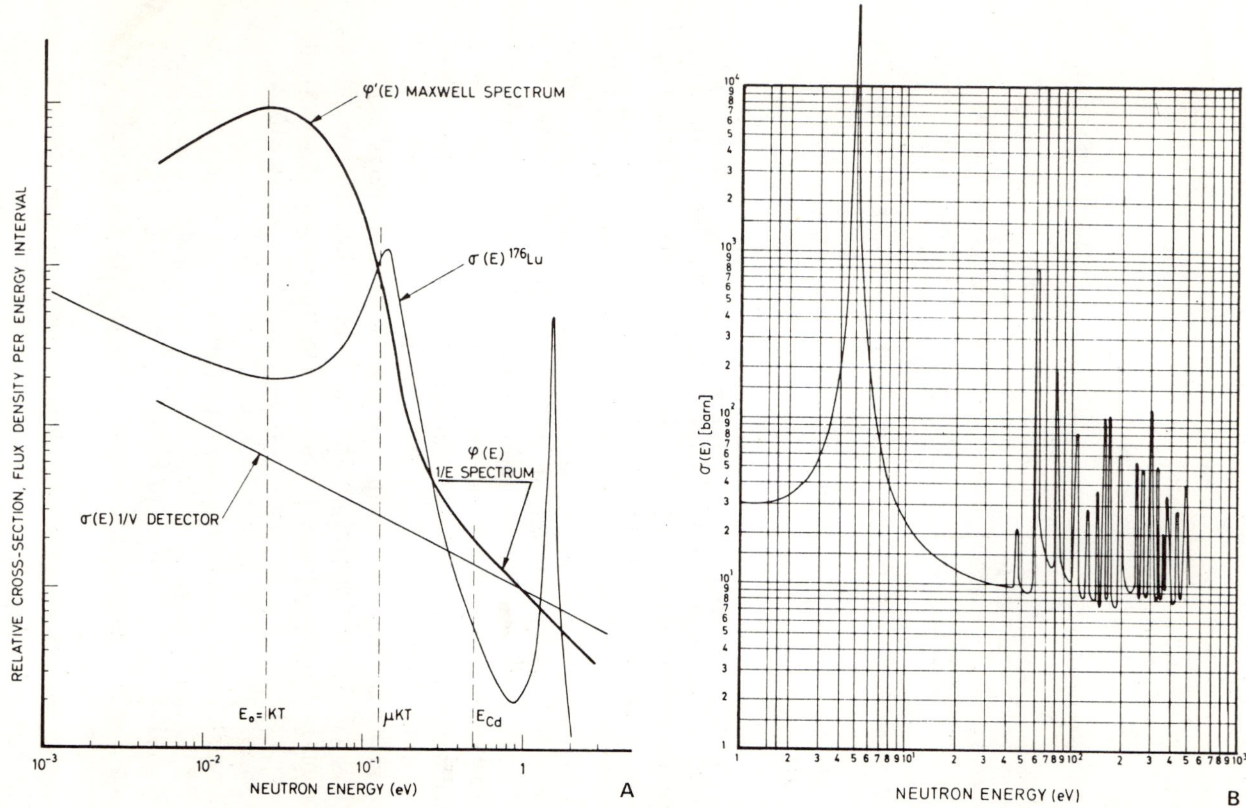

Figure 2.16 Typical neutron cross section curve (Au): (a) total scattering; (b) capture, thermal and resonance regions.

If the compound nucleus remains excited, de-excitation may result in the emission of another particle, either immediately or after some delay. The capture process is then referred to as *activation,* with a half-life and decay mode that may be characteristic of the activated nuclide. The sequence of events may be depicted as:

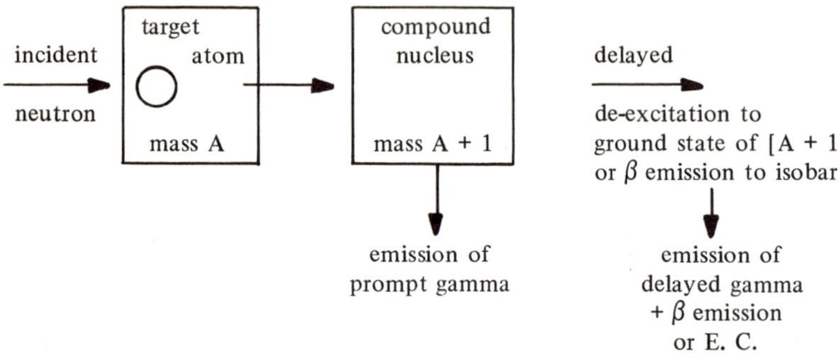

The process of activation, involving target materials with high-capture cross sections and product nuclides of convenient half-lives and emissions, is an important technique in neutron measurement methodology.

Two materials, indium and cadmium, are of particular importance in the measurement of neutron spectra and intensities because their capture cross sections show an unusually sharp "cut-off". This is shown in Figure 2.17. As one can see, the cross section curve for ^{115}Cd (n,γ) ^{116}Cd exhibits a sharp drop at 0.144 eV; below that energy, cadmium above a few microns in thickness is virtually opaque to thermal neutrons; above it the neutron capture is minimal. This property is the basis for a convenient means of classifying neutrons as thermal, if their energy lies essentially below the cadmium cut-off, and as epithermal if above. In this way cadmium foil can serve as a filter of epithermal neutrons and as an effective shield for thermal neutrons as σ_{th} = 2,440 barns. Cd-116 is stable so that there is no induced delayed activity.

Indium has a sharp resonance absorption peak at 1.8 eV for the ^{115}In (n,γ) ^{116m}In reaction, which permits similar use of indium foils to separate the energy region above and below that energy. A Cd-In foil sandwich can also be used to define a neutron energy window in the low epithermal range.

At higher epithermal neutron energies, nuclides exhibit pronounced resonances in their cross-section curves, permitting their use to detect preferentially neutrons in specific energy ranges. Table 2.4 lists some materials that can be employed as resonance foil activation detectors for

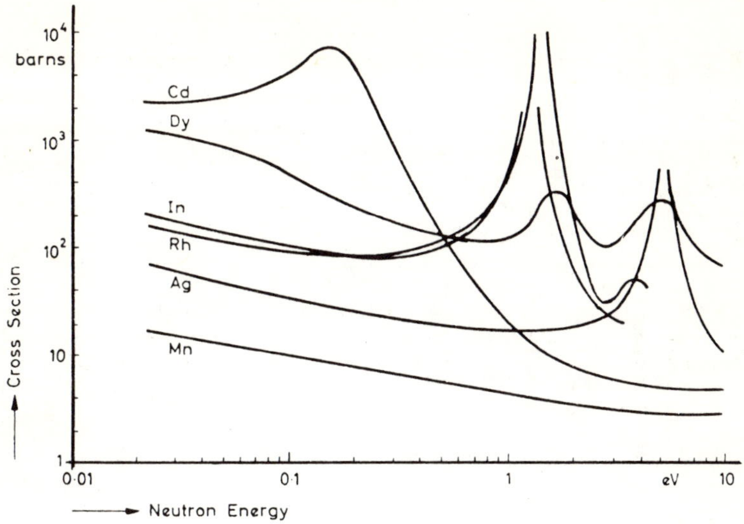

Figure 2.17a Cross section of various resonance detector materials as a function of energy (Beckurts and Wirtz, 1964).

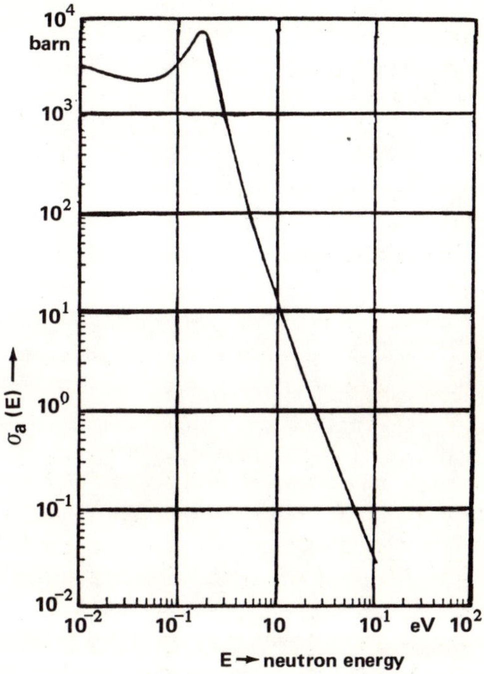

Figure 2.17b Absorption cross section of cadmium.

Table 2.4 Characteristics of Various Probes for the Detection of Thermal and Intermediate Neutrons [after Dennis]

Nuclear Reaction	Isotope Fraction (%)	Cross-section (barn): Thermal (0.025 eV)	Resonance Integral: Total	Resonance Integral: 1/E Fraction	Resonance Energy (eV)	Half-life of Induced Activity	Emission per Disintegration	Activation (disintegrations/min g rad): 1/E Neutrons	Activation (disintegrations/min g rad): Thermal Neutrons
In:									
$^{115}In(n,\gamma)^{116m}In$	96	150	2,640	87	1.45	54 min	0.6 MeV β^-, 100%	2.4×10^7	1.5×10^8
							1.27 MeV γ, 75%	1.8×10^7	1.2×10^8
$^{115}In(n,\gamma)^{116}In$	96	52	—	—	—	13 sec	3.3 MeV β^-, 100%	—	1.3×10^{10}
$^{113}In(n,\gamma)^{114m}In$	4	56	1,050	26	2.4	50 d	0.2 MeV γ, 97%	3.8×10^2	1.7×10^3
$^{113}In(n,\gamma)^{114}In$	4	2.0	—	—	—	72 sec	>0.7 MeV β^-, 100%	—	4.0×10^6
Au:									
$^{197}Au(n,\gamma)^{198}Au$	100	98.9	1,556	45	4.9	2.7 d	0.96 MeV β^-, 99%	1.47×10^5	8.5×10^5
							0.41 MeV γ, 100%	1.47×10^5	8.5×10^5
Ag:									
$^{107}Ag(n,\gamma)^{108}Ag$	51	45	74	13	16.5	2.3 min	1.8 MeV β^-, 97%	1.1×10^7	6.0×10^8
$^{109}Ag(n,\gamma)^{110m}Ag$	49	3.2	—	—	—	253 d	0.1 MeV β^-, 100%	—	2.6×10^2
							0.66 MeV γ, 100%	—	2.6×10^2
$^{109}Ag(n,\gamma)^{110}Ag$	49	87	1,160	37	5.2	24 sec	2.2 MeV β^-, 99%	9.5×10^8	6.0×10^9
							0.66 MeV γ, 60%	5.7×10^8	3.6×10^9
Zn:									
$^{64}Zn(n,\gamma)^{65}Zn$	49	0.44	—	—	—	245 d	1.1 MeV γ, 44%	—	28
$^{68}Zn(n,\gamma)^{69}Zn$	19	0.097	—	—	—	13.8 h	0.4 MeV γ, 100%	—	2.3×10^3
$^{68}Zn(n,\gamma)^{69}Zn$	19	1.0	—	—	—	57 min	0.9 MeV β^-, 100%	—	3.4×10^5
$^{70}Zn(n,\gamma)^{71}Zn$	0.6	0.085	—	—	—	2.2 min	2.4 MeV β^-, 100%	—	2.4×10^4
							0.5 MeV γ, 100%		2.4×10^4
Cu:									
$^{63}Cu(n,\gamma)^{64}Cu$	69	4.5	4.4	1.9	590	12.8 h	0.6 MeV β^+, 19%	8.1×10^2	8.1×10^4
$^{65}Cu(n,\gamma)^{66}Cu$	31	1.8	2.2	0.9	240	5.2 min	>1.5 MeV β^-, 99%	1.5×10^5	1.1×10^7
							1.0 MeV γ, 9%	1.4×10^4	1.0×10^7
							>0.65 MeV β^-, 100%	1.0×10^5	1.0×10^7
Mn:									
$^{55}Mn(n,\gamma)^{56}Mn$	100	13.2	11.8	5.9	337	2.6 h	0.85 MeV γ, 100%	1.0×10^5	1.0×10^7

this purpose. To separate the portion of the activation activities that are due to thermal and to resonance-energy neutrons, two identical detectors are used, one of which is covered by cadmium to remove the thermal neutrons, whereas the other is activated "bare." Corrections must be made for the residual contribution by cadmium to absorption in the resonance region.

Whereas activation cross sections for (n,γ) reactions may be significant at all energies, other capture processes, resulting in the emission of heavier particles, are possible if the incident neutron supplies enough energy to overcome the binding energy for the escaping particle. Thus, for fast neutrons, with thresholds above 1 MeV, (n,2n), (n,p), (n,α), (n,αt) etc., reactions become possible leading to de-excitation of the compound nucleus with ionizing particle emission. Three important exceptions should be noted here: the following reactions occur without threshold and have significant cross sections at thermal neutron energies (E_n = 0.024 eV).

$^{6}Li\ (n,\alpha)\ ^{3}H$	σ_{th} = 940b	abundance 7.5%
$^{10}B\ (n,\alpha)\ ^{7}Li$	σ_{th} = 3836b	abundance 20%
$^{14}N\ (n,p)\ ^{14}C$	σ_{th} = 1.81b	abundance 99.6%

These reactions are widely employed for thermal neutron detection, since they lead to heavily ionizing particles. The nitrogen reaction can be an annoying background effect in gas-filled detectors and is of significance in tissue-dose determinations, where the 0.58-MeV proton produced loses its energy over a range of less than 10 μm. The other two reactions find use in gas-filled and solid-state detectors.

For measurements on fast neutrons which frequently must be carried out against a background of slow neutrons from reactor operations or from scattered neutrons, the use of threshold-activation foil detectors is often convenient. These reactions have an effective threshold slightly above the minimum possible energy, to permit the product particle to escape, and have cross sections of the order of tens of millibarns. By selection of convenient threshold energies and product half-lives, a sequence of foils can be established that will cover the fast neutron region above 1 MeV in MeV intervals. Table 2.5 lists some of the more frequently used reactions. For convenience, these materials can also be incorporated into scintillator materials (Chapter 5).

In addition to the above threshold-capture reactions, many fission reactions fall into that category. In fission reactions, the incident neutron is absorbed and the consequent nuclear excitation leads to a splitting of the compound nucleus into two or more nuclear fragments with the additional emission of prompt gamma radiation. In the best-known case, the

Table 2.5 Threshold Reactions for the Detection of Fast Neutrons (NBS Handbook 63)

Nuclear Reaction	Isotope Fraction	Threshold Energy (MeV)	Average Cross-Section for Fission Neutrons (barn)	Half-Life	Emission per Disintegration	Activation (Disintegrations per min-g-rad) for Fission Neutrons
$^{31}P(n,p)^{31}Si$	100%	2.7	0.030	2.6 h	1.5 MeV β, 100%	7.5×10^{2}
$^{32}S(n,p)^{32}P$	95%	2.8	0.065	14.2 d	1.7 MeV β, 100%	10.9
$^{24}Mg(n,p)^{24}Na$	79%	7.2	0.0013	15 h	1.4 MeV γ, 100%	5.44
					2.8 MeV γ, 100%	
$^{27}Al(n,\alpha)^{24}Na$	100%	7.3	0.0006	15 h	1.4 MeV γ, 100%	2.84
					2.8 MeV γ, 100%	
$^{27}Al(n,p)^{27}Mg$	100%	4.7	0.032	9.5 min	1.0 MeV γ, 30%	—
					0.84 MeV γ, 70%	1.44×10^{4}
$^{58}Ni(n,p)^{58}Co$	67%	2.8	0.111	71 d	0.8 MeV γ, 99.5%	1.37
$^{115}In(n,n')^{115m}In$	96%	1.2	0.174	4.5 h	0.34 MeV γ, 50%	3.09×10^{2}

Neutron Threshold Detectors in Scintillators (after Cowan and O'Brien)

Nuclear Reaction	Threshold Energy (MeV)	Detector Material	Minimum Detectable Neutron Flux Density (n/cm^2 sec)
$^{238}U(n,f)$	1.1	Uranium nitrate	6.6
$^{31}P(n,p)^{31}Si$	2.0	$NH_4H_2PO_4$ in "phoswich"	200
$^{32}S(n,p)^{32}P$	2.0	Sulfur	230
$^{107}Ag(n,2n)^{106}Ag$	9.6	Silver orthophosphate in "phoswich"	350
$^{127}I(n,2n)^{126}I$	10	NaI(Tl) crystal	20
$^{12}C(n,2n)^{11}C$	20	Anthracene	2.3
$^{209}Bi(n,f)$	50	Bismuth nitrate	

fission of uranium-235, the fission results in the production of two heavy fission products of unequal mass, heavily ionized and usually decaying down an isobaric chain, with beta emission, to the most stable isobar. Also produced are 2 or 3 fast neutrons with an average energy of 2.6 MeV. Whereas the $^{235}U(n,f)$ reaction will take place with thermal neutrons, most fission reactions require a minimum energy to occur. Table 2.4 also lists some of the fission reactions that are useful in neutron-energy determinations and in energy-selective measurements. In many applications the fissionable materials are used as "converters" or "radiators" where neutrons are detected by observing the ions produced by the ejected fission products.

Spectrum Hardening

For electromagnetic radiations the reduction in mean energy through the Compton effect leads to a gradual reduction in linear energy transfer by all mechanisms. A similar situation exists for neutrons, but here, because of the existence of absorption resonances and the small cross sections for fast neutron absorption, the effective energy spectrum within a scattering medium can vary in a very complex fashion. Once a neutron has been slowed by one or two collisions it tends to get thermalized or captured rather rapidly, whereas an uncollided neutron retains its high energy and its low probability of being moderated. As a result the spectrum tends to get "hardened", *i.e.*, of the original Maxwellian energy distribution that one may obtain from fission, the more energetic portion may remain, though reduced in number, with the balance being degraded all the way to thermal energies. This spectral hardening has important consequences for shielding purposes and for precise energy and cross-section determinations.

The interactions of neutral particles encountered in high-energy physics research usually have very low cross sections. The properties of most of these particles have to be derived from those of their charged, paired decay products which must then be detected, often very rapidly, through their ionizing properties and track characteristics.

CHEMICAL AND BIOLOGICAL EFFECTS OF ENERGY DEPOSITION

In the process of losing energy to an absorbing medium, a variety of direct and indirect effects occur that depend not only on the atomic composition of the absorber but also on its molecular makeup, its crystal structure, and the nature of surrounding material. Ionization of atoms and molecules may lead to the breaking of chemical bonds or the formation of new ones, the energy deposited in a complex molecule may lead to excitation to higher vibrational or rotational states, and the resulting rise in temperature in a larger assembly of irradiated molecules may result in more rapid chemical reaction rates or in faster molecular diffusion. Thus, a whole branch of chemistry has grown around such radiation-induced or radiation-affected reactions; it is called *radiation chemistry*. Several radiation-chemical reactions, such as those leading to color changes or the formation of new compounds, that were absent from the initial target material, have found applications in radiation detection and monitoring. Some of them will be discussed in Chapter 8.

Many of the bombardment reactions lead to the production of secondary atoms, produced by the recoil of atoms of the target material in scattering events. In other cases, such as the capture of neutrons, energetic heavy charged ions are introduced into the medium causing a variety of secondary events, but always resulting in a transfer of energy from the incident particle to the target medium, with a steady, often indirect, degradation of the initial kinetic energy to heat, *i.e.*, vibrational energy of the target atoms. The amount of energy thus transferred to the medium and converted to chemical energy or heat is usually called the *dose* to that medium. More formally, the absorbed dose D is defined as

$$\text{absorbed dose D} = \frac{\Delta E}{\Delta m} \qquad \text{(2-45)}$$

i.e., the dose equals the change in kinetic energy per unit mass observed as the radiation traverses the material. Since most chemical, physical and biological effects in the target material depend on this quantity, the determination of dose is a major field of radiation measurement and various instruments have been designed specifically for this purpose. They are called *dosemeters* or *dosimeters.*

One of the major applications of dosimetry is for the determination of dose in human tissue. Although the chemical composition of tissue, made up as it is of hydrogen, carbon, oxygen and a variety of lesser components, appears to vary little from one form of tissue to another, the highly specialized nature of the molecules forming the various organs and tissues makes them susceptible to radiation to a varying degree, with more highly specialized material and cells with high mitotic activity being particularly susceptible to damage.

Since the mode of energy transfer [(dE/dx) or LET] depends both on the nature, charge and energy of the incident radiation and on the nature and vulnerability of the target material, any actual health effects will not necessarily be proportional to the measured dose. Instead it is customary to express the effective dose, *i.e.*, the relative effectiveness of the absorbed energy in causing a certain biological effect, in terms of the dose equivalent, H.

It is defined by

$$H = D \times Q \times N \tag{2-46}$$

where D = absorbed dose
Q = "quality factor"
N = "distribution factor"

The quality factor, Q, also related to the "relative biological effectiveness (RBE)," compares the relative efficiency of different types of radiation in causing comparable health effects: Q = 1 for 200 kV X-rays. The distribution factor N is a product of various modifying factors to allow for non-uniformity of distribution of a given radiation source in relation to the organ under observation and other geometry-sensitive effects. Table 2.6 compares these factors for the principal radiation types.

Table 2.6 Representative LET and RBE Values

Radiation	Energy (MeV)	Av. LET (keV/μ)	RBE	Quality Factor
X-Rays, 200 kVp	0.01-0.2	3.0	1.00	1
Gamma Rays	1.25	0.3	0.7	1
	4	0.3	0.6	1
Electrons (β)	0.1	0.42	1.0	1
	0.6	0.3	1.3	1
	1.0	0.25	1.4	–
Protons	0.1	90	–	6
	2.0	16	2	10
	5.0	8	2	10
Alpha Particle	0.1	260	–	–
	5.0	95	10-20	10
Heavy Ions	10-30	~150	~25	20
Neutrons	thermal		4-5	3
	1.0	20	2-10	10

These values are general and approximate. RBE and QF values vary widely with different measures of biological injury.

The absorbed dose is measured in grays or rads,

$$1 \text{ gray (Gy)} = 1 \text{ joule/kg} = 100 \text{ rads}$$

$$1 \text{ rad} = 100 \text{ ergs/g tissue}$$

The unit of dose equivalent was defined originally as

$$1 \text{ rem} = 1 \text{ rad} \times \text{QF}$$

It clearly has the same dimensions as the rad but now is determined by use of Equation 2-46. For X- and gamma-rays the rad and the rem are close, numerically. The SI unit is the *sievert* (Sv) which is related to the dose in grays by Equation 2-46. The dose to tissue arising from an X-ray exposure of 1 roentgen (1 R) is about 93 ergs/g tissue, not much below one rad. Since for X-rays this also equals the dose equivalent of 1 rem, there has been a tendency to use exposure measurement, in roentgens, as dose indications in rems. Although convenient, this should be considered improper, since both the dose and the dose-equivalent vary strongly with energy; hence, dosimeter calibrations are valid only for a specific radiation type and energy range.

As both dose and dose-equivalent depend on the nature of the tissue irradiated and on target geometry, it is often found that the introduction of the dosimeter may affect the measurement by additional scattering or absorption effects in the instrument materials. For this reason much effort has been devoted to the development of "tissue-equivalent" dosimeters, whose composition and structure do not change any of the dose and exposure conditions. Typically a tissue-equivalent detector should be so small as to cause negligible disturbance or it should be constructed of materials whose atomic composition closely matches that of the surrounding tissue. Such instruments are discussed in Chapter 12. It should be remembered that tissue equivalence for X- or γ-radiation, which depends mainly on electron density, is not the same as tissue equivalence for neutrons, which depends on capture and scattering cross sections.

A device which simulates the human body to permit dose measurement and dose planning in a given radiation field is called a "phantom." Phantoms are widely used in medical therapy applications.

Radiation exposure may be *chronic* or *acute*. Chronic exposure implies continued exposure over long periods of time leading to a given total dose value. Acute exposure refers to sudden, perhaps massive, radiation exposure, such as might arise in radiation therapy or a nuclear explosion. In the same way, dosimeters have to be designed to operate at low or high dose levels, to have rapid response times or to integrate and hold the reading over long periods. No one instrument, in general, can be expected to meet all requirements and a careful selection must be made to obtain the desired sensitivity and performance. Some instruments must be capable of recording dose accurately in mixed fields, *i.e.,* in the presence of both neutrons and gamma-rays, while others are expected to discriminate, with only low sensitivity to the interfering radiations. In some cases different absorbers may be introduced to distinguish between effects from different radiations, such as in the "film badge" shown in Figure 2.18. The lead foils filter out X-rays and γ-rays of various energy ranges, while the indium foil discriminates against thermal neutrons and the plastic sheet slows down fast neutrons.

Calibration of a dosimeter must be done under appropriate conditions and for the correct energy range. Dose-equivalent readings themselves are no direct indication of any health effects resulting from exposure; it is important to maintain the distinction between exposure measurements and dose readings.

RADIATION DAMAGE

In solid materials, where the atoms are arranged in a three-dimensional crystal structure, reversible and irreversible changes in the physical structure may be introduced by the deposition of energy by high-LET particles. Most of these changes arise through the physical displacement of lattice atoms as energetic particles hurtle through the material and by the motion of recoiling atoms from elastic or inelastic collisions. These collisions, which can be initiated by fast neutrons, energetic protons, alpha particles and other charged heavy particles, such as fission products, will impart to the hit atom the kinetic energy given by Equation 2-42. If that energy is

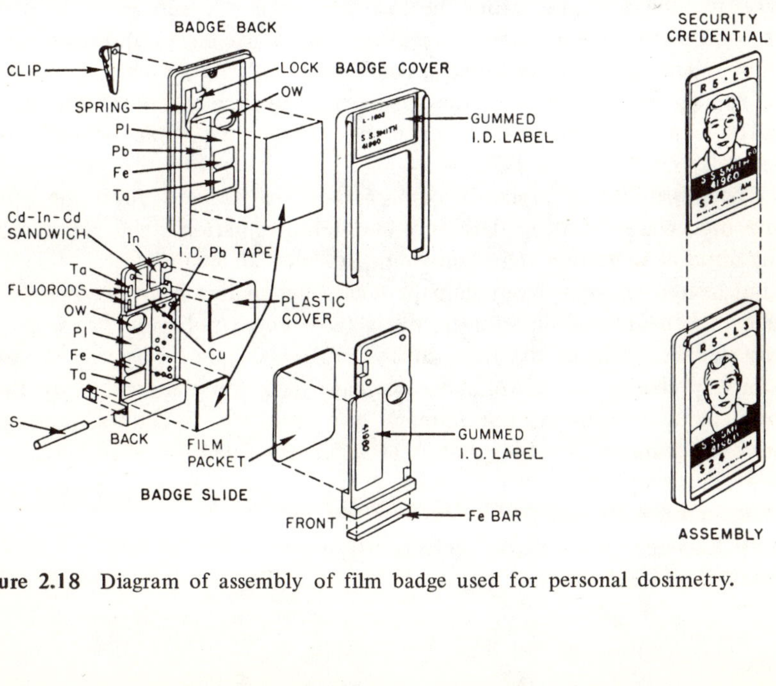

Figure 2.18 Diagram of assembly of film badge used for personal dosimetry.

in excess of the displacement energy, typically > 20-30 eV, the recoiling atom itself can move through the lattice undergoing further collisions and creating secondary recoils of its own. If the initial energy transfer is high enough, especially in low-Z materials, a cascade of secondary, tertiary and higher-order collision events will occur that may involve thousands of atoms in a relatively small volume. If only a few atoms are displaced, they can usually be induced to find a vacant lattice position by heating the material; this is called "annealing" the defect. Radiation damage may be caused by the presence of a large number of irregular, interstitial atoms and represents a reversible defect if, in fact, all the displaced atoms are annealed out. If, on the other hand, a very large number of atoms are hit and displaced, the original lattice structure may be lost and the resultant defect may be irreversible. Such a region whose properties have been severely modified by radiation-induced displacements is called a *displacement spike*. Displacement spikes can be observed in mica, some plastics, glass and other transparent materials along the path of fission products, energetic cosmic rays and alpha-particles, and in some minerals give rise to so-called "pleochroic haloes."

In some materials, especially those with low melting points, the highly localized deposition of energy by strongly interacting particles over a short particle range can, through the dissipation of this energy through the slowing down of the cascade atoms, inject enough heat into that lattice volume that the temperature may exceed the melting point or a phase transition temperature before the heat can be conducted away. This may result in a irreversible phase change, perhaps local recrystallization on cooling; such a region is called a *thermal spike*. The formation of irreversible damage regions, such as the spikes, represents another interaction and energy transfer mechanism that may be utilized in certain detection methods (see Chapter 7).

The possibility of radiation damage may also severely limit the lifetime of certain radiation detectors in intense radiation fields. Many glasses will discolor in strong radiation fields and not all materials can tolerate being heated to anneal out damage. Glass dosimeters have been developed where the degree of discoloration is a measure of radiation exposure. In most semiconductors, displaced atoms introduce hole or electron traps; in high-level irradiations, radiation damage may completely destroy the initial detector properties. For many semiconductor detectors, the maximum exposure to fast neutrons is at a fluence (nvt) value of 10^{14} – 10^{15}, *i.e.,* such a detector would last only about 100 sec if placed into a fast neutron flux of 10^{13} neutrons/cm^2-sec. Similar problems arise in many high-energy research applications and must be carefully guarded against.

REFERENCES FOR FURTHER READING

Attix, F. H., and W. C. Roesch, Eds. *Radiation Dosimetry,* Vol. 1, 2nd ed., (New York: Academic Press, Inc., 1968).

Beckurts, K. H., and K. Wirtz. *Neutron Physics* (New York: Springer Verlag, 1964).

Evans, R. D. *The Atomic Nucleus* (New York: McGraw-Hill Book Co., 1955).

Friedlander, G., J. W. Kennedy and J. M. Miller. *Nuclear and Radiochemistry,* 2nd ed., (New York: John Wiley & Sons, Inc., 1964).

Grodstein, G. W. "X-ray Attenuation Coefficients from 10 keV to 100 MeV," NBS Circular 583, National Bureau of Standards, Washington, DC (1957).

Kiefer, H. and R. Maushart. *Radiation Protection Measurement* (Oxford: Pergamon Press, 1972).

Protection Against Neutron Radiation up to 30 Million Electron Volts. NBS Handbook 63. National Bureau of Standards, Washington, DC (1957).

PROBLEMS

2.1 Determine the maximum range in air and in aluminum of the following 6-MeV particles: alpha particles, protons, deuterons and electrons.

2.2 Explain the term "infinite thickness." Calculate the thickness of aluminum that would be infinitely thick for 2-MeV β particles and 5-MeV α particles.

2.3 A 500-keV photon is scattered through a 60° angle during a Compton interaction. What are the frequency, wavelength and energy of the scattered photon?

2.4 Calculate the maximum energy loss during Compton scattering of (a) a 50-keV X-ray photon; (b) a 2-MeV γ-ray photon.

2.5 Estimate the rate of energy loss (LET) as a 1-MeV proton passes through a gas-filled chamber. If W = 35 eV for that gas, how many ion pairs may be produced?

2.6 Repeat the previous calculation for a 1-MeV α particle.

2.7 Calculate the attenuation to a collimated beam of 0.8-MeV photons in passing through 3 cm of iron. (μ_a = 0.54 cm^{-1}).

2.8 The half-value thickness of concrete for 150 kVp X-rays is 2.3 cm. What fraction of a collimated beam of such X-rays remains after passage through a concrete thickness of 11.5 cm and 23 cm?

2.9 Iodine-122 is a 3.2-MeV positron emitter. What would be the energy and source intensity of the gamma-ray emission from a thick 10-kBq source of this material?

2.10 Calculate the thickness of natural cadmium metal that would be effectively opaque to thermal neutrons.

2.11 Compute the energy loss of a 2-MeV beta particle in passing through a 2 mm thick slice of intrinsic silicon. Assuming a W of 3.65 eV per carrier pair, how many pairs would be produced per incident particle? How many charge carriers would be produced by a beam current of 1 mA or 2-MeV electrons under those conditions assuming a carrier lifetime of 500 μsec?

2.12 The Roentgen was defined as that amount of radiation that would produce one e.s.u of charge in 1 cm^3 of air at 0°C and 760 torr. Given that W = 34 eV per ion pair in air, show how the metric equivalent definition for the Roentgen is obtained.

2.13 Calculate the rate of consumption (burnup) of the boron in a 50 cm^3 BF_3 counter tube, maintained at 300 torr, when exposed to a thermal neutron flux of 10^{14} n/cm^2-sec. (Cross section for $^{10}B(n,\alpha)$ is 3,840 barns; boron-10 abundance 19.6%.)

CHAPTER 3
COUNTING STATISTICS AND ERROR DETERMINATIONS

INTRODUCTION

In reporting the outcome of any laboratory measurements, it is incumbent upon the experimenter to report not only the results of his measurements, but also the accuracy with which they were made or, more appropriately, the confidence with which they are held to be valid. Consequently, it is important that the experimentalist be aware of the various sources of error in his work and able to assess their significance. The mathematical framework for this assessment is the science of statistics.

It is particularly important that statistics be applied to any laboratory work which includes radiation detection as part of its data collection, since the very process of radioactive decay is probabilistic in nature and the result of any counting experiment is perforce a random quantity. Since the result of any physical measurement can be considered as a random variable, the basic ideas developed in this chapter are not limited to measurements of radiation. They are applicable to all situations in which measurements are subject to some source of random error.

UNCERTAINTY IN THE MEASUREMENT PROCESS

The errors that accompany most laboratory measurements can be divided into two classes: systematic errors and random errors. Systematic errors result from faults in the measurement process or its interpretation that lead to a bias in the results. For example, if one measured the length of a table with a tape measure that had been stretched through misuse, the result would consistently be less than the actual length of the table. The measured length would be biased toward low values by the incorrect tape; the results would have a systematic error.

On the other hand, the proper use of accurately calibrated instruments is no guarantee that the result of a measurement will be errorless. For example, even if perfectly accurate tape measures were used, it is unlikely that forty people would agree to the nth significant figure on the length of a table. Individual variation in holding the tape, reading the scale, etc. would produce a range of results for the measured length, regardless of how carefully the measurements were made. However, it is unlikely that there would be any bias in this set of measurements. A result chosen from this group is likely to contain an error, but the error will be random, not systematic.

In the example just mentioned, the source of the error is the randomness inherent in the measurement process. It is also possible for a result to contain an error because the phenomenon measured is itself a random process. Radioactive decay is an obvious example. As a measurement process, counting the pulses that are generated in a radiation detector in a given length of time can be virtually error-free, but several consecutive counts taken in identical situations will generally be different, though they will be clustered around the "true value." Since any one of these counts might be taken as a measure of the activity of the source, it is apparent that the results of the counting process contain a random error, the source of which lies in the probabilistic nature of the decay process.

In the discussion that follows, it will be assumed that the error associated with a laboratory measurement is random in nature. However, it is immaterial whether the source of the error is the measurement process or the phenomenon being examined.

For the purpose of illustration, consider the problem of determining the half life of a specified radioisotope. So stated, the problem incorporates the assumption that there exists a "true" albeit unknown value for the half-life, τ_*. However, due to various sources of random error, any single measurement will in general deviate from the true value, so the task becomes one of providing *best estimates* of the true value and the size of the potential error.

Any single measurement of the half-life is one of a virtually infinite number of potential measurements. These potential measurements, if realized, could provide results which would be distributed in some fashion about the true value of the half-life, τ_*, as shown in Figure 3.1. This distribution may be characterized in a very crude fashion by specifying its *mean*, μ, and its *variance*, σ^2. The mean is a measure of the central tendency of the distribution and is defined as

$$\mu \equiv \lim_{N \to \infty} \frac{1}{N} \sum_{i=1}^{N} x_i \tag{3-1}$$

where x_i is the result of a single measurement of the half-life, and N is the large number of measurements possible. If the sources of error that generate

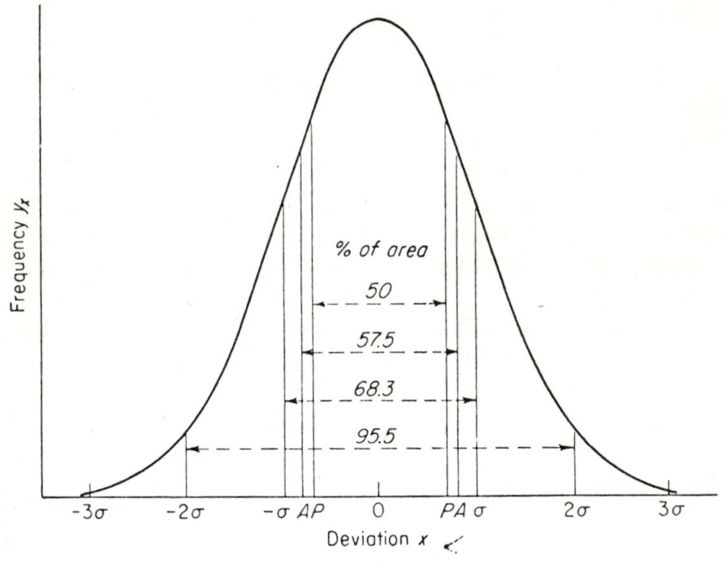

Figure 3.1 Example of a normal distribution of results.

this distribution are truely random, *i.e.*, there are no systematic errors, then it is reasonable to identify the mean, μ, with the true value of the half-life, τ_*.

The variance is a measure of the dispersion of the distribution about the mean and is defined by the relation

$$\sigma^2 \equiv \lim_{N \to \infty} \frac{1}{N} \sum_{i=1}^{N} (x_i - \mu)^2 \tag{3-2}$$

If the population of potential measurements has a *normal* distribution about the mean, μ, with variance, σ^2, *i.e.,* the probability of obtaining a result between x and (x + dx) is given

$$p(x)dx = \frac{1}{\sqrt{2\pi\sigma^2}} \exp\left[-\frac{(x-\mu)^2}{2\sigma^2}\right] dx \tag{3-3}$$

then the variance, or more appropriately, the *standard deviation,* σ, the square root of the variance, is a measure of the "width" of the distribution. In particular, the probability of obtaining a result between $(\mu - \sigma)$ and $(\mu + \sigma)$ is 0.6745. This range is indicated in Figure 3.1.

Therefore, the task of providing a best estimate of τ_*, the true value of the half-life sought, is accomplished by determining a best estimate of μ, the mean of the population of all potential measurements of the half-life. Furthermore, in order to provide a measure of confidence in the estimate of the mean, one must have a best estimate of σ^2, the variance of the population.

To determine μ and σ^2 directly by use of the defining Equations 3-1 and 3-2 is obviously impractical since it would require an infinite number of measurements. The trick is to determine estimates of μ and σ^2 from a limited number of measurements such as the half-life results listed in Table 3.1. Such a collection is called a *sample* and the estimates obtained from them are termed *sample statistics.* For a sample of size N, statistical theory indicates that the best, unbiased estimates of the population mean, μ, and variance, σ^2, are the *sample mean,* M, and the *sample variance,* s^2, defined as

$$M = \frac{1}{N} \sum_{i=1}^{N} X_i \tag{3-4}$$

and

$$s^2 = \frac{1}{(N-1)} \sum_{i=1}^{N} (X_i - m)^2 \tag{3-5}$$

Table 3.1 A Sample of Half-Life Measurements

Measurement Number	Half-Life Measured
1	5.26 yr
2	5.24
3	5.20
4	5.21
5	5.27

$$M = \frac{1}{N} \sum_{i=1}^{N} \tau_i = 5.236 \text{ yr}$$

$$s^2 = \frac{1}{(N-1)} \sum_{i=1}^{N} (\tau_i - M)^2$$

$$= \frac{1}{(N-1)} \left[\sum_{i=1}^{N} (\tau_i^2) - NM^2 \right] = 0.00093 \text{ (yr)}^2$$

$$S = \sqrt{S^2} = 0.030 \text{ yr}$$

$$S_M = \frac{S}{\sqrt{N}} = 0.014 \text{ yr}$$

respectively. Therefore, in the example of measuring a half-life, the sample mean calculated in Table 3.1 is the desired result. Given the limited knowledge contained in the sample of N measurement, each of which contains some random error, the sample mean is the best possible estimate of the true value of the half-life.

One cannot, however, infer that the sample variance, or its square root, s, the *sample standard deviation,* is the error associated with this result. This fact can be seen quite readily if the sample number N is allowed to get larger and larger. In that case the sample mean will be an increasingly better estimate of the true half-life and the error associated with this result should decrease uniformly. However, the sample variance will not decrease but will simply become a better estimate of the constant σ^2, the population variance, which might be quite large. The resolution of this problem lies in the interpretation of σ, the population standard deviation. Recall that σ is a measure of the width of the population distribution and, consequently, is an indicator of the error associated with a *single* measurement drawn from that distribution. Therefore, to determine the error associated with m, the mean of a sample of size N, one needs to know the width of the distribution of the population of such sample means, *i.e.,* the population of all the potential means calculated from random samples of N measurements drawn from the parent population. Such a distribution is illustrated in Figure 3.2.

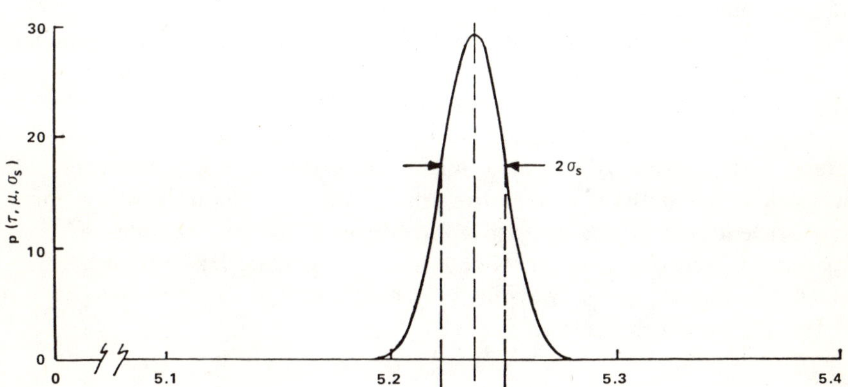

Figure 3.2 Distribution curve for sample means.

The theory of errors predicts that the resulting distribution will be characterized by a mean, μ_μ, equal to the mean of the parent population, and a variance, σ_μ^2, given by

$$\sigma_\mu^2 = \frac{\sigma^2}{N} \qquad (3\text{-}6)$$

The square root of this variance, σ_μ, is called the *standard error of the mean,* and represents the error associated with the mean, M. The best estimate of σ_μ is the quantity S_M, given by

$$S_M = \frac{S}{\sqrt{N}} \qquad (3\text{-}7)$$

where S is the sample standard deviation and N is the sample size.

Therefore, to report the result of an experiment, one takes several similar measurements, calculates (using Equations 3-4 and 3-5) the mean, M, and standard deviation, S, for this sample, and gives the result as $M \pm (S/\sqrt{N})$.

Two comments are in order concerning the procedure just described. First, the method is correct if applied to either a single measurement such as the reading of a dial or a whole experiment. However, when an experiment is comprised of many individual measurements, as they generally are, it is more appropriate to apply the method to each individual measurement and to determine the final error by the propagation of error techniques that will be presented in a subsequent section of this chapter. Such a procedure will maximize the information obtained from the raw data.

Secondly, the procedure calls for at least two separate and independent measurements of each quantity desired. In practice, the accuracy of a measurement is often limited by the precision of the measuring instruments as opposed to any source of random error. When such a case occurs, the several measurements of a sample might all be the same within the precision of the instruments used to record them. Instead of reporting zero error, it is conventional in this case to report the error as half the value of the smallest instrumental division. For example, a reading taken from a clock which has seconds as its finest division should not be reported with an error of less than one-half second.

THE STATISTICS OF COUNTING

Radiation detection instruments are used to measure processes involving radioactive decay and therefore such measurements are subject to the random fluctuations inherent in the decay process. In their simplest form, these measurements are called counting experiments since the data collected are merely the number of detection events occurring in a fixed time interval. When the random nature of the decay process is overwhelmingly the largest source of uncertainty in the measurement, the procedure for determining the best estimate of the true value of the quantity sought and its associated error can often be simplified because the population of all potential counts meets a particular statistical condition: it has a *Poisson distribution.* This property permits even a sample of *one* to be used to obtain a best estimate of both the mean and variance of the population.

In order to fully appreciate this result, consider a simple counting experiment in which a detection system records in a fixed time, t, the number of decays that occur in a certain amount of a single radioisotope characterized by an unknown decay constant λ. The objective of the experiment will be to determine the *expected* number of counts in the time t. This number might then be used in a variety of ways, *e.g.,* to calculate the activity of the source.

As discussed in the previous section, the procedure for obtaining the expected count is to take several counts of duration t, calculate the mean and variance of the sample, and report the results as described. However, the fact that the distribution of all potential counts is Poisson means that exactly the same information about the expected count and its associated error can be obtained from a single count of duration N x t as from N counts within the time interval t. Data acquisition is thereby simplified (in this case). A detailed examination of this simple experiment will reveal the mathematics underlying this rather surprising conclusion and will identify those situations for which it is valid.

First, assume that at the beginning of the count the samples of radioactive material consists of n nuclei of a single isotope characterized by a decay constant λ. The decay of any one of the nuclei during the count is totally independent of the fate of any other nuclei in the sample. The probability, q, that any given nucleus will survive the counting time, t, is easily seen to be

$$q = e^{-\lambda t} \tag{3-8}$$

Since one of two fates, decay or survival, must befall every nucleus, the probability, p, that a given nucleus will decay during the time, t, is obviously given by

$$p = 1 - q$$

or

$$p = (1-e^{-\lambda t}) \tag{3-9}$$

Hence, if the detection system is characterized by an efficiency, ϵ, less than one, where ϵ is defined as the probability that a decay will be registered as a count, then the probability of a nucleus producing a count is found to be

$$p = \epsilon(1 - e^{-\lambda t})$$

and the probability of a nucleus failing to produce a count becomes

$$q = 1 - \epsilon(1 - e^{-\lambda t})$$

Whether p and q refer to counts or decays, the following arguments are applicable.

This binary characteristic of the decay process coupled with the independence of decay events means that the probability P(k;n,p) of exactly k nuclei out of the original n decaying in the time t is given by the *binomial distribution,*

$$P(k;n,p) = \frac{n!}{k!(n-k)!} p^k (1-p)^{(n-k)} \qquad (3\text{-}10)$$

This expression is simply the product of the probability of k nuclei decaying, p^k, times the probability of (n-k) nuclei surviving, $q^{(n-k)}$ times the number of ways that this event can occur; *i.e.*, the number of different combinations of n nuclei divided into two groups of k and (n-k). The binomial distribution is applicable to any situation in which independent events have two potential outcomes. For example, Equation 3-10 is also the expression one would use to calculate the probability that exactly k heads would turn up if one tossed n coins, each of which had a probability p of coming up heads. Notice that the probability given in Equation 3-10 is just the kth term of the binomial expansion of the expression $(p+q)^n$; *i.e.*,

$$(p+q)^n = \sum_{k=o}^{n} P(k;n,p) \qquad (3\text{-}11)$$

Therefore, since (p+q) = 1, the sum of the probabilities over all possible outcomes gives unity as it should. The binomial distribution is illustrated in Figure 3.3 for several values of its two parameters n and p.

The above analysis indicates that the binomial distribution is the distribution of all the potential counts that could be taken during a time t. Therefore, the mean of this distribution is the result sought, the expected number of counts. An expression for the mean can be derived by use of the formula

$$\mu = \sum_{k=o}^{n} k\ P(k;n,p) \qquad (3\text{-}12)$$

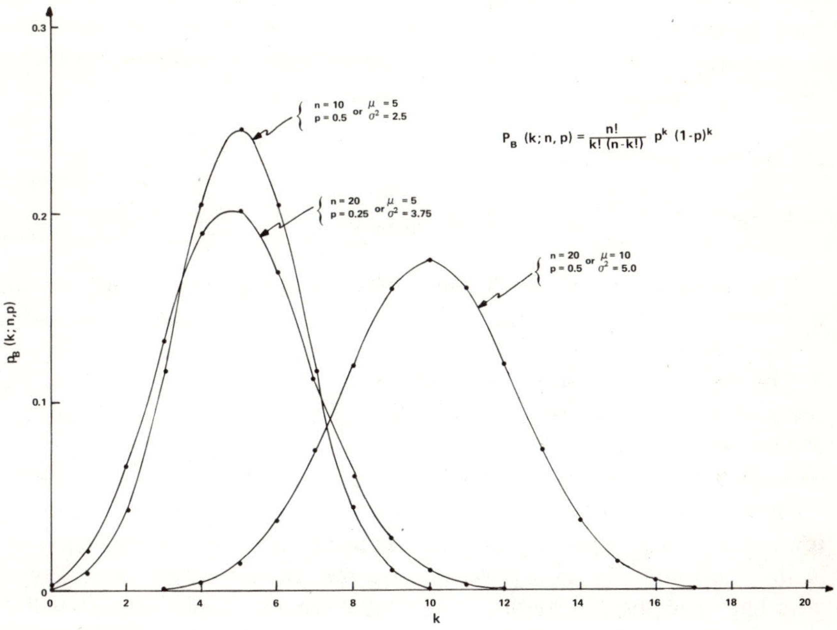

Figure 3.3 Binomial distribution curves for various values of n, μ, p and σ.

which follows from the definition of the mean (Equation 3-1), for the case in which the distribution is known. The substitution of the expression for P(k;n,p) from Equation 3-10 gives

$$\mu = \sum_{k=o}^{n} \frac{n!}{k!(n-k)!} kp^k q^{n-k} \qquad (3\text{-}13)$$

or

$$\mu = \left\{ x \frac{\partial}{\partial x} \left[\sum_{k=o}^{n} \frac{n!}{k!(n-k)!} x^k q^{n-k} \right] \right\}_{x=p} \qquad (3\text{-}14)$$

The sum is easily evaluated by use of Equation 3-11. Therefore,

$$\mu = \left[x \frac{\partial}{\partial x} (x+q)^n \right]_{x=p} \qquad (3\text{-}15)$$

and performance of the indicated differentiation gives

$$\mu = \left[xn(x+q)^n \right]_{x=p}$$

$$\mu = pn \qquad (3\text{-}16)$$

By a similar procedure, the expression for the variance of the population from Equation 3-2,

$$\sigma^2 = \sum_{k=o}^{n} (x - \mu)^2 \, P(k;n,p) \qquad (3\text{-}17)$$

can be evaluated to give

$$\sigma^2 = p(1 - p)n \qquad (3\text{-}18)$$

The constants μ and σ^2 can be used to characterize the binomial distribution instead of p and n. The curves in Figure 3.3 are labeled for both sets of parameters.

The binomial distribution then is the exact distribution of all potential counts for the simple experiment under consideration. It is a two-parameter distribution, depending on either the set p and n or the set μ and σ^2, and consequently a best estimate of these two parameters cannot be made from a sample of less than two separate counts. This situation is no different from that in which the distribution was assumed to be *normal.* However, in the usual counting experiment the number of radioactive atoms present, n, is huge and the probability, p, that any specific single nucleus will decay during the counting time is virtually zero. For example, in one microcurie of Co-60, there are about 8.88×10^{12} radioactive nuclei, while the probability of decay during a counting interval of one minute is about 2×10^{-7} for each nucleus. In a case such as this, the expressions 3-13 and 3-15 indicate that the mean and the variance of the distribution are *essentially the same number*:

$$\sigma^2 \simeq \mu = np \qquad (3\text{-}19)$$

This conclusion is the key result of the above analysis. For when p is so small that σ^2 and μ are essentially equal, the distribution of all potential counts can be characterized by a *single* parameter, μ, and a single sample count can be used to make a best estimate of that parameter. This estimate then serves as the best estimate of the variance as well as the mean.

In this limit of very small p and very large n, it is conventional to express the distribution of all possible counts, P(k;n,p), as a function of the single parameter, μ, instead of the set n and p. To this end, consider the limit of the binomial distribution, Equation 3-10, when p approaches zero and n approaches infinity such that the product np = μ remains finite. Since values of k near the mean, μ, are the ones of interest, it will be appropriate to assume that k is finite, *i.e.*, $k \ll n$, as well. The factor of Equation 3-10 that gives the probability of (n-k) nuclei surviving can be written:

$$(1 - p)^{(n-k)} = (1 - p)^n / (1 - p)^k = \left[(1 - p)^{\frac{1}{p}}\right]^{(np)} /(1 - p)^k \qquad (3\text{-}20)$$

and in the limit that $p \rightarrow 0$, the denominator goes to one, and the quantity in the square brackets becomes

$$\lim_{p \rightarrow 0} (1 - p)^{\frac{1}{p}} = \frac{1}{e}$$

Therefore, one obtains

$$(1 - p)^{(n-k)} \rightarrow e^{-(np)} \qquad (3\text{-}21)$$

In the limit that $n \rightarrow \infty$, the number of ways in which n objects can be divided into two groups of k and (n-k) becomes

$$\frac{n!}{k!(n-k)!} = \frac{n \quad (n-1) \ . \ . \ . \ (n - k + 1)}{k!} \qquad (3\text{-}22)$$

$$\rightarrow \frac{n^k}{k!}$$

Therefore, the binomial distribution approaches the form:

$$P(k;n,p) \approx \frac{n^k}{k!} p^k e^{-(np)} \qquad (3\text{-}23)$$

or, since $np = \mu$, the limiting distribution is

$$P(k;\mu) = \frac{\mu^k}{k!} e^{-\mu} \qquad (3\text{-}24)$$

This is the expression for the *Poisson distribution.* In the typical case where the number of nuclei in the sample of radioactive material is very large and the probability of any one of them producing a count during the counting time is very small, the distribution of all possible counts is a Poisson distribution, characterized by the single parameter, μ, which is both the mean and the variance of the distribution. Some examples of the Poisson distribution are shown in Figure 3.4.

If the mean, μ, is large–*e.g.,* greater than 10–then the Poisson distribution can be approximated very closely by a normal distribution with the variance set equal to the mean. Figure 3.5 illustrates the close correspondence between the binomial, Poisson and normal distributions for the typical counting situation. Each of the distributions shown has a mean and variance equal to 100. The values of n and p for the binomial distribution are 10^3 and 0.10, respectively.

It should be noted that these distributions are *not* identical; the Poisson and normal distributions can only be considered as approximations to the exact binomial distribution. Furthermore, the approximations are only

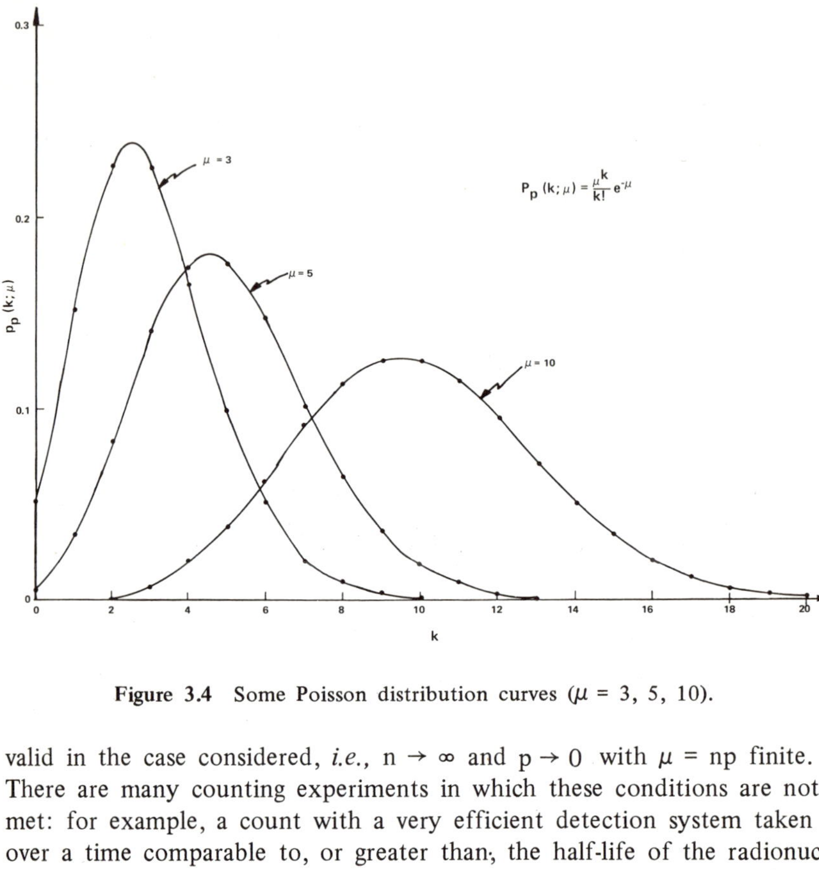

Figure 3.4 Some Poisson distribution curves (μ = 3, 5, 10).

valid in the case considered, *i.e.*, $n \to \infty$ and $p \to 0$ with $\mu = np$ finite. There are many counting experiments in which these conditions are not met: for example, a count with a very efficient detection system taken over a time comparable to, or greater than, the half-life of the radionuclide being counted. In such a case it may be inappropriate to assume that p is very small and, consequently, one *cannot* assume that the distribution of all possible counts will have a Poisson distribution depending on a single parameter, μ.

Note also that the normal distribution, Equation 3-3 is continuous *i.e.*, it is defined for *any* real value of the variable x, and is therefore qualitatively different from the discrete binomial and Poisson distributions which have meaning only when the variable k is a nonnegative integer. The normal distribution function can be interpreted as a probability *only* when it appears under an integral sign.

If the distribution of a given process is known and can be represented equally well by a normal and a Poisson distribution, the normal distribution will generally be more useful for calculating various probabilities. For example, consider the situation shown in Figure 3.5, for which the mean

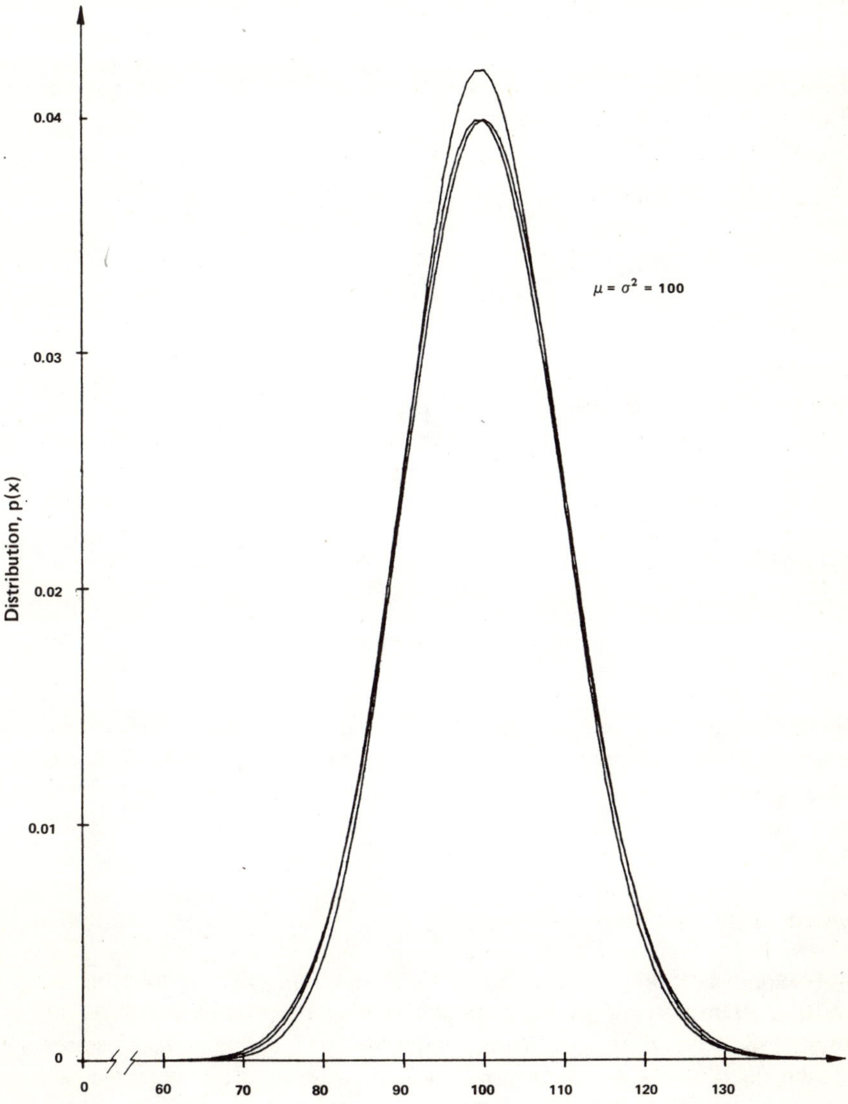

Figure 3.5 Effect of detector efficiency on counting times required to obtain 5% accuracy for various sample activities for a certain detector.

and variance of a population are 100. To calculate the probability that the outcome of a trial will be exactly 110 is tedious using the Poisson distribution (Equation 3-17):

$$P(110;100) = \frac{100^{110}}{110!} e^{-100} \qquad (3\text{-}25)$$

because of the large numbers involved. On the other hand, the normal distribution (Equation 3-3) taken over an interval of one, is quite easily evaluated:

$$\begin{aligned} p(x)\Delta x &\simeq \frac{1}{\sqrt{2\pi(100)}} \exp\left[-\frac{(100-110)^2}{2(100)}\right] \\ &\simeq \frac{1}{\sqrt{628}} \exp(-\tfrac{1}{2}) \\ &\simeq 0.0242 \end{aligned} \qquad (3\text{-}26)$$

This result is within three percent of the result of the Poisson calculation.

The normal distribution is even more useful if one needs to calculate the probability of an event occurring within a given range of values. Of particular interest is the probability that a result will lie within one standard deviation, σ, of the mean. For a Poisson distribution, the probability is given by

$$P = \sum_{k = <\mu - \sqrt{\mu} + 1>}^{<\mu + \sqrt{\mu}>} \frac{\mu^k}{k!} e^{-\mu} \qquad (3\text{-}27)$$

where $<n>$ is the largest integer in n. If a normal distribution can be used to approximate the Poisson distribution, the probability sought in the relatively simple integral expression:

$$\begin{aligned} P &= \frac{1}{\sqrt{2\pi\mu}} \int_{\mu - \sqrt{\mu}}^{\mu + \sqrt{\mu}} e^{-\frac{(x-\mu)^2}{2\mu}} dx \\ &= 0.6827 \end{aligned} \qquad (3\text{-}28)$$

To appreciate the utility of the Poisson distribution in simplifying sample statistics, return to the problem of determining the expected number of counts, $<C_t>$, in time t for a radioactive source. If one *cannot* assume a Poisson distribution, the appropriate way to determine the expected counts is to take several counts of equal time, t, such as those listed in Table 3.2, and to calculate the sample mean and variance as indicated. Since the sample size is ten in this example, the result of the experiment is

$$\langle C_t \rangle = \bar{C}_t \pm \frac{\sigma_{C_t}}{\sqrt{N}}$$
$$= 283.1 \pm \frac{17.3}{\sqrt{10}}$$
$$= 283.1 \pm 5.5 \text{ counts} \qquad (3\text{-}29)$$

Table 3.2 A Sample of Counts

Count Number	Count
1	285
2	256
3	279
4	318
5	270
6	274
7	303
8	288
9	281
10	277

$$\overline{C_t} = \frac{1}{N} \sum_{i=1}^{N} C_i = 283.1$$

$$\sigma^2_{C_t} = \frac{1}{N-1} \sum_{i=1}^{N} (C_i - C_t) = 298.8$$

$$\sigma_{C_t} = \sqrt{\sigma^2_{C_t}} = 17.3$$

$$\sigma_{\bar{C}_t} = \frac{\sigma_{C_t}}{\sqrt{N}} = 5.5$$

If the process is known to obey a Poisson distribution, the population mean and variance must be equal, and a single sample may be used to estimate both. In the example of Table 3.2, the *total* number of counts in the ten trials, which could have been acquired in a single measurement, can be considered a sample of one for the determination of the expected number of counts in an interval of 10 x t. Since this number is the best estimate of both the mean and variance of the population of all such counts of duration 10 x t and the sample consists of one measurement, the expected number is given as

$$\langle C_{10t} \rangle = C_{10t} \pm \sqrt{C_{10t}}$$

$$= 2{,}831 \pm 53 \qquad (3\text{-}30)$$

Obviously, for an interval of time t, these values are reduced by a factor of ten:

$$\langle C_t \rangle = 283.1 \pm 5.3 \quad \text{counts} \qquad (3\text{-}31)$$

This result is essentially identical to the one presented in Equation 3-29. However, data acquisition was considerably simplified since only a single count was necessary, even though the total counting time was equal, and the calculations were reduced to the taking of a single square root. Note that using a sample of one count which is the same size as individual counts in the larger sample does not constitute an equivalent measurement. For example, if the third entry in Table 3.2 is used as a sample of one, the resulting expected count in time t is

$$\langle C_t \rangle = C_t \pm \sqrt{C_t}$$

$$= 279 \pm 17 \quad \text{counts} \qquad (3\text{-}32)$$

This result is obviously less accurate, as indicated by the larger error, than either of the results in Equations 3-29 or 3-30, and this is as it should be since only a tenth as much information has been used. In fact, one can see that the *relative error* in the measurements, defined as the ratio of the standard error of the sample mean to the sample mean is inversely proportional to the square root of the total number of counts taken in the measurement, regardless of the duration of the count.

There are, in general, two reasons for not using single samples in most counting experiments. First, if there is any doubt about the validity of the assumptions which allow the binomial distribution to be treated as Poisson, it is better to base error estimates on the more credible foundation of a large sample. Secondly, a mistake in the measurement process can easily be hidden in a single sample, whereas it might be detected if the measurement were repeated. Therefore, samples of at least two are often taken as a check on the measurement process even when the assumption of Poisson statistics is justified.

ERROR PROPAGATION

To this point, the discussion of experimental error has dealt with the problem of reporting the result of a single measurement. However, the end result of most laboratory work is a number determined from many individual data each of which may have an error associated with it. There arises the question of how to report the error of this resultant quantity when the errors in the individual data are known.

In mathematical terms, one wishes to find the statistics μ_y and σ_y^2 of a random variable y which is a function of other random variables,

$$y = y(x_1, x_2, \ldots, x_n) \tag{3-33}$$

where the applicable statistics of the x_i's are known. For example, y might represent the count rate for a given source while x_1 and x_2 are the counts in an interval and the duration of the interval, each of which is a measured quantity and consequently subject to error. This problem is generally very involved, but a relatively simple result can be obtained under the following two conditions:

1. The various input data, x_i, are measured independently, *i.e.,* the magnitude or sign of the error in any one input quantity x_i in no way affects the magnitude or sign of the error in any other input quantity, x_j. This condition is rarely satisfied rigorously, but is approximately true in many cases of interest.

2. The distributions of the errors in the input quantities are unbiased normal distributions. Again, while seldom exactly true, this assumption is justified in enough instances to be useful.

If both conditions are met, it is legitimate to assert that: (a) the best estimate of the mean of the random variable y is the function y evaluated at the means of the random variables $x_1, x_2, \ldots, x_n$;

$$\bar{\mu}_y = y(\bar{x}_1, \bar{x}_2, \ldots, \bar{x}_n), \tag{3-34}$$

and

(b) the variance of y is given by the expression

$$\sigma_y^2 = \sum_{i=1}^{n} \left(\frac{\partial y}{\partial x_i}\right)^2 \sigma_{x_i}^2 \tag{3-35}$$

where the partial derivatives are evaluated at the means of the variables x_i, and $\sigma_{x_i}^2$ is the variance of the x_i.

The simplest examples of error propagation are those for which the desired quantity, y, is a function of only a single random variable. For instance, it is often useful to know the variance associated with the logarithm of a count. In this case,

$$y = ln\ x$$

where x is a count and can be assumed to come from a Poisson distribution, *i.e.*,

$$\overline{x} = C \tag{3-36}$$

and

$$\sigma_x = \sqrt{C}$$

From expressions 3-34 and 3-35

$$\overline{y} = ln\ C$$

and

$$\sigma_y^2 = \left(\frac{1}{C}\right)^2 \left(\sqrt{C}\right)^2$$

$$= \frac{1}{C} \tag{3-37}$$

Therefore, the result of this calculation would be reported as

$$y = ln\ C \pm \frac{1}{\sqrt{C}} \tag{3-38}$$

A common situation in which the result desired is a function of multiple variables is the correction of a count for background. Assume that a count, C, taken over time t_C is the result of the decay of a source plus background. With the source removed, the background alone produces a count B in time t_B. The count rate, R, due to the source alone, is

$$R = \frac{C}{t_C} - \frac{B}{t_B} \tag{3-39}$$

The variance in the rate is determined with the aid of Equation 3-35:

$$\sigma_R^2 = \left(\frac{1}{t_C}\right)^2 \left(\sqrt{C}\right)^2 + \left(-\frac{1}{t_B}\right)^2 \left(\sqrt{B}\right)^2 + \left(-\frac{C}{t_C^2}\right)^2 \sigma_{t_C}^2 + \left(\frac{B}{t_B^2}\right)^2 \sigma_{t_B}^2 \tag{3-40}$$

where $\sigma_{t_C}^2$ and $\sigma_{t_B}^2$ are the variances in the measured counting times. Table 3.3 consists of some particularly useful error propagation formulas derived from Equation 3-35 for some simple, but common, situations.

In conclusion, three comments are in order concerning error propagation calculations using experimental data. First, an awareness of error propagation is essential to good experiment design. A lot of wasted time and effort can be avoided if the experimenter anticipates the use of his data in

Table 3.3 Error Propagation Formulas

Function	Variance
$y = ax_1 \pm bx_2$	$\sigma_y^2 = a^2\sigma_1^2 + b^2\sigma_2^2$
$y = \begin{cases} a(x_1x_2) \\ a(x_1/x_2) \end{cases}$	$\dfrac{\sigma_y^2}{y^2} = \dfrac{\sigma_1^2}{x_1^2} + \dfrac{\sigma_2^2}{x_2^2}$
$y = ax_1^b$	$\dfrac{\sigma_y^2}{y^2} = b^2 \dfrac{\sigma_1^2}{x_1^2}$
$y = ae^{bx_1}$	$\dfrac{\sigma_y^2}{y^2} = b^2\sigma_1^2$
$y = a\ln(bx_1)$	$\sigma_y^2 = a^2 \dfrac{\sigma_1^2}{x_1^2}$

Note: the quantities a and b are constant parameters and the variables x_1 and x_2 have the variances σ_1^2 and σ_2^2, respectively.

arriving at the desired result. One objective of a good experiment is to minimize the variance in the result through a judicious choice of experimental parameters. A very simple example of this process is the division of a fixed amount of time between the counting of a source and the background. Let the time available, T, be divided between source counting, t_c, and background counting, t_b, *i.e.*,

$$t_b = T - t_c$$

If the counting rates of the source and the background are approximately known to be R_c and R_b, respectively, and the errors in the counting times are negligible, then expression 3-35 gives the variance in the corrected rate as:

$$\sigma_R^2 \simeq \left(\frac{1}{t_c}\right)^2 \left(\sqrt{R_c t_c}\right)^2 + \left(\frac{1}{t_b}\right)^2 \left(\sqrt{R_b t_b}\right)^2$$

$$\simeq \frac{R_c}{t_c} + \frac{R_b}{T-t_c} \qquad (3\text{-}41)$$

Minimizing this quantity with respect to t_c gives

$$0 = -\frac{R_c}{t_c^2} + \frac{R_b}{(T-t_c)^2} \qquad (3\text{-}42)$$

or, the ratio of source count to background count time, t_c/t_b, should be given by

$$\frac{t_c}{t_b} \simeq \sqrt{\frac{R_c}{R_b}} \tag{3-43}$$

This procedure, then, gives a rational way to choose the division of counting times in this simple case. This general technique is also useful in more complicated situations.

The second comment concerns the need to use error propagation in every case in which there is any calculation. The assignment of an error or variance to a quantity on the basis of sample statistics is valid only when that quantity is actually sampled. For example, in the simple case of measuring a counting rate, R, the final answer is the result of at least two separate measurements: the counts, C, in an interval and the duration of the interval, T. The assignment of error on the basis of sampling can only be made to the two pieces of data, C and T. The error in R must be calculated using the error propagation formulas, even if there is only one piece of the data, C!. In that case, since

$$R = \frac{C}{T}$$

the error propagation formula gives

$$\sigma_R = \frac{\sigma_c}{T} \tag{3-44}$$

If C is assumed to come from a Poisson distribution, Equation 3-44 reduces to

$$\sigma_R = \frac{\sqrt{C}}{T} \tag{3-45}$$

Since R is not directly sampled but calculated, its error cannot be simply assigned, it must be calculated.

Finally, it is important to remember that the deviation from the mean for any individual event has a finite probability to be large; that is, a counting result that does not seem to fit the assumed general trend cannot by itself be rejected as incorrect unless there is evidence of a systematic error or unrelated source of interference. Many important scientific discussions have been invalidated because the experimenter did not allow pure chance free play and instead "selected" those data points that fit preconceived notions of what the results should be.

Experimental data points may be rejected only if firm and consistently applied criteria are used for this purpose. These may include the following:

1. any point obtained during visible or evident malfunction of equipment, fluctuation in voltage or interference from other radiation sources;
2. any point whose individual variance from the mean, determined from all other points without considering it, exceeds 5σ;

3. any point, for which application of the statistical "chi-square" test gives a value of χ^2 for which the probability is less than one percent.

In drawing curves through experimental points care should be taken to give all points a weight equal to their σ^2 value; it is not usually satisfactory to adjust curves to pass through the highest and lowest point disregarding intermediate values.

As Equation 3-35 shows, the uncertainty in a given gross count rate, C, depends on the proportional contribution of the net count, N, and the background count rate, B, *i.e.*,

$$\sigma_C^2 = \sigma_N^2 + \sigma_B^2 \tag{3-46}$$

Since the standard deviation σ is proportional to the count rate, $\sigma_N \propto \sqrt{N}$ and the percent error is

$$\text{p.e.} \propto \frac{\sqrt{N}}{N} \times 100 = \frac{1}{\sqrt{N}} \times 100 \tag{3-47}$$

The percent error can be reduced by increasing the total number of counts recorded; for a steady count rate this means counting for a longer time.

Since the background count also increases in the same proportion, not much is gained unless the signal to background ratio $[C/B = 1 + (N/B)]$ is large, *i.e.*, $N >> B$.

For "good statistics," to obtain a low percent error, both C and C/B should be maximized, that is (C^2/B) should be as large as possible. For a given source and detector type this can be accomplished only in two ways: reduction in background by improved shielding, by guard counter or anticoincidence arrangements and/or by increases in the net count rate through improvements in the sample-detector geometry. This can sometimes be accomplished by changing the sample configuration, going from a planchet source to a well-type configuration, or by going to an internal counting arrangement.

Equation 3-47 also determines the minimum source strength that can be detected, for instance, in radiotracer experiments in a given detector system. Depending on the minimum acceptable net count rate considered significant, a certain total count rate must be exceeded for the test to be meaningful. For example, assume a detector is available with a background count rate of 200 cpm. If the count results must be determined with an accuracy of 10%, then

$$\frac{C - B}{\sqrt{N^2 + B^2}} < 0.1 \tag{3-48}$$

For a 5-min count B = 1,000 counts and from Equation 3-48, if $N >> B$, so that $(C/N) - (B/N) < 0.1$, $N > 9B$ approximately; *i.e.*, $N > 1{,}800$ cpm.

Longer counts can improve the accuracy in N, but the basic condition that N >> B remains. In practice, a net count rate equal to four to five times the background rate is usually acceptable; much lower net counts introduce rather large errors. A different approach must be taken when both source counts and background are extremely low as, for instance, in carbon dating or neutrino research, where even very long counting times result in very few detectable events and chance interference must be taken into account.

Another practical consideration, then, is what constitutes the minimum detectable sample activity. It is usually defined, as in the example above, as the least activity in a given geometry that results in a statistically significant net count, taking into account all background events and possible systematic errors. Altshuler and Pasternack, in 1963, defined detector sensitivity in terms of two other quantities, the "minimum significant measured activity" representing the smallest measured count which can be interpreted as indicating the presence of radioactivity in the sample, and the "minimum detectable true activity," which is the smallest amount of radioactivity required to be in a sample so that a measurement can imply correctly the presence of activity with a predetermined degree of confidence. Below this limit of detection two types of error can occur: a Type I error, concluding there is sample activity if there is none in reality, or a Type II error, concluding that there is no activity in the sample when there really is some.

If the background count B is stable and predictable, and the probability of making a Type I error is denoted by α and the probability of making a Type II error by β, then with β minimized it can be shown that the minimum significant sample count $C_S = k_\alpha \sqrt{B}$ and the minimum measured activity

$$A_I = \gamma k_\alpha \sqrt{B} \tag{3-49}$$

where γ = calibration constant, converting counts to activity, and the probability constant k_α, for normal distributions can be obtained from Table 3.4. The minimum detectable true activity, determined by the condition that the probability of a Type II error is not greater than the given value β, can be shown to be

$$A_{II} \approx \gamma(k_\alpha + k_\beta)\sqrt{B}$$

when

$$\frac{k_\alpha + k_\beta}{\sqrt{B}} \ll 1. \tag{3-50}$$

Table 3.4 Error Probability Parameters

α, β	k_α, k_β
0.005	2.576
0.010	2.326
0.025	1.960
0.050	1.645
0.100	1.282

Although the percent error can be reduced steadily by longer and longer counting times, the quantity $\sqrt{N}/N$ shrinks only slowly for large values of N and there are practical limits on system stability and equipment availability that set effective lower bounds to the minimum activity that may be detected to a given degree of confidence.

Figures 3.6 and 3.7 indicate the relationship between sample activity, efficiency, counting time and accuracy for a particular scintillation detector system. It is evident that there is a practical lower limit to the activity that can be determined in that system regardless of counting time. In most cases any counting period over three to four hours would introduce additional uncertainties unless very special measures are adopted.

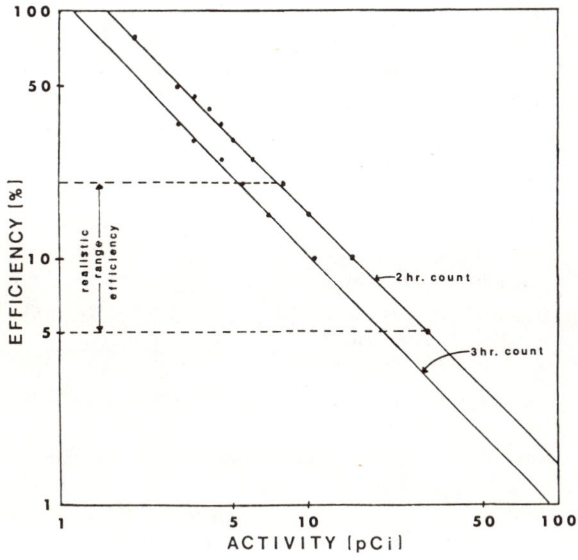

Figure 3.6 Relation between probable error and sample activity for various counting times (same detector).

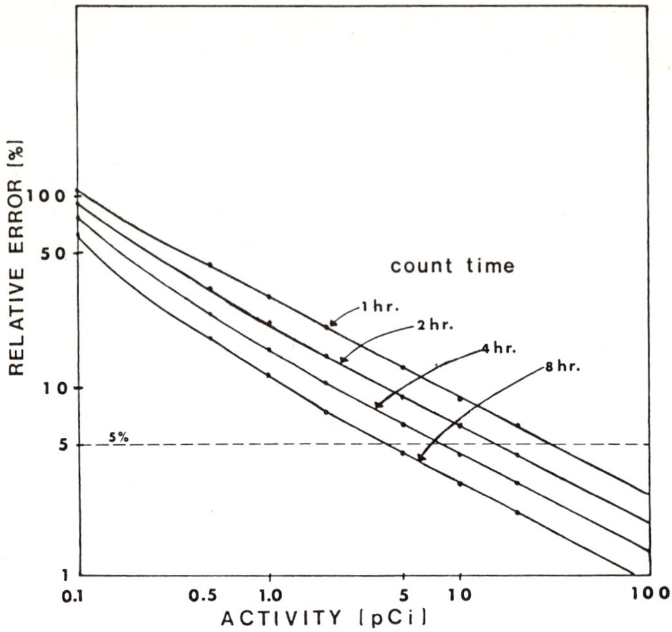

Figure 3.7 Relation between probable error and sample activity for various counting times (I-131 samples).

REFERENCES FOR FURTHER READING

Altshuler, B., and B. Pasternack. "Statistical Measures of the Lower Limit of Detection of a Radioactivity Counter," *Health Physics* 9:293 (1963).

Barford, N. C. *Experimental Measurements: Precision, Error and Truth* (Reading, MA: Addison-Wesley Publishing Co., Inc., 1967).

Beers, Y. *Introduction to the Theory of Error*, 2nd ed. (Reading, MA: Addison-Wesley Publishing Co., Inc., 1957).

Bevington, P. R. *Data Reduction and Error Analysis for the Physical Sciences* (New York: McGraw-Hill Book Company, 1969).

Lyon, A. J. *Dealing with Data* (New York: Pergamon Press, Inc., 1970).

Parratt, L. G. *Probability and Experimental Errors in Science* (New York: John Wiley & Sons., Inc., 1961).

Schiff, L. J., and R. D. Evans. "Statistical Analysis of the Counting Rate Meter," *Rev. Sci. Inst.* 7:456 (1936)

Whittaker, E., and G. Robinson. *The Calculus of Observations*, 4th ed. (London: Blackie & Sons, 1944) [New York: Dover Publications, Inc., 1961].

PROBLEMS

3.1 The following counts were obtained in successive one-minute runs with a certain detector: 68, 75, 62, 59, 77, 71, 80, 64, 67, 72, 67, 70, 65, 79, 84, 66, 55, 59, 68, 90, 78, 75, 59, 45, 75, 58, 92, 71, 67, 89. Calculate the mean, variance and standard deviation for that sample. Calculate the average count and standard deviation for the first 12 data points. Are they greatly different?

3.2 A one-minute count of a sample gives a count of 100, whereas a one-minute count of background alone gives a count of 36. What is the standard error of each of the counts?

What is the best estimate, and its associated error for the counts due to the source?

3.3 How many counts must be recorded to have a probable error (± one standard deviation) of 5% of the count?

3.4 Repeated measurements have been made on a sample yielding an average count of 880. Within what range should 90% of the measurements fall?

3.5 In the data series in Problem 3.1:
(a) Was the counter operating properly? Why do you think so?
(b) What is the maximum variation in this sequence that would be acceptable; at what point could a result be rejected?
(c) What should be reported as the sample count at the 95% confidence level?

3.6 A given sample has count rates of 35 cpm and a background of 5 cpm. How long does the sample need to be counted for a 2% accuracy in the sample activity?

If only one hour is available for a count-rate determination, including background, what is the best attainable accuracy for this sample?

3.7 Carbon-14, half-life 5,730 yr, is used to date archaeological samples. In a particular experiment the total count rate was about 10 counts/min and the background 6.4 counts/min. How long would this sample have to be counted to determine its activity to an accuracy of 5%? If the above sample was 10,000 years old, what is the oldest sample whose age could be determined in that detector to any significant accuracy, assuming a maximum counting time of one day per sample?

3.8 A sample was counted for 10 min and the average count rate was 28 cpm. The background was counted for 20 min and the average count rate was 17 cpm. What is the probable error in the sample activity measured?

CHAPTER 4

GAS-FILLED DETECTORS

INTRODUCTION

Gas-filled detectors represent probably the most widely used class of radiation detectors. All detectors in this class are based on the collection of ions produced in the sensitive volume of the detector due to the passage of ionizing radiation. Ionization chambers, proportional counters and Geiger-Müller counters are the primary detectors in this class. However, ionization is also important in a number of other detector systems that will be discussed in subsequent chapters.

The popularity of gas-filled detectors stems from a number of important factors. The detectors are relatively simple to construct, are easy to operate and require only a minimum of equipment. This is especially true for some types of ionization chambers. A large number of gases may be used to fill the detectors, even air. Finally, detectors in this class may be constructed in a variety of shapes and sizes to fit virtually any application and to provide a wide range of sensitivities. Most of these detectors are available commercially.

GENERAL CONSIDERATIONS

Radiation interacting in a gas-filled volume produces ion pairs in the volume through the process called ionization. Excitation of the molecules also occurs; both processes were discussed in detail in Chapter 2. The ion pairs (electrons and positive ions) are the result of interactions with orbital electrons of the gas molecules. Radiation that produces ionization directly through interactions with the gas is called *directly ionizing radiation*. This class of radiation is comprised of charged particles (*e.g.*, electrons, alpha particles, beta particles, fission fragments, etc.).

Gamma radiation and neutrons are common examples of indirectly ionizing radiations. These radiations interact with molecules of the gas or those composing the chamber wall, producing charged particles that result in ionization and excitation of the gas.

Gamma-ray interactions, especially the photoelectric and Compton processes, are usually classed as ionizing events. However, in radiation detectors, and especially gas-filled detectors, gamma-ray interactions occur with a much higher probability in the dense wall of the detector than in the less dense, gas-filled volume. Thus, the primary source of charged particles in the gas-filled volume due to gamma-ray interactions are electrons released from (or knocked out of) the wall by these interactions.

Neutrons may be detected by supplying the radiation detector with a material which has a high neutron absorption cross section. In this case, the neutron interaction produces an excited nucleus which emits a charged particle that ionizes the gas. This technique, as well as others, will be discussed in more detail later in the chapter.

Charged particle interactions in the gas produce a region around the particle track in which the ion pairs exist. The ion pair density is dependent on the specific ionization of the particle. For a heavy charged particle, *e.g.*, an alpha particle, the track is very straight and the density of ion pairs along the track is quite high. The specific ionization of an alpha particle may exceed 10^4 ion pairs/cm. Electron tracks are not straight due to the large number of scattering interactions these particles may experience. In addition, the specific ionization is much less (perhaps 100 ion pairs/cm) and thus the ion pair density is less than that for alpha particles or protons.

Electrons produced in ionizing events may make many collisions with the gas molecules as they move through the detector volume. For most common gases the *mean free path, i.e.,* the average distance between collisions, is in the range of 10^{-4} to 10^{-5} cm. Often, electron collisions with gas molecules result in electron attachment to the molecule, forming a negative ion. The probability of attachment to a neutral molecule per electron collision is called the *electron attachment coefficient*. The value of the coefficient is dependent on the electron energy and the type of gas. Values range from 10^{-6} for gases such as argon to 10^{-3} for the halogens. The best gases for use in radiation detectors are those which have a low electron attachment coefficient.

Since the positive and negative ions and electrons exist in close proximity to each other, recombination is a common occurrence; that is, the ions simply recombine to form neutral molecules. Of course, the number of recombinations is proportional to the density of positive and negative charges present. Therefore, recombination along the track of an alpha particle is much higher than along a beta particle or electron track. If no electric field is present in the detector, the above effects predominate and tend to erase the effects of the ionizing radiation.

IONIZATION CHAMBER REGIME

Consider a radiation detector shown schematically in Figure 4.1. The diagram shows a gas-filled volume with a central electrode which is insulated from the outer walls of the chamber. A voltage is applied, through a resistor shunted by a capacitor, between the outer wall and the central electrode.

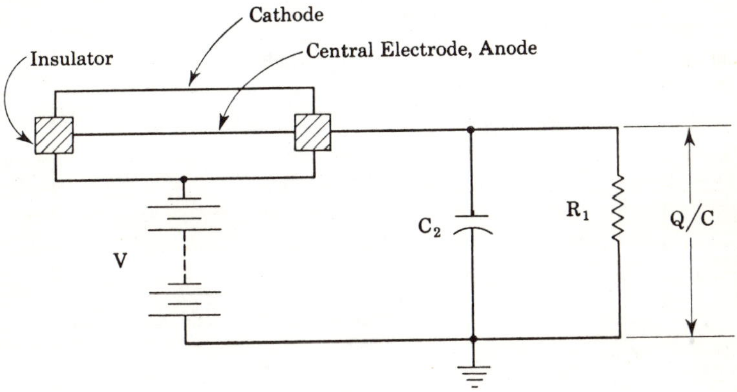

Figure 4.1 Diagram of gas-filled detector.

Ionizing radiation passing through the sensitive volume of the detector will produce ion pairs within the volume due to interactions either in the walls or the gas filling. The positive and negative charges, under the influence of the applied electric field, will move toward their respective electrodes. The number of ion pairs reaching the collecting electrodes is dependent on the chamber design, the gas-filling and the applied voltage. A typical plot of the relationship between the number of ion pairs collected and the applied voltage is shown in Figure 4.2.

In this figure there are four very different regions. In Region I, at low applied voltages, several mechanisms are active and compete for the ion pairs produced. The primary factors are a loss of ion pairs due to recombination and the removal of the charges from the volume by the collecting electrodes. At low voltages there is a net drift of the charges in a direction parallel to field lines of force. The drift velocity is a complex relation

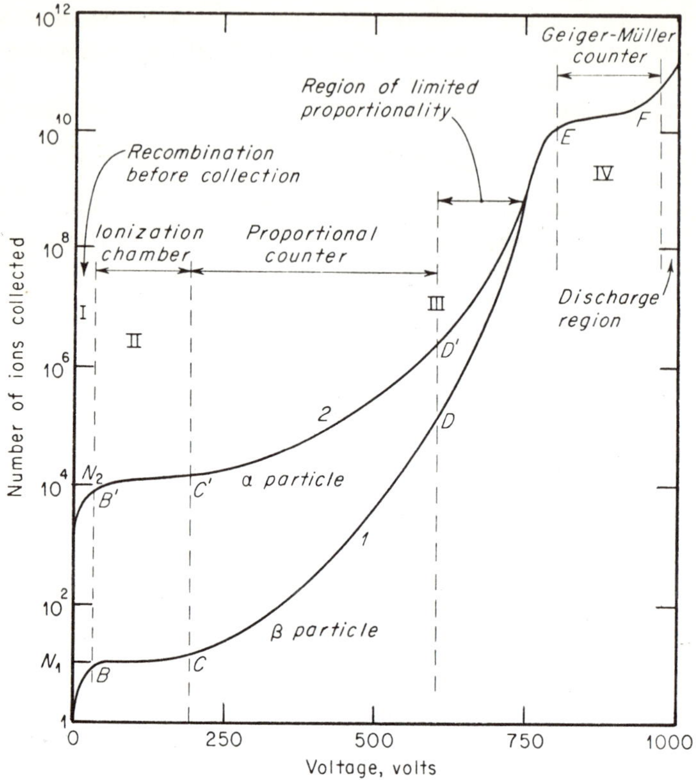

Figure 4.2 Counting characteristics of gas-filled detectors.

dependent on the type of gas. However, the drift velocity is directly proportional to the electric field strength E′ and inversely proportional to the gas pressure, *i.e.*, for a parallel plate chamber

$$w = \frac{dx}{dt} = \mu \frac{E'}{p} = \frac{\mu}{p} \cdot \frac{V}{d} \tag{4-1}$$

where μ = mobility
d = plate distance
p = gas pressure
V = applied voltage

For cylindrical chambers with electrodes of radii a and b

$$w = - \frac{dr}{dt} = \frac{\mu}{p} \frac{V_0}{r \log b/a} \tag{4-2}$$

As the applied voltage is increased the drift velocity of the ions increases significantly and there is a corresponding decrease in the time available for recombination. Thus, the number of ion pairs collected increases with increasing voltage. In addition, the force acting on the ions accelerates the electrons faster than the heavier positive ions and hence the drift velocity of the electrons is much higher than that of the positive ions. Figure 4.3 shows the variation in electron drift velocity with field strength in various gases.

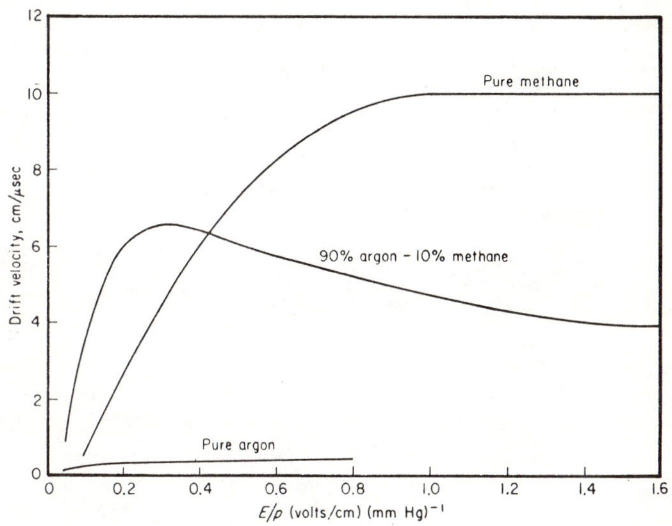

Figure 4.3 Electron drift velocity as a function of E'/p for argon, methane and methane-argon mixture (from Bortner *et al.*, 1957).

The number of ion pairs collected increases with increasing voltage. In this situation, because the probability of reaching the collecting electrodes significantly exceeds that of meeting an oppositely charged ion, all ion pairs produced in the volume by the primary ionizing event are swept from the volume and collected. At this point, further increases in applied voltage result in no increase in the number of ion pairs collected. The voltage at which this occurs is called the *saturation voltage* and the region, *i.e.,* Region II in Figure 4.2, is called the ionization region. Gas-filled detectors which operate in this region are called ionization chambers. The voltage required to produce saturation in the ionization region depends on the type of gas used, the gas pressure and chamber dimensions.

The charge per ion, e = 1.6×10^{-19} coulombs; thus the ionization current caused by primary ion pairs produced in the chamber is small. This results in a current or a pulse from the chamber that is very small. Usually an electrometer or an electrometer tube with amplification stages is required to detect this small current. The events which occur in an ionization chamber are summarized schematically in Figure 4.4.

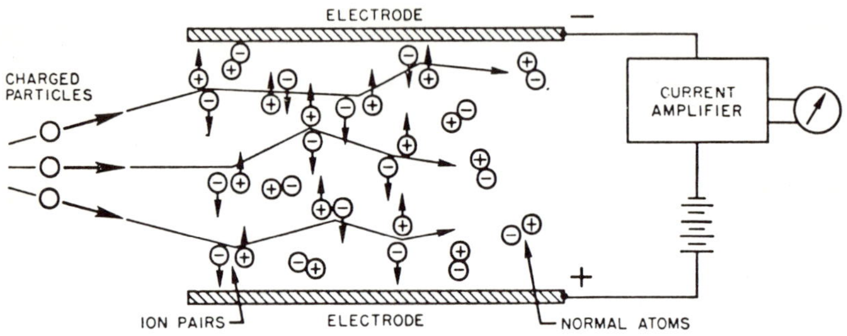

Figure 4.4 Operation of ionization chamber.

THE PROPORTIONAL COUNTER REGION

In Region II a voltage was reached at which all primary ions were collected, that is, the collecting voltage was sufficient to sweep all ions from the detector volume before significant recombination occurred. In this situation the drift velocity of the electrons may reach 10^5 to 10^7 cm/sec as they acquire kinetic energy due to the accelerating force of the electric field. As these ions move toward the electrodes they collide with gas molecules. In each collision a fraction of the kinetic energy is transferred to the molecules, exciting but not ionizing them. Energy is also lost in each scattering collision but more energy is acquired as each ion moves on under the influence of the electric field.

The broad area of Figure 4.2 marked Region III presents a regime where a phenomenon called *gas multiplication* begins to predominate. In this regime the voltage has been increased so much that the ions acquire sufficient energy to create further ionization in the gas themselves.

Under normal conditions, electrons produced by a primary ionizing event will drift under the influence of the electric field toward the anode, while positive ions move outwards. In the arrangement shown in Figure 4.1 the electric field intensity is strongest surrounding the central electrode. When electrons enter this steeply increasing electric field, which is perhaps only a few mean free paths away from the central electrode, the result of their interactions with the gas molecules changes significantly. In this region the electrons receive sufficient kinetic energy to produce secondary ionizations in the gas; that is, the electrons, which were the result of a primary ionizing event, become directly ionizing particles. The electrons released in these ionizing events are called secondary electrons. These electrons are also under the influence of the electric field and ultimately may possess enough kinetic energy to produce additional ionizations and so on. (The electrons produced in these tertiary ionizing events may also produce ionization.) Thus, each primary ionizing event produces an avalanche of electrons all moving toward the anode. This is the phenomenon of *gas multiplication* which is common to both proportional counters and Geiger-Müller counters.

As the applied voltage is increased, the gas volume in which gas multiplication can occur will expand. In a cylindrical chamber such an increase in applied voltage implies an increase in the radius of a cylindrical "multiplication" region around the anode which results in a growth in the number of electrons produced per primary ionizing event. Under suitable conditions the effect of gas multiplication will become quite significant and the charge collected may be increased by several orders of magnitude (*e.g.*, a gas multiplication factor, the number of ions collected per primary event, of 10^4 or more is typical).

In the detector itself there may be several primary ionization events produced by the passage of a single ionizing particle. The electron avalanche, often called a *Townsend avalanche*, produced by each primary interaction moving parallel to the collector is confined to a small length of the wire. If there are only a few ionizations produced in the detector volume, there will be no interaction between the avalanches caused by the primary electrons. Within certain limitations, all the avalanches produced by individual electrons (from primary ionizations) are approximately the same size, that is, the total charge collected per primary event is uniform and independent of location within the chamber of the passing particle. Thus, the resultant output pulse, which is the sum of all these avalanches, is *proportional* to the number of primary electrons produced by the passage of the ionizing particle. The region designated Region III in Figure 4.2 is called the *proportional region* and radiation detectors that operate in this region are called proportional counters. This process is illustrated schematically in Figure 4.5.

Electrons are more mobile and hence are collected much more quickly than the positive ions. In many cases the avalanche region may extend only a fraction of a millimeter into the gas. This rapid collection of electrons from the gas volume leaves a "positive ion sheath" around the central electrode which, in the time required for electron collection, appears to be

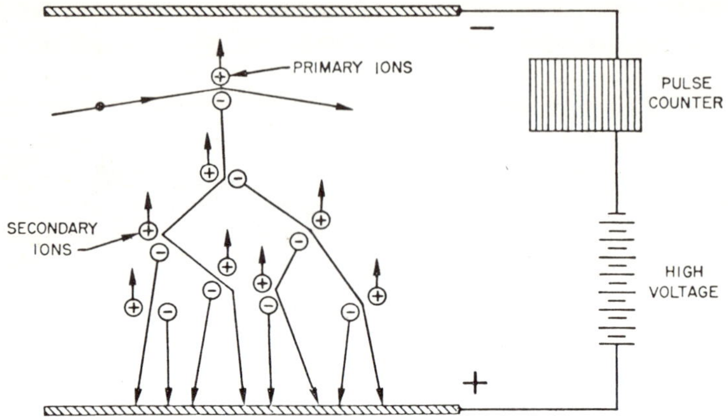

Figure 4.5 Operation of proportional counter.

essentially stationary. In the proportional counter, these positive ion sheaths remain localized. The counter may receive another pulse at another position on the electrode even while the ion sheath remains in position in one place. Thus, the ability to resolve the passage of many ionizing particles is quite good in a proportional counter. That means that the counter is responsive to new ionizing events, as soon as existing ions have been swept out of the main counter volume by the applied voltage.

If the electrode potential is further increased, there is a corresponding increase in the gas multiplication factor and the positive ion sheath spreads along the electrode. This spreading causes a corresponding increase in the resolving time of the detector. Since the avalanche in a proportional counter is terminated by the collection at the anode of the liberated electrons, the rise time of the pulse is governed primarily by the time required for electrons produced at the extreme end of the primary ion track to travel to the electrode. Usually this time is very short. In some applications a resolving time of 0.2 to 0.5 μsec may be obtained. However, such applications usually involve only the detection of radiation. In many other applications the actual resolving time may be limited, not by the detector, but by the external electronic circuits associated with the detector system.

In the region in which gas multiplication occurs, the multiplication for a given applied voltage is independent of the initial ionization event since it only depends on the energy of the positive and negative ions drifting through the gas volume. In this way, the proportionality of pulse sizes is maintained, *i.e.*, the total number of ions is still proportional to the total energy lost by the incident particle in the gas volume. As the voltage is increased further, the strict proportionality no longer holds because not all primary ions generate an identical number of secondaries. In the upper portion of Region III, the pulse size is independent of the initial ionization and this is called the region of limited proportionality.

Before considering the last region, note that Figure 4.2 also illustrates the difference in the number of ion pairs produced by densely ionizing particles (curve 2 for alpha particles) and more sparsely ionizing radiation (curve 1 for beta particles). In the region of limited proportionality it is still possible to distinguish between the two types of particles but the ratio of the pulse heights provides no useful information. As the voltage is increased further even this ability disappears and in Region IV the ability to distinguish between particles with different ionizing power does not exist.

THE GEIGER-MÜLLER REGIME

The last region, Region IV, represents one in which the charge collected is independent of the initial ionization because each primary event causes an "avalanche" of secondary ions extending throughout the whole counter volume, so that only the number of events is detected, not the energy transferred. Actually, the degree of gas multiplication is what differentiates this region from the proportional region. In Region IV gas multiplication is limited only by the characteristics of the detector and the external circuit. This regime is commonly called the Geiger-Müller or simply the GM mode of operation.

In a Geiger-Müller counter the gas-multiplication process produces a pulse of a uniform size regardless of the number of ion pairs formed by the primary ionization. The avalanche continues until a certain number of ion pairs are produced in the detector volume; typically on the order of 10^9 electrons.

In the GM counter the initial Townsend avalanche builds up rapidly. It is terminated when all the electrons produced in the avalanche reach the central electrode because the positive ion sheath causes a reduction in the field strength near the wire. The initial ionization avalanche is followed by successive avalanches, each one triggered by the preceding one. In a GM counter the effects of the Townsend avalanche are thought to be propagated by the excitation of neutral atoms within the avalanche region. These atoms may deexcite by the emission of photons in the ultraviolet region that may initiate further avalanches. In counters filled with a mixture of gases, avalanches are propagated by ionization of other atoms that have a lower ionization potential than the original excited atom.

The ultraviolet radiation may also initiate the emission of photoelectrons by interactions in the cathode. The time sequence of these secondary avalanches may be delayed significantly when compared to the original discharge. This is due to metastable excitation states of the noble gases which serve to delay the emission of the ultraviolet radiation, and the subsequent production of photoelectrons and another avalanche.

In pure gases, positive ions may initiate photoelectrons as they are neutralized at the cathode. Figure 4.6 illustrates schematically this process.

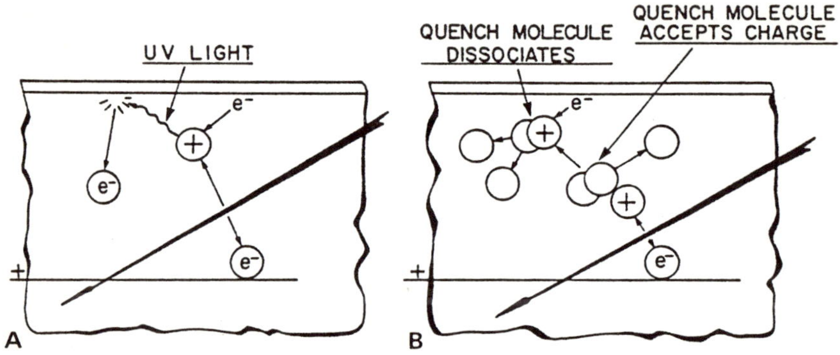

Figure 4.6 Operation of Geiger-Muller counter: (a) formation of initial ion pair; (b) effect of quench molecules.

Positive ions migrate to the cathode and are neutralized by combining with electrons from the wall. This process (*i.e.*, neutralization) may produce an additional electron by two separate mechanisms. First, a photon may be radiated from the wall and may produce more photoelectrons. The energy of this photon is initially the energy difference between the ionization potential of the ion and the work function of the cathode material.

The second mechanism which produces an additional electron occurs if the positive ion has an ionization potential that is at least twice the work function of the cathode. In this case, a second electron may be ejected from the cathode, rather than a photon. In either case, the emission of this excess energy results ultimately in another avalanche.

The latter two processes are obviously governed by the transit time of the positive ions to the cathode. This implies a significant delay after the initial discharge before additional discharges are produced. In a typical situation the time interval may approach 200 μsec; this, in turn, introduces a long "dead time" during which the detector will not respond to new ionizing particles.

In all cases discussed above, the production of another electron, by whatever process, results in another discharge in the detector. This propagation of avalanches will continue unless steps are taken to prevent these occurrences. This is called *quenching* the discharge. Quenching consists of the introduction of an electronegative impurity which can absorb some of the excitons without further ionization. Thus, it limits further spreading of the avalanche region and permits early recovery of the ionized gas.

Quenching is not important in proportional counters because the number of excited atoms and ions formed is small. Thus, successive avalanches will get smaller and result in no further charge generation. The need for quenching is another characteristic of the difference between proportional and Geiger-Müller counters.

A number of methods have been used to quench the self-perpetuating avalanches in GM counters. External circuits were used with older GM counting systems. These circuits were designed to reduce the voltage across the detector below the value required to maintain the discharge. The voltage is reduced only momentarily but this is sufficient to terminate the discharge of the counter. External quenching results in long resolving times and for this reason "self-quenching" GM counters are commonly used.

Self-quenched GM counters contain a quenching gas which is added as an impurity to the major gas filling. Two types of self-quenching tubes are available, those quenched by the addition of a halogen and those which are organically quenched. In self-quenched GM counters, as positive ions move toward the cathode, there is a transfer of charge to the molecules of the quenching gas. When the charged molecules reach the cathode they dislodge electrons from the chamber wall, but this process results in the dissociation of the quenching gas molecules rather than the production of additional electrons (see Figure 4.6b).

Typical organic quenching gases are polyatomic gases, including ethyl alcohol, ethyl formate and amyl acetate. For organic quenching gases the dissociation is irreversible and, therefore, the life of an organically quenched GM counter is limited. The useful lifetime is typically 10^8 to 10^{10} pulses. Bromine and chlorine gases are commonly used in halogen-quenched GM counters. In this type of counter the quench gas molecules recombine after a finite period and, thus, the counters have essentially an unlimited useful lifetime. However, all gas-filled counters have a practical limitation on their lifetime set by leakage of counter seals.

Geiger-Müller counters are typically cylindrical in shape and filled with a mixture of argon and quenching gas ($< 1\%$) at a pressure of 40 torr ($\sim$ 5.3 kPa). The collecting electrode is usually a tungsten wire; the outer electrode, a thin stiffened steel or aluminum shell coated with graphite to ensure uniform field distribution. For alpha or beta detection, counters have been designed with thin mica windows or very thin walls. Such counters require very careful handling and the window should never be touched by hand. Figure 4.7 illustrates the design or appearance of some typical GM tubes.

Since the Geiger regime is such that all ions produced by a single initiating event are collected as a single pulse event, the output pulses have a constant amplitude regardless of the energy of the initiating event. Hence, a GM counter is a very sensitive radiation monitor whose large output pulses may be of the order of volts in amplitude, making it convenient to count them directly. Another important feature is its relative insensitivity to minor variations in applied voltage as long as sufficient voltage is applied to maintain avalanche conditions. That minimum voltage is called the threshold voltage. The region over which the GM counter will function with almost constant output when exposed to a fixed source is called its "plateau." This is shown in the detector characteristic plot, Figure 4.8, which represents the relation between count rate and applied voltage. The plateau extends from the threshold voltage to the sharp upper knee where dielectric breakdown of the ionized gas begins to occur. The quality of a GM detector is usually judged by the length of the plateau and its "slope," *i.e.*, the percentage change in count rate per 100 volt increase in operating voltage near the center of the plateau. For a halogen-quenched GM tube, a plateau of 200 volts and a slope of 5-6% may be considered reasonable. As the counter tube ages and deteriorates, this will be reflected in an increase in the slope and a shortening of the plateau. A GM tube with a very short steep plateau will usually be considered unacceptable and the counter characteristic should be checked at intervals.

IONIZATION CHAMBERS

Two broad classifications of ionization chambers are usually assigned in the discussion of radiation detectors of this type. These are passive detectors and active detectors. The term passive implies a situation in which several steps must be taken to utilize the detector. That is, the detector must be prepared for use, exposed and evaluated. The information obtained from the detector may be obtained after the end of the radiation exposure and/or after the detector is removed from the field. Therefore, passive detectors have to be integrating devices since the reading is for the total exposure accumulated over the entire exposure period, unless the detector is overexposed.

Active detectors are those which provide an immediate indication of the exposure received. This information is obtained through the use of an external electronic circuit which is connected directly to the detector,

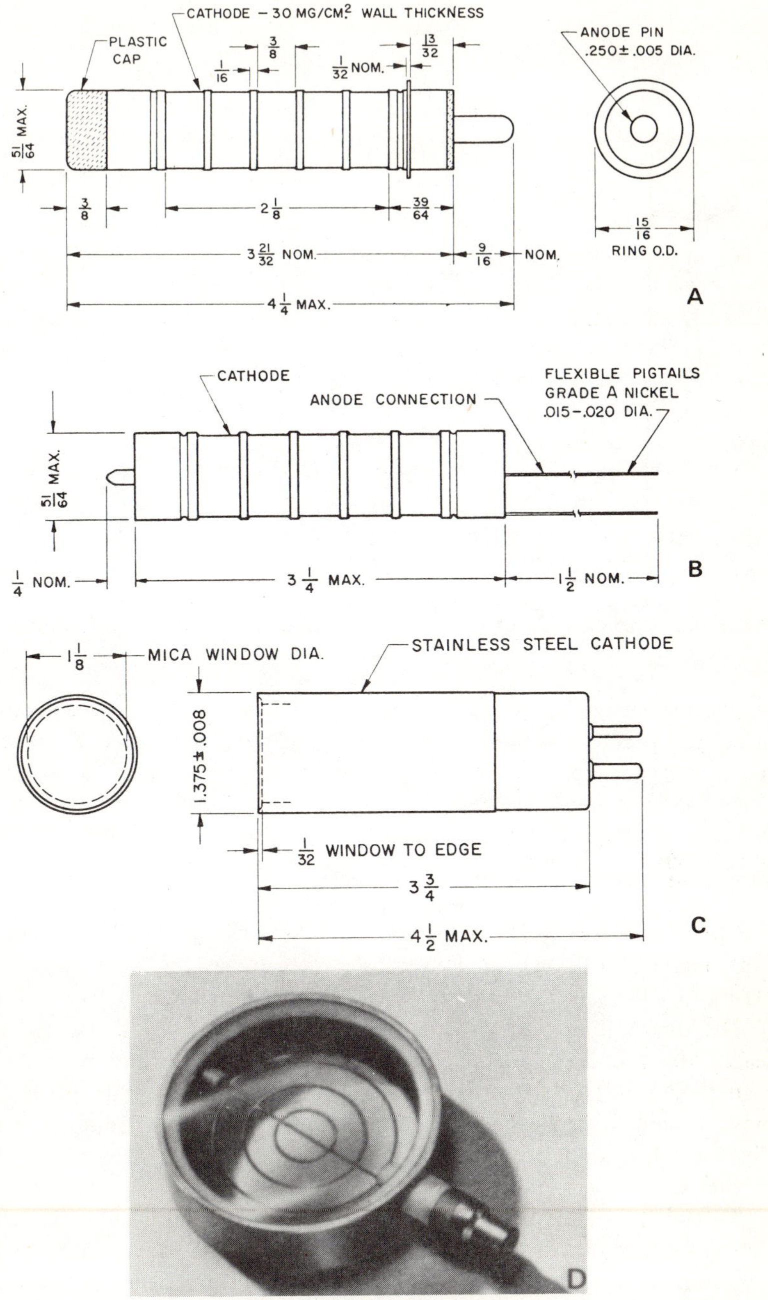

Figure 4.7 Typical Geiger counter configurations: (a) single-ended; (b) concentric thin-walled; (c) mica end window; (d) pancake type.

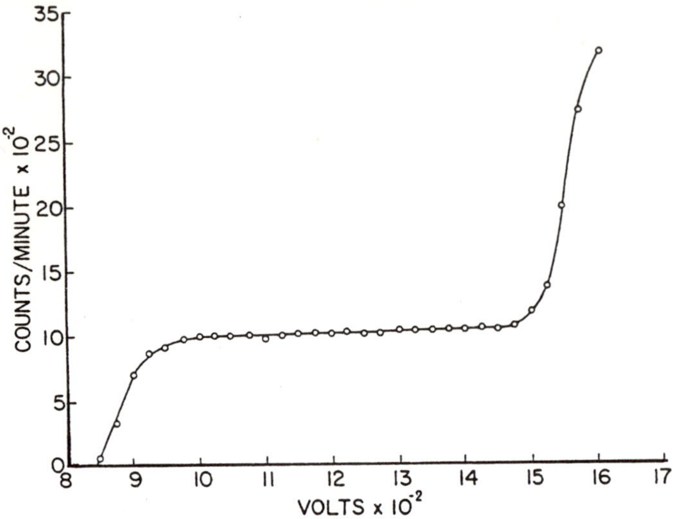

Figure 4.8 Characteristic curve for a Geiger counter (organic quench agent).

semipermanently. The electronic system may be close to the detector, as in certain portable radiation detectors, or it may be located at some distance from the detector, as in the case of radiation detectors used in the operation of a nuclear reactor.

Passive Ionization Chambers

One of the earliest radiation detectors was the gold-leaf electroscope. This detector, although no longer used, remains as the basis for some radiation detectors used in personnel monitoring.

The quartz-fiber electroscope, developed originally by Lauritsen, replaced the gold-leaf design because it is more compact and portable (see Figure 4.9). In this detector there are two metallized quartz fibers, one movable and the other stationary, mounted on a central support. These fibers are insulated from the outer case and represent the sensitive element of the detector.

To use the detector, it is charged through an external circuit; usually 200 volts is required. Since the two fibers have the same polarity, the repulsion of like charges displaces the movable fiber from the stationary fiber. The electroscope is equipped with an eyepiece, a transparent scale and a window to illuminate the fiber. When fully charged, the movable quartz fiber casts a shadow at the zero position on the scale. When the

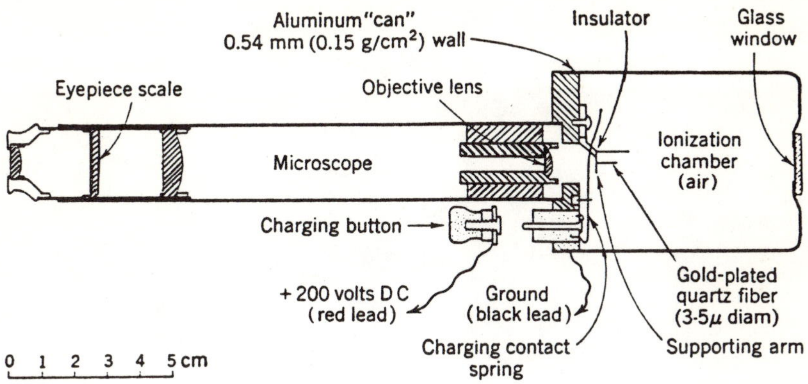

Figure 4.9 Lauritsen electroscope.

electroscope is exposed to radiation, ionization of the gas in the volume reduces the charge on the electroscope and the repulsive force between the fibers is reduced. The movement of the fiber toward the fixed fiber is reflected on the scale as the total amount of radiation exposure.

The above principle is applied in the self-reading pocket ionization chamber. This detector is shown schematically in Figure 4.10. In this device the fibers are in the form of loops and all components are enclosed in a rugged metal case. The chamber is small and can be carried easily in a shirt pocket. Typically the range of the chambers is 200 mR but others are available with full-scale readings up to 50 R.

The second major type of passive ionization chambers are those called "condenser-type" chambers. These devices are simpler than the electroscope type but require an external circuit not only for charging but also for evaluation. The detector (shown in Figure 4.10b) consists of a cylindrical outer case constructed of a material, such as bakelite, and a central metallic electrode coaxial with, but insulated from, the case. The detector is charged from a battery pack, removed from the charger and exposed to radiation. The exposure is evaluated by replacing the detector in the charger, which also serves as a reader, and evaluating the reduction in total charge which is proportional to the total ionization produced in the chamber.

Usually the charger-reader is calibrated to read directly in units of exposure, *i.e.*, roentgens. However, the exposure can be evaluated by the relation:

$$C \times \Delta V = \Delta Q \qquad (4\text{-}3)$$

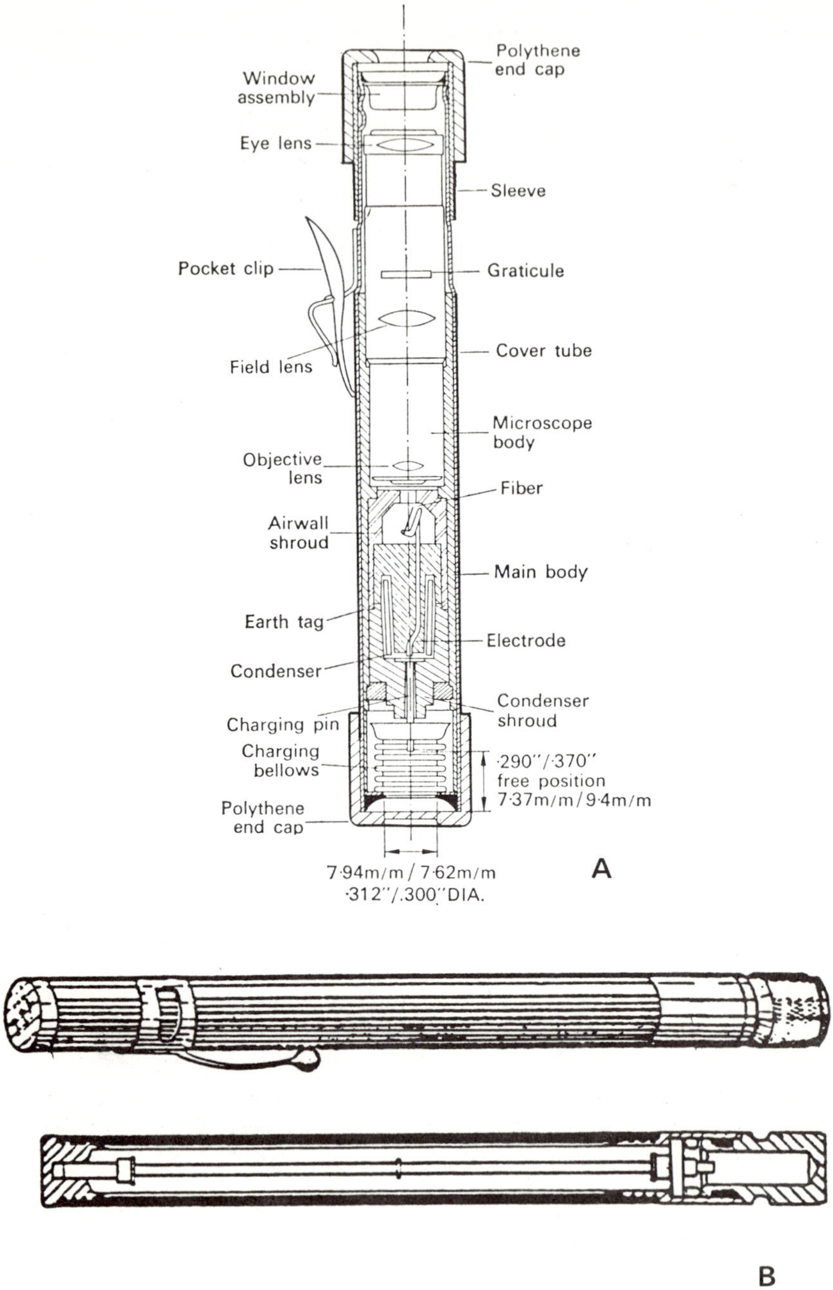

Figure 4.10 Portable ion chambers: (a) direct reading quartz-fiber electroscope chamber; (b) condenser-type chamber.

In the above relation, C is the electrical capacitance of the chamber, ΔV represents the change in voltage before and after exposure, and ΔQ is the charge collected during the exposure. If the chamber volume is known, the exposure can be calculated based on the definition of exposure.

Certain types of condenser chambers may serve as secondary standard devices for use in calibrating the output of X-ray machines, radioisotope sources, etc. These detectors are often referred to as R-chambers. These chambers are manufactured with walls of different thicknesses and compositions for use in a wide range of photon fields. Usually, the wall material is air-equivalent and the thickness of the wall is adjusted for a particular photon energy. For example, R-chambers are available for use in the energy range 6-35 keV and also for photons in the energy range 0.25-1.3 MeV. In addition, the total exposure range may be selected by choosing a detector of a certain volume. Typical detectors in use have sensitivities from 0.025 R to 250 R. Other sensitivities are available (*e.g.*, 0.001 R) for use in special exposure situations. However, it must be remembered for all direct-reading chambers that the roentgen calibration strictly is valid only for energies near that of the calibration source.

The R-chambers require a charger-reader for use. This charger-reader gives an output directly in units of exposure. The complete system, usually called an R-meter, is normally purchased with calibration certificates from the NBS or from a regional calibration laboratory.

The simplicity of passive ionization chambers is such that there are a variety of detector types and arrangements which have been designed for special applications. Usually, the walls of the chambers are constructed of material which is "air-equivalent," *i.e.*, the material has similar photon absorption properties to those of air. Materials, such as bakelite, graphite, polystyrene, nylon, aluminum or any other material that for a selected photon energy has similar characteristics are commonly used.

For other applications, chambers may need a wall material that is *tissue-equivalent* or *muscle-equivalent* (Table 4.1). Chambers have also been manufactured with "bone-equivalent" materials. A large number of equivalent materials are available for the construction of the chambers (Figure 4.11). The choice is dependent on the application, the materials available to the designer-builder and the expected accuracy of the intended use. For example, small condenser chambers can be made sensitive to thermal neutrons by the addition of a material such as boron, either as a liner or as part of the wall material. Condenser pocket chambers used as neutron dosimeters are commercially available and are used widely for personnel monitoring. However, since the chamber responds to photon irradiation as well as thermal neutrons, the user obtains no information as to the relative contribution of each radiation to the measured exposure when both are present.

Table 4.1 Tissue-Equivalent Materials (McGinley, 1971)

(a) Composition of Muscle and Muscle-Equivalent Materials in Weight Percent

Element	Typical Muscle	TE Gas[a] I	TE Gel	TE Plastic A-150	TE Gas II	TE Liquid
H	10.2	9.7	10.0	9.9	10.1	10.2
C	12.3	15.7	14.9	77.1	45.7	12.0
N	3.5	3.5	3.5	4.4	3.5	3.6
O	72.9	71.1	71.6	5.0	40.7	74.2
Other Elements	1.1	–	–	3.6	–	–

(b) Composition of Bone and Bone-Equivalent Plastic in Weight Percent

Element	Bone	Bone-Equivalent Plastic B-100
H	6.4	6.39
C	27.8	53.41
N	2.7	2.67
O	41.0	3.06
Ca	14.7	17.69
P	7.0	—
F	—	16.77
Other Elements	0.4	—

[a]Explosive mixture.

This disadvantage can be overcome by the use of two chambers, one sensitive only to photons and the other sensitive to neutrons as well as photons. Both detectors must be properly calibrated since even the detector which responds only to photons has a finite (but small) response to thermal neutrons. Although this presents no real problems, the simple example illustrates the complexities encountered with measurements in mixed radiation fields, even with the simplest radiation detectors.

Active Ionization Detectors

In an active (or current-type) ionization chamber, electrons collected at the anode compose a direct current which can be amplified in an electrometer tube and measured with a microammeter. For situations requiring high accuracy the small current may be measured with a vibrating reed electrometer which transforms a small direct current into pulses that can be amplified. Essentially all ionization chambers are evaluated in one of the above ways.

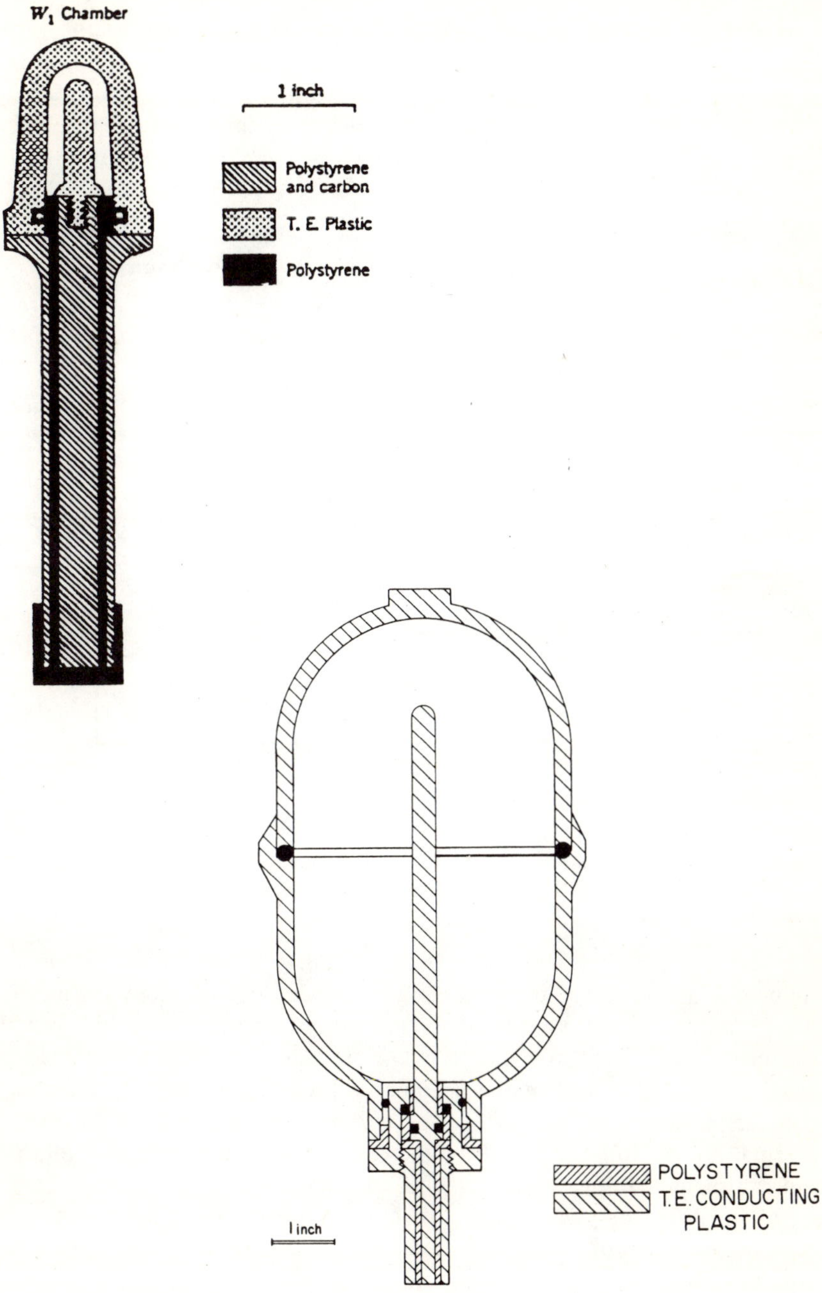

Figure 4.11 Two tissue-equivalent ionization chambers.

Probably the simplest of the active ionization chambers, yet the type least familiar in terms of widespread use, is the free-air ionization chamber. This detector is used as a primary standard in national standardization laboratories throughout the world. The chamber, shown schematically in Figure 4.12 is a parallel-plate design which satisfies the operational definition of exposure in units of roentgens. The photon beam is collimated as

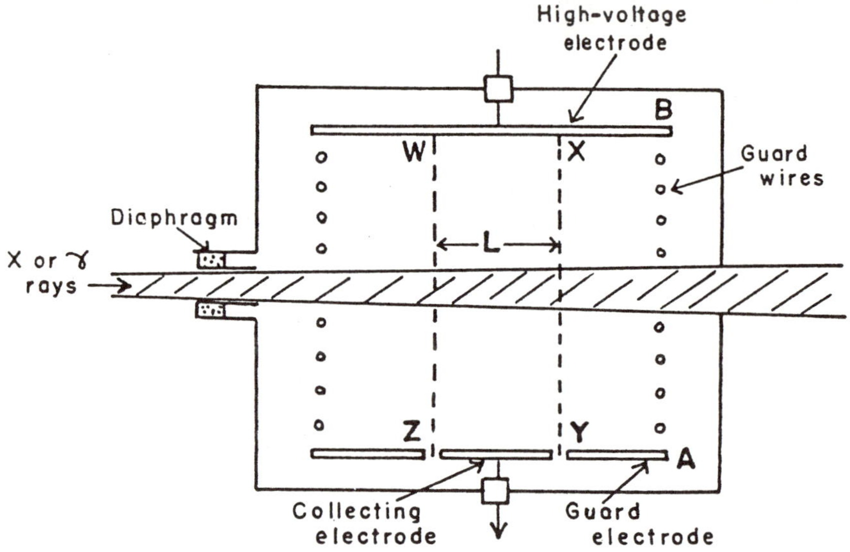

Figure 4.12 Free-air ionization chamber. Dotted lines indicate collecting volume of length L; hatched area depicts incident beam.

it enters the chamber and interacts in a volume of air defined by the collimator aperture and the collecting electrode. The chamber features a guard ring and guard wires to maintain straight lines of force between the two electrodes. The entire device is enclosed, usually with a lead-lined material. Ions produced in the chamber volume due to photon interactions are collected at the plates. The current flow is measured by an external circuit and from it the number of ions produced in the volume and, ultimately, the exposure can be calculated.

In order for this measurement to be valid, electronic equilibrium must exist in the detector. In other words, all the energy of primary electrons produced in the sensitive volume of the chamber must be dissipated in the chamber. Obviously, many electrons produced in the detector volume by photon interactions will leave the sensitive volume. Electronic equilibrium is maintained by making the entire chamber larger than the maximum range of the primary electrons in air. In this situation, primary electrons produced in the sensitive volume that leave the volume are replaced by primary electrons which were produced outside the sensitive volume but enter it. Thus, electronic equilibrium is obtained as an electron of equal energy enters into the sensitive volume for every electron that leaves.

The thickness of air between the entrance port and the collecting volume needed to provide electronic equilibrium increases with increasing photon energy. For example, 9 cm of air is required for highly filtered X-rays, whereas for 500 kV X-rays the air thickness is 40 cm. This fundamental requirement limits the use of free-air chambers since the size of the chamber for higher photon energies is extremely large. For example, the National Bureau of Standards (NBS) has three free-air ionization chambers for use in their activities. The chambers are intended to cover the X-ray generating potentials of 10 to 60 kV, 20 to 100 kV, and 60 to 250 kV. These chambers were manufactured at the NBS but similar chambers are commercially available with a useful range up to about 300 keV. Above this photon energy, operational difficulties, chamber size and so on limit this detector's usefulness even in a standards laboratory.

The surface dose due to beta emitters may be determined by use of the extrapolation chamber. This special ionization chamber (see Figure 4.13) is a parallel-plate detector similar in design to the free-air chamber described above. However, in this design the distance between the plates can be varied. Usually, one plate which acts as a thin window is placed as close as possible to the source to be measured. A series of measurements are obtained while the spacing between the plates is decreased. The results of these measurements are plotted and extrapolated to zero spacing. This gives the dose at the surface of the beta source and eliminates secondary gas or wall effects.

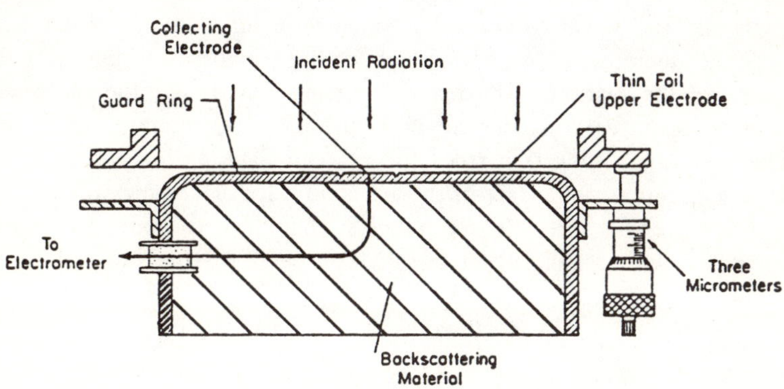

Figure 4.13 Failla extrapolation chamber.

The use of an extrapolation chamber is a good example of the application of the Bragg-Gray principle to the measurement of absorbed dose. The chamber, introduced by Failla, is quite useful because it recognizes the fundamental requirement that the detector cavity be small compared with the electron ranges. The chamber has also been utilized for measurements in areas where no electronic equilibrium exists, *e.g.*, at interfaces between two dissimilar materials.

Ionization chambers have found wide use in surveys for radiation protection purposes. The ionization chamber is the only gas-filled detector that allows the direct determination of the absorbed dose. This is because the measured current is directly proportional to the ionization produced in the sensitive volume and that in turn is directly proportional to the energy deposited in the detector.

One of the most popular survey detectors using an ionization chamber is the "Cutie pie." This instrument is a portable large-volume air-filled ionization chamber. The detector is a right-circular cylinder, 7.3 cm in diameter and 14.0 cm long, with a total volume of about 600 cm^3. One end of the detector is usually covered with thin aluminized mylar to allow the detection of alpha and beta radiation. The detector is sensitive to gamma-rays with energy greater than 7 keV, beta particles with energy greater than 70 keV and alpha particles more energetic than 3.5 MeV. The energy dependence is $\pm$ 10% over the range 7 keV to 2 MeV for photons.

As with the passive detectors, many designs of ionization detectors are available commercially. The user has a wide variety of detectors from which to select depending on the particular application.

A number of active detectors have been designed with characteristics similar to the condenser R-chambers discussed in the previous section. One such system is a precision instrument designed specifically for the measurement of ionizing radiation used in medical diagnostic and therapeutic procedures. The individual chambers have walls constructed of "air-equivalent" materials and the sensitive volume is filled with air. A preamplifier located close to the detector allows a reasonably long run of cable between the detector and the readout. The readout system functions either as a ratemeter or as an integrating device. In addition, a high voltage supply for the chamber is an integral part of the readout. The entire system is very stable and accurate, and state-of-the-art solid-state electronics makes it easy to operate. As with the condenser R-chambers, the detectors may be purchased with a calibration traceable directly to the National Bureau of Standards.

Pulsed Ionization Chambers

Pulsed ion chambers offer the advantage of a fast response relative to the current-type chambers. In these detectors the applied voltage is pulsed and the output pulse is due only to the collection of the electrons at the anode. This type of detector has been used primarily for the study of alpha particles. Alpha particles and neutrons produce relatively large pulses in these detectors which can be amplified easily and measured.

Two special considerations must be observed in the use of pulsed ionization chambers. First, the pulse size is sensitive to the interaction site in the detector volume. The influence of this effect may be reduced by using a wire-mesh grid around the anode. The grid does not interfere significantly with the movement of the electrons, yet it reduces the influence on pulse size caused by the positive ion space charges.

Secondly, since this detector is designed to produce voltage pulses by the rapid collection of electrons, gases with a high electron attachment coefficient such as the halogens, oxygen and water vapor should not be used.

PROPORTIONAL COUNTERS

Proportional counters in common use are usually of two types; gas-flow and sealed tubes. In the gas-flow device, the counting gas is continuously circulated at a slow rate through the detector volume. These detectors have essentially an infinite life because the counting gas is constantly replenished and molecules degraded by the ion formation process are continually removed. Sealed proportional counters have a finite life since there is usually no mechanism for replenishing the filling gas which will accumulate impurities due to corrosion of sealing materials in bombardment effects.

Gas-Flow Counters

A schematic diagram of a gas-flow counter is shown in Figure 4.14. This detector will be used as an example in the general discussion of proportional counter characteristics. The chamber is cylindrical or sometimes hemispherical and the anode is usually a thin loop of wire (*e.g.,* 0.003 in. in diameter) supported in the center of the detector. A thin window made of aluminized mylar or other suitable material allows radiation to enter the sensitive volume of the detector. The counting gas, typically a mixture containing 10% methane and 90% argon called "P-10 gas," flows through the detector continuously. Before operation the detector volume is usually purged with a high volume of gas, but during use the flowrate of counting gas is greatly reduced to save gas and avoid unnecessary disturbance of the counting gas volume. "Windowless" proportional counters are also in

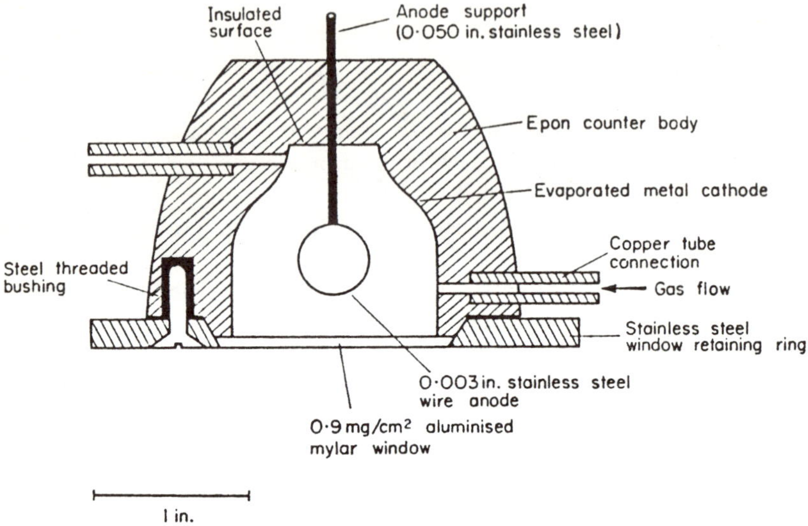

Figure 4.14 Low background end-window counter, made of metallized plastic (after Softky and Nather).

common use; in these the radioactive material is placed inside the sensitive volume of the detector. These windowless detectors may be hemispherical or spherical in shape and have a 2π or 4π geometry, respectively. Such windowless operation is particularly important when alpha and soft beta emitters are counted since otherwise an appreciable portion of the radiation would be lost in passing through a wall or window to reach the sensitive gas volume. Flow counters are particularly suitable for such internal counting since any air introduced when the sample source is changed will be swept out rapidly by the gas flow. Although the design of various flow detectors may vary, the principle of the gas-flow detector remains essentially the same and is discussed below.

For a given type of radiation the count rate versus operating voltage characteristics of these detectors possess apparent plateaus. In the proportion counter, separate plateaus are obtained for the various particle types which exhibit large differences in primary ionization, *e.g.*, alpha and beta radiation (see Figure 4.15). This figure was obtained by routing pulses produced by a mixed alpha-beta source from the detector to a scaler and counting those with heights sufficient to pass the discriminator voltage setting (called the starting voltage). At low operating voltages, only the most energetic alpha particles will produce such a pulse. As the voltage is increased, more and more pulses pass the discriminator level. Finally all the alpha pulses are large enough to pass and a plateau results. Operation of the

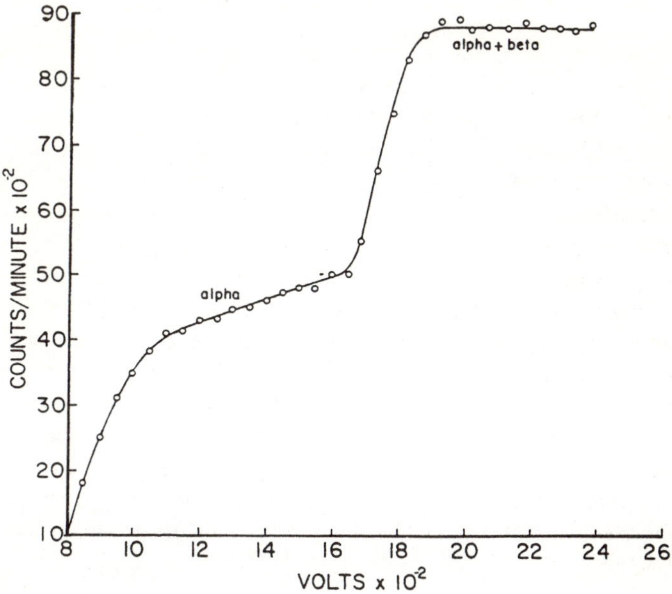

Figure 4.15 Characteristic curve for a proportional counter (P-10 filling gas).

detector on the plateau leads to a condition where small changes in operating voltage, *e.g.*, due to line fluctuations, do not significantly affect the operating efficiency of the detector. In normal use the operating voltage is selected in a region located on the first one-third to one-half of the plateau.

If the voltage is increased even further, an increase in the observed count rate will result. In this situation, a voltage has been reached at which gas multiplication produces additional pulses. These pulses are the result of ionizations produced by beta particles which have been amplified sufficiently to pass the discriminator. At some higher operating voltage, another plateau is reached which represents a condition where both alpha and beta radiations cause pulses of sufficient amplitude to pass the discriminator. The beta plateau (or more properly the combined alpha-beta plateau) is not as flat as the alpha plateau. This is due primarily to the wide distribution of beta energies. In a typical gas-flow proportional counter, the slope of the beta plateau will be less than 0.2% per 100 volts. The detector operating voltage will depend on the type of counting gas used. In general, P-10 gas results in a lower operating voltage than other standard gases.

Windowless gas-flow counters are used primarily for the detection of alpha particles. They offer the advantage that the source is part of the detector and thus the alpha particles do not dissipate a portion of their energy in reaching the detector window and passing through it into the sensitive volume. In addition, because of the large difference in specific ionization, alpha particles may be counted efficiently in the presence of other radiations, *e.g.,* beta or gamma radiation. These radiations can be easily discriminated against due to their low specific ionization. Finally, proportional counters operated on the alpha plateau exhibit a low background count rate, often less than one count per minute, making it possible to assay alpha sources with low activities.

One application of gas-flow proportional counters to the detection of alpha particles outside the normal laboratory situation is the use of the gas-flow alpha survey meter. The detector features a flat detector with 50 cm^2 active area. The window is aluminized mylar 0.85 mg/cm^2 thick and the counting gas is propane. The detector is designed to operate in the presence of high-humidity, volatile solvent vapors, inert or other atmospheres, and other types of radiation. The detector may be used at count rates up to 500,000 counts per minute. A discriminator adjustment is included as part of the electronics package which, if adjusted properly, will allow alpha surveys to be made in the presence of gamma radiation levels up to 50 R/hr.

Sealed Proportional Counters

The widest use of sealed-tube proportional counters is in the detection of thermal neutrons. A material with a high neutron cross section, *e.g.,* boron, is introduced into the chamber and the counter functions by detecting the charged particles liberated by the (n,α) reaction in boron-10. The reaction, ${}^{10}_{5}B(n,\alpha){}^{7}_{3}Li$, has a high thermal neutron cross section (~ 3,800 barns) and exhibits a simple-energy dependence over a wide neutron energy range, from the thermal energy to about 30 keV. The reaction is exothermic and an energy release of up to 2.78 MeV is shared between the alpha particle and the recoil lithium nucleus. There is also an excited state of ${}^{7}Li$ in which a 0.48 MeV gamma-ray is emitted and, in this case, the shared energy is 2.30 MeV. This latter reaction is the most probable.

The boron may be incorporated in the detector in two ways. First, a thin layer of boron may be used to coat the inside of the cathode. The boron layers are typically in the range 0.5 mg/cm^2 or less because coatings thicker than the maximum range of the alpha particles (about 0.75 mg/cm^2) would actually reduce the sensitivity of the detector. This type of detector has the advantage that a conventional counting gas can be used. However, alpha particles produced in the boron lining lose some of their energy in escaping the wall and produce a wide distribution of pulse heights. The count rate versus operating voltage curve obtained has a plateau with a slope greater than that obtained in other proportional counters.

An alternative method of introducing the neutron absorber is to combine it with the counting gas. In this detector the gas filling is boron trifluoride ($^{10}BF_3$), typically at a pressure of 120-600 torr (16.3-81.6 kPa). This detector has the advantage that the (n,α) reaction is produced in the gaseous volume and, therefore, the entire energy of the alpha particle will be utilized in producing ionization in the sensitive volume. Usually the plateau in the operating curve is very flat.

Proportional counters designed to detect neutrons have found wide use in mixed neutron-gamma radiation fields that may exist around reactors, accelerators or certain isotopic neutron sources. The detectors not only have good discrimination characteristics against gamma radiation but may also be used to detect both thermal and fast neutrons. This is accomplished by the use of an additional moderator. For example, in normal use a $^{10}BF_3$ proportional counter is sensitive only to thermal neutrons. However, if a thick moderator, such as paraffin or polyethylene, is placed around the detector it becomes sensitive to fast neutrons. Fast neutrons incident on the moderator are thermalized and are detected by the proportional counter as thermal neutrons. The moderator may be any convenient shape; spheres, right-cylinders and rectangular moderators are in common use. Thus, with the BF_3 counter outside the moderator the thermal neutron component of the radiation field is detected. Inside the moderator the fast neutron component of the field is detected. In practice, these detectors must be carefully calibrated before use because often no precise information is available on the sensitivity of a specific detector, the energy spectrum of the fast neutron field or the percentage of the fast neutron component which is thermalized. In addition, the response of the detector system depends on the thickness of the moderator materials and the incident neutron energy. This is illustrated in Figure 4.16 which presents the relative count rate of a BF_3 counter as a function of moderator thickness for several neutron energies.

Another method for detecting thermal neutrons with proportional counters is to utilize a "fission counter." This detector uses a thin coating of a fissionable material, *e.g.*, ^{235}U, on the electrodes to detect thermal neutrons. Thermal neutron capture in the ^{235}U results in a fission event and the fission fragments produce a high density of ion pairs in the detector gas. The advantages of a fission counter are its low sensitivity to very high gamma radiation fields and the large amplitude of pulses due to the fission fragments, allowing easy discrimination against other radiations, including alpha particles emitted in the normal radioactive decay of ^{235}U. Fission chambers are used widely as part of the startup instrumentation on nuclear reactors. This is further discussed in Chapter 9. Proportional counters have also been

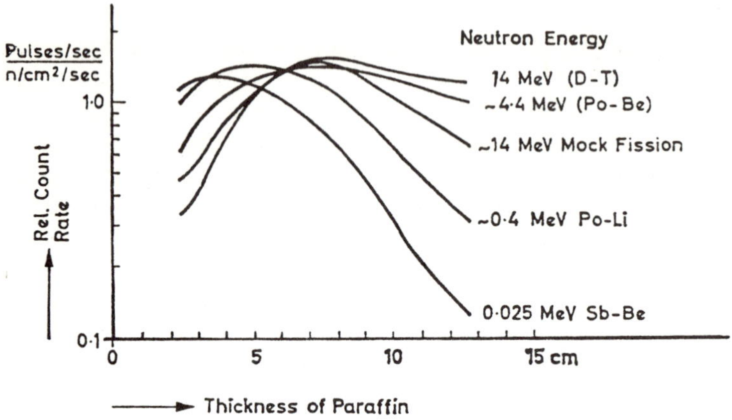

Figure 4.16 Relative count rates for a paraffin-covered BF_3 counter for different neutron energies. The paraffin is covered with cadmium to remove slow neutron background (after Wallace, 1960).

lined with hydrogenous material, paraffin or polyethylene, and used as fast neutron detectors. One such detector is the Hurst absolute fast neutron dosimeter. This cylindrical proportional counter is lined with polyethylene and the gas filling is cyclopropane (ethylene has also been used). The gas pressure is typically 500 torr (68 kPa) for cyclopropane and 750 torr (102 kPa) for ethylene. In either case the gas and the walls have the same atomic composition, satisfying one of the requirements of the Bragg-Gray relationship. In this case, ionization produced by proton recoils produces relatively large pulses. Using this detector it is possible to measure the energy absorbed per unit mass of the gas. From this measurement the energy absorbed per gram of tissue can be determined. Over the energy range 0.01 to 20 MeV the ratio of the absorbed dose in ethylene to that in tissue is about 1.45.

The Hurst counter has a low response to gamma radiation due to the differences in pulse heights between those produced by proton recoils and secondary electrons. In addition, the associated electronics usually feature a discriminator which allows the measurement of a fast neutron dose of 0.001 rad/hr in a ^{60}Co gamma radiation field of 50-100 R/hr.

Many other specialized proportional counters have been designed and used in radiation measurements. One of these is the Rossi LET chamber. This proportional counter consists of a sphere of tissue-equivalent material featuring a helical "field-defining" wire around the central electrode. The intent of the detector is to simulate small volumes of tissue, by varying the filling pressure of the tissue-equivalent gas, measure the pulse-height distributions from the detector, and calculate the LET distribution from this data. These detectors are available commercially, in sizes up to about 2 in. (5 cm) in diameter, with quick-disconnect fittings for gas-filling and built-in preamplifiers. However, after an initial flurry of interest, this type of proportional counter has not found wide use.

Multiwire proportional counters (MWPC) have found wide use in many high-energy physics and nuclear physics experiments. The interest in this detector has grown rapidly since its introduction in the late 1960s and the applications have expanded into astrophysics, crystallography and nuclear medicine, as well as other fields.

Multiwire counters are suitable for internal counting of low-energy events and may be considered as windowless detectors for low-energy X-rays. The wires must be spaced far enough to minimize interferences due to capacitative coupling or photoelectric effects. A typical MWPC is shown in Figure 4.17.

Multiwire drift chambers are made of planes of anode wires sandwiched between two cathode planes. The wires may be spaced as close as 0.5 mm although the practical limit in a large MWPC is about 2 mm. In many cases each wire is connected to an elaborate computer circuit for online computer analysis of the pulse. Some detectors with nearly 10^5 wires are in operation.

The properties of multiwire detectors which have contributed to their study and use are

1. The arrays of anode wires constitute independent detectors, *i.e.*, it is possible to localize ionizing radiation within an error equal to the wire spacing.
2. Signals induced on the cathode permit the localization of the avalanches along the wires.
3. The time lag between the production of ionization electrons in the gas filling and the pulse can be used to determine the position of the particle trajectory.
4. The time resolution of the detectors is quite good. In some detectors the time resolution is on the order 25 μsec, a significant factor in the success of these devices.

Because sequential nuclear reactions within the detector volume can be followed from wire to wire, large multiwire chambers called "drift chambers" have been built with elaborate online data evaluation. Thus, the primary use of these detectors is in experiments in which the accurate localization of ionizing particles is essential. This use has been extended to the detection of low-energy gamma rays (30-80 keV) as well as the development of two-dimensional neutron detectors.

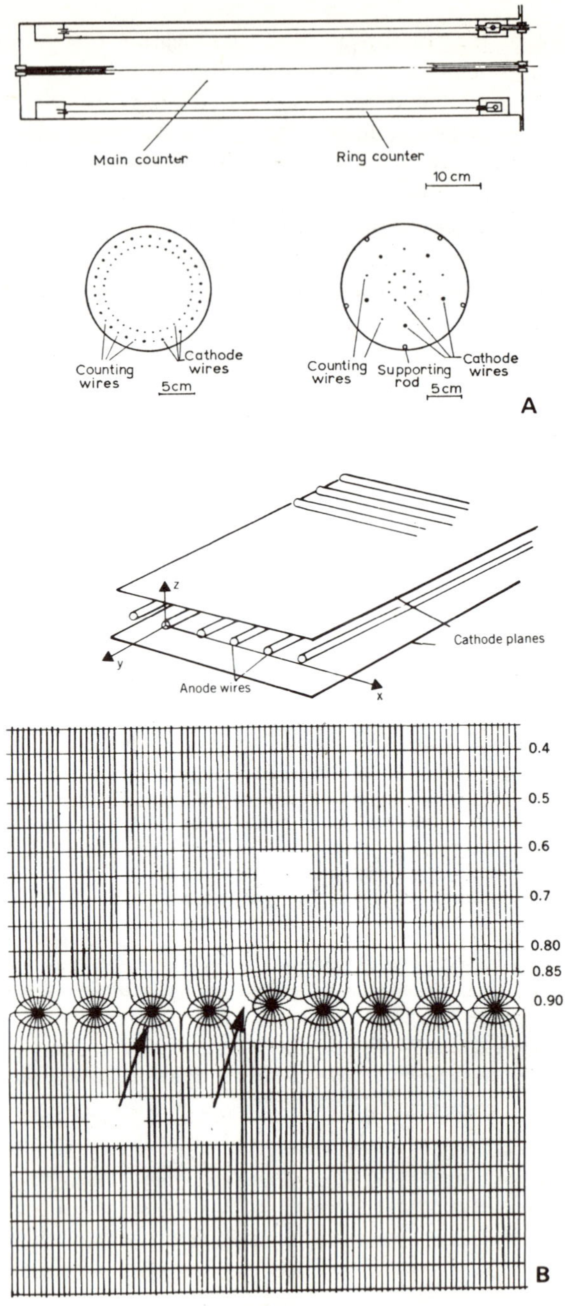

Figure 4.17 Operation of multiwire proportional counter.

Portable Geiger-Müller Counters

Although proportional counters and Geiger-Müller counters are often discussed together because of their similarities, there is one fundamental difference which must not be overlooked. The pulse amplitude from a proportional counter is closely related to the number of ionizing particles entering the gas volume, whereas the GM counter delivers essentially the same size pulse for any directly ionizing particle entering the detector volume. That is, as long as the ionizing particle generates even one primary ion pair in the gas volume, the GM counter will produce a pulse of a uniform height. For this reason GM counters are used simply to indicate the presence of radiation rather than as precise instruments to measure absolute activity, absorbed dose, etc. Nevertheless, there are many applications for which GM counters are quite suitable in many research, industrial and medcal situations, for example, because of their simplicity and reliability.

Usually GM tubes are cylindrical but the exact dimensions vary widely. Tubes are available commercially from 2.5 to over 100 cm in length and 0.32 to 5.4 cm in diameter. Typical construction materials of modern detectors are either stainless steel or aluminum. Many detectors are available with thin windows covered with mica (0.5-0.85 mg/cm^2) (Figure 4.7). The usual gas filling is Ne with a halogen quenching gas added. Ar-, Kr- and He-filled detectors are also available and some tubes have organic quenching gases added. Because of the wide selection on the market it is easy to select a GM counter to meet most applications.

One of the major uses of GM counters is in radiation surveys for protection purposes. The counter provides a fast reliable indication of the presence of radiation. GM portable detectors can be used to detect alpha, beta and gamma radiation. For example, a typical portable beta-gamma survey meter would use a halogen-filled 30 mg/cm^2 stainless steel wall GM tube, 3.5 cm in diameter and 15.2 cm long. The detector has an operating voltage of about 900 volts with a minimum plateau length of 100 volts. The plateau slope is less than 0.1% per volt and the dead time is approximately 100 μsec. The GM tube features an energy-compensating shield to limit the characteristic overresponse of GM tubes in the photon energy range 30-300 keV. The adjusted energy response is about ± 20% from 40 keV to 1.25 MeV. In some cases, a GM tube with essentially identical characteristics but with a thin mica end window (1.4 to 2.0 mg/cm^2 and about 9 cm^2 area) can be substituted. This tube would be suitable for measuring β-contamination as well as low-energy photon radiation. In either case, the maximum count rate is about 50,000 cpm.

Another interesting type of survey detector is a GM counter that can be used in high gamma radiation fields. The detector is enclosed in a hemispherical tungsten alloy shield which allows the user to obtain directional information on the radiation field or to monitor a specific object while rejecting background influences. The front-to-back ratio of the shield is 9:1 for ^{60}Co radiation. This halogen-filled GM tube is constructed of chrome-plated steel (0.25 cm thick) and may be used in radiation fields in the range 2 mR/hr to 20 R/hr. The operating voltage is about 600 volts with a plateau length of at least 100 volts (slope less than 0.25% per volt) and a dead time of about 30 μsec.

For monitoring α- and β-radiation a "pancake" detector is used. This detector is a flat cylindrical GM tube with a large area mica window (see Figure 4.7). Typical dimensions of such a window are 4.5 cm in diameter with a total area of more than 15.5 cm^2. The window thickness is 1.4 to 2.0 mg/cm^2 providing a sensitivity to alpha particles which impinge on the window with energies greater than 3 MeV. The detector is sensitive to β-radiation with energies down to 40 keV. However, the detection efficiency drops rapidly as the beta energy decreases. For example, a typical detector would have a sensitivity of about 45% of the 2π emission rate for ^{90}Sr - ^{90}Y (E_{max} = 2.2 MeV), whereas the efficiency for ^{14}C (E_{max} = 0.15 MeV) is only about 10%. The normal operating voltage of such a detector is about 900 volts and the plateau length is at least 100 volts. The maximum plateau slope is 0.1% per volt and the dead time is less than 50 μsec. In addition, the detector is shielded on all sides except the window by a tungsten shield to reduce the influence of gamma radiation; a wire screen may be provided to protect the mica window.

Special medical counting tubes are also available that are designed to be inserted in body cavities during procedures utilizing radioactive material in diagnosis. These tubes are constructed with thin stainless steel walls and some feature mica windows (1.5 to 2.0 mg/cm^2). The overall diameter of these detectors is about 8 mm and the length of the device depends on the particular application (typically 7 to 22 cm). Plateaus for these Ne-filled tubes are in the range 650-800 volts and the dead time ranges from 15 to 75 μsec.

GENERAL CONSIDERATIONS IN THE USE OF GAS–FILLED DETECTORS

The fundamental circuit diagram for a gas-filled detector was given in Figure 4.1 and the characteristics of such detectors have been discussed. In the figure, capacitor C and resistor R were shown to indicate all the capacitance and resistance of the circuit including that of the detector. When the detector is exposed to radiation, the gas within the volume becomes electrically conducting due to the production of ion pairs in the gas. The relation between the ionization current i and the voltage V is

$$RC \frac{dV}{dt} + V = Ri \qquad (4\text{-}4)$$

If the ionization current changes from i_1 to i_2, the time-dependence of V is

$$V = Ri_2 + R(i_1 - i_2)e^{-t/RC} \qquad (4\text{-}5)$$

The quantity RC in the exponent is called the *time constant* of the circuit.

If the time constant of the detector circuit is larger than the collection time of the ions produced by the passage of an ionizing particle through the volume, then a voltage pulse is obtained corresponding to curve a in Figure 4.18. The magnitude V′ of the pulse is given by

$$V' = Q/C \qquad (4\text{-}6)$$

where Q is the total charge collected and C is the capacitance of the circuit.

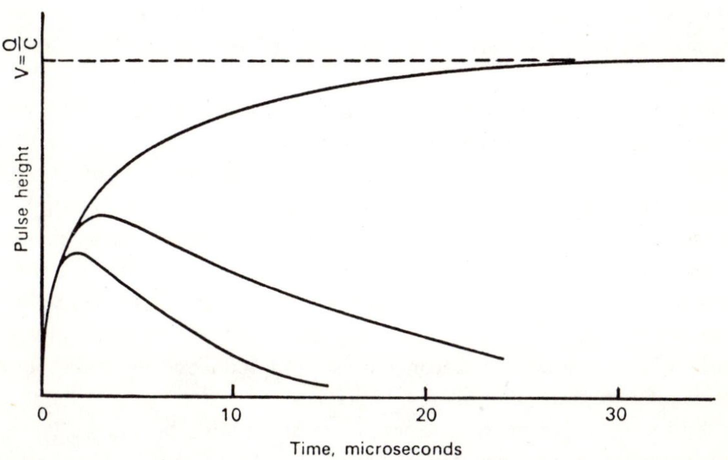

Figure 4.18 Dependence of pulse shape on time constant of the detector circuit. Top curve (a) for RC = ∞; middle curve for RC < ion collection time τ; bottom curve RC ≪ τ.

Pulses similar to those shown in curve (a) would be difficult to separate from successive pulses. Usually the time constant of the circuit is less than the time required to collect all the ions. In that case the pulse shape developed at the output is similar to curves b and c in

Figure 4.18. These pulses are narrower and it is relatively easy to separate and count the individual pulses.

The ionization current from ion chambers may be measured in several different ways. Measurements requiring high accuracy normally employ a vibrating-reed electrometer. This electrometer is of the dynamic capacitor type, *i.e.*, it consists of a capacitor with one moving plate. The capacitor, which oscillates at a frequency of 200-500 Hz, converts the ionization current into an alternating voltage. The resultant ac signal can be amplified with an ac amplifier and rectified to a dc indication. Conventional measuring techniques allow the signal to be measured with a high precision of the order of 0.1% or better.

Two other methods are used to measure the ionization chamber signal. These are the rate-of-charge method and the voltage-drop method. In the former, the ionization chamber current is used to charge a capacitor and the increase in voltage across the capacitor serves as input to the vibrating-reed electrometer. This method is quite sensitive and is normally used to measure very low radiation levels. For example, ionization currents in the range of 10^{-16} to 10^{-17} amp may be measured.

In the voltage-drop method, the charge on the capacitor is allowed to leak through a precision resistance (typically 10-10,000 megohms or greater). The voltage drop across the resistance is used as input to a vibrating-reed electrometer. The electrometer unit produces an ac signal and also provides the necessary amplification, rectification and measurement functions. This method is in many ways easier to use, but it does not provide the sensitivity of the rate-of-charge method.

Pulse counters are normally used to observe the signals from proportional and Geiger-Müller detectors. The signals from these detectors have larger amplitudes than those from ionization chambers and a pulse-height discriminator is often used to select only those pulses above a certain amplitude. This eliminates electronic noise and reduces the background count rate of the system.

The ability of a detector to separate or distinguish between two sequential pulses is an important factor that must be considered in the use of proportional and GM counters. The *resolving time* of a detector is the minimum time between two pulses required so that the pulses are registered as two separate events. Detectors with short resolving times may be used for measurements on high activity levels or in radiation fields of high intensity.

To determine the resolving time, usually a multiple-source method called the "split-source" method is used. This method employs a radiation source which is fabricated in two identical halves. The counting rate when both sources are counted together is compared with that obtained from a count of one or the other of the individual sources alone. The resolving time τ is given by

$$\tau = \frac{R_1 + R_2 - R_{12}}{(R_{12})^2 - (R_1)^2 - (R_2)^2} \tag{4-7}$$

where R is the gross count rate and the symbols 1, 2 and 12 correspond to the counting rates of source 1, source 2, and both sources 1 and 2 combined, respectively. In this equation it is assumed that any measured count rate R is the sum of the source counting rate and the detector background counting rate.

In the GM counter the resolving time is considered to be composed of two separate phases. These are the *dead time* and the *recovery time*. The electric field in a GM counter, immediately following the discharge, is reduced due to the presence of the positive-ion sheath. Additional ionizing events during this period will result in pulses similar to those shown in Figure 4.19. In the early period after a discharge, any additional ionizing

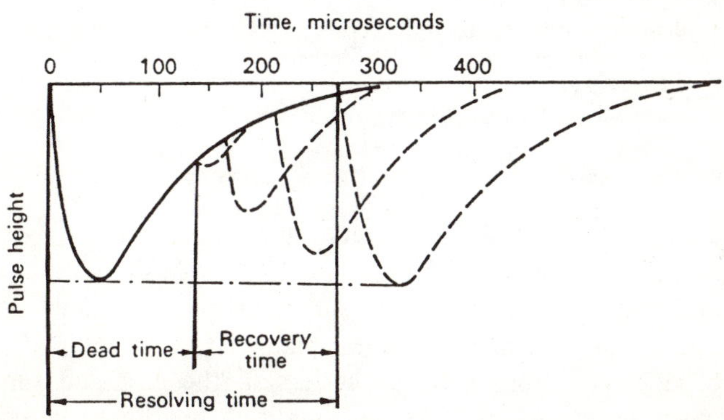

Figure 4.19 Relation between dead time, recovery time and resolving time.

event will not be able to trigger another discharge. This time is called the dead time. Later, small pulses will be seen to occur and the amplitude will increase until at some point a full-size pulse (discharge) occurs. This period is called the recovery time and is determined by the clearing rate of positive ions in the sensitive volume. In many cases the dead time sets the lower limit on the interval of time required for the resolution of individual ionizing particles, but since most counting systems have a built-in pulse-height discriminator which requires a certain minimum pulse height, the resolving time is usually the sum of the dead time and the recovery time.

The overall efficiency of a gas-filled detector is an important parameter in most applications. In its simplest form, overall efficiency of any system is the quotient of the counting rate and the source strength (emission rate). This statement oversimplifies the actual case because the counter efficiency is the product of many parameters. Consider the typical counting arrangement shown in Figure 4.14, which shows a gas-flow proportional counter, though, within certain limitations the following discussion is applicable to many types of detectors.

The overall counting efficiency η in this arrangement is given by

$$\eta = \frac{C_R}{S} = G\,\epsilon\, f_m f_t\, f_w\, f_b\, f_s\, f_o \qquad (4\text{-}8)$$

where C_R = observed net count rate (minus background)
S = source strength
G = geometry factor for the counting setup
ϵ = efficiency of counter for a particular radiation
f_m = multiple-count factor
f_t = dead-time correction
f_w = correction for attenuation in detector window, air, sample covering, etc.
f_b = source-mount backscattering factor
f_s = source self-absorption factor
f_o = other scattering corrections, *e.g.*, side scatter

In precise work all of these parameters must be evaluated to arrive at the overall counting efficiency of the system. Much of this work can be avoided by using standard radiation sources of known activity which are prepared, mounted, etc. in an identical manner (see Chapter 10). The number of different samples to be evaluated, the cost and time involved must be evaluated in order to select the proper course.

SUMMARY

Gas-filled detectors are widely used in radiation detection. The size and shape of the detectors vary and may be selected according to the desired use.

Ionization detectors provide very small currents and for very precise measurements sensitive instrumentation must be used. Many small portable chambers are available and are used commonly in radiation protection for surveys and personnel monitoring.

Proportional counters and Geiger-Müller counters operate in the pulse mode with pulses large enough to allow the use of pulse-height discriminators, scalers and, in some cases, very sophisticated electronics. These detectors also are widely used in radiation protection, radiochemical laboratories, reactor startup and high-energy physics, to name only a few. Table 4-2 summarizes typical characteristics of the three detector systems.

Table 4.2 Typical Output Signals, Amplification Factors and Resolving Times for Gas-Filled Detectors

Detector	Output Signal	Amplification Factor	Resolving Time
Pulse Ionization Chamber	0 - 1 mV	1	1 μsec
Proportional Counter	0 - 100 mV	10^2-10^4	5-50 μsec
G-M Tube	0 - 10 V	10^6-10^8	10-1,000 μsec

REFERENCES FOR FURTHER READING

Attix, F. H., and W. C. Roesch. *Radiation Dosimetry, Vol. 2, Instrumentation* (New York: Academic Press, Inc., 1966).

Charpak, G. "Multiwire and Drift Proportional Chambers," *Physics Today* 31(10):23-30 (1978).

Hendee, W. R. *Radiologic Physics, Equipment and Quality Control* (Chicago: Yearbook Medical Publishers, 1977).

Rice-Evans, P. *Spark, Streamer, Proportional and Drift Chambers* (London: The Richelieu Press, 1974).

Rossi, B. B., and H. H. Staub. *Ionization Chambers and Counters* (New York: McGraw-Hill Book Company, 1949).

PROBLEMS

4.1 A 4-MeV alpha particle produced a track parallel to the axis of a cylindrical ionization chamber. Assuming that a saturation voltage has been applied and that all the energy of the alpha particle is absorbed in the chamber, compute the charge deposited on the collecting electrode. Calculate the voltage rise if the total capacity of the collector electrode and associated circuit is 20 pF, neglecting leakage.

4.2 A standard thimble chamber is exposed to a radiation field and loses a charge of 3×10^{-11} coulombs in 10 sec. If the volume of the chamber is 1.2 cm^3, what is the exposure dose rate in R per second?

4.3 A vibrating-reed electrometer with a input capacity of 10 pF is used to measure currents in an ionization chamber of 100-cm^3 volume by the rate drift method. If the minimum rate of voltage change which can be reliably detected is 10 millivolts per minute, calculate the sensitivity of this method for the assay of tritiated gases in μCi/cc.

4.4 Draw a semi-logarithmic plot of counter tube pulse height versus voltage for two initial ionizing events differing in magnitude by a factor of 10. Cover the voltage range from zero to above the point of continuous discharge. Label the various counting regions. Describe the actions which are determining pulse height in each region.

4.5 A 2-MeV gamma-ray beam falls on a thin-window Geiger counter. The counter has an efficiency of 1.5% for this radiation. The flat plate area of the counter is 5 cm^2. How many counts per minute would you expect to correspond to 1 mR/hr?

4.6 Compute the pulse height produced by a 5.1-MeV alpha particle in a parallel-plate gridded ionization chamber of 15 pF capacitance and a time constant RC = 0.5 x maximum transit time.

4.7 A pocket dosimeter has a volume of 2.5 cm^3, a capacitance of 150 pF and is charged by a 200-V potential. If full scale is 5 R, what will be the voltage change ΔV when the instrument is discharged to read 3 R?

4.8 A proportional counter consists of a cylindrical cathode, 2 cm i.d., with a 0.1-mm-thick tungsten anode wire. It is filled with a 90:10 argon/methane mixture. Calculate the collection time for an uncollided electron generated 1 mm from the cathode if the applied voltage is 300 V, the pressure 100 torr, the ionization potential of methane 14.5 eV.

4.9 In the previous detector, estimate the distance that an electron would travel before it was capable of producing a secondary ion in the gas mixture.

4.10 Estimate the length of the plateau for the GM characteristics in Figure 4.8. If the tube deteriorated to a plateau slope of 4%, what would be the count rate at 1,000 V applied potential in the same source-detector system? What would happen to the plateau length?

4.11 A boron trifluoride counter is used to monitor the fast-neutron flux around a critical assembly and a tokamak fusion device. If it has been calibrated with a 5-cm paraffin layer and a Pu-Be source, what corrections must be made for the above measurements, or what changes in moderator thickness would be desirable? Explain.

4.12 Compare the dead times and resolving times of ionization chambers and Geiger tubes of comparable design.

What is the minimum pulse height and pulse length required to "clear" a Geiger counter in pulsed operation?

4.13 Calculate the minimum sensitivity (lowest number of ion pairs detected) of a proportional counter with an amplification factor of 10^4, a capacitance 20 pF and an associated amplifier gain of 6×10^4 if the minimum signal amplitude accepted by a counting system is 5 V.

CHAPTER 5

SCINTILLATION DETECTION SYSTEMS

INTRODUCTION

While gas-filled detectors fulfill many detection tasks quite satisfactorily, they have a number of inherent shortcomings for many applications: the low density of the gas medium makes the interaction with incident radiations inherently inefficient, especially for photons and high-energy radiations, where large detector volumes or high gas pressures would be required for good sensitivity; also, the finite drift mobility of the ions produced in the gas results in a rather slow response time, of the order of milliseconds in most cases, which leads to an appreciable dead time, during which the detector may not respond to fresh incident radiation, and to a slow rise time in the charge collection pulses, again limiting the rate of detection. One way of overcoming these limitations is to utilize condensed-state detectors, in which signal generation processes occur more rapidly, and where the high atomic density results in a high probability of interaction over a relatively short range. There are several phenomena that meet these conditions, though again no one detection mechanism is ideal for all applications.

The first one of the condensed-state detection processes to be discussed is known as *scintillation detection.* It was one of the earliest to be used, if only in a rather primitive form, when Rutherford and his collaborators observed the emission of alpha particles by means of a "spinthariscope" in the 1920s. The spinthariscope consisted of a screen of zinc sulfide which could fluoresce or "scintillate" when struck by an alpha particle. By visual observation of the screen by an observer sitting in a fully darkened room, after some hours of adaptation of the eye to the dark, these scintillations could be counted and used to estimate alpha activities. Modern scintillation counting dates from 1947 when Kallmann and Broser combined the scintillating material with a photomultiplier, enabling one to obtain more efficient light detection while eliminating problems of eye fatigue.

The general principle of scintillation counting depends on interaction of the incident radiation with a suitable fluorescent material, called the "scintillator" or "phosphor". On absorbing energy from the incident radiation the phosphor undergoes excitation to a higher electron state, followed by a prompt or delayed return to the ground state, accompanied by emission of electromagnetic radiation (light) of a wavelength appropriate to the difference in energy levels. Provided the material is transparent to light of that wavelength, this light may be observed outside the phosphor material; otherwise only light emission from the surface layer can be seen. In many materials the emitting transition also corresponds to an absorption transition, so that the energy cannot escape, *i.e.*, the material is opaque at that frequency. For this reason it is usually necessary to introduce an impurity into the scintillator to which the exciton can be transferred. The impurity gives rise to trapping levels from which transfer to the ground state can occur at a wavelength significantly different from the absorption wave length so that light is emitted in a region for which the material is transparent. This is illustrated in Figure 5.1.

Another reason for introducing impurities into phosphor materials is to serve as "wavelength shifters." Many materials, especially organic materials, will fluoresce readily when excited, but the transitions associated with light emission tend to give rise to wavelengths in the near or far ultraviolet regions where detection of this light is relatively inefficient and may require replacement of glass envelopes by quartz. By introducing small concentrations of impurities into the scintillator one can provide enough characteristic trapping centers to ensure a high probability for exciton transfer leading to a high proportion of fluorescence emission from these now lower energy levels. The net effect of this is that emission then may occur in the blue or green regions of the visible spectrum where sensitive photomultiplier detectors are easily available, and at a much lower cost.

The fluorescent light emitted by the excited scintillator is guided through a suitable optical medium to a *photomultiplier.* In the photomultiplier each incident light photon is photoelectrically converted to one or more electrons, which are then accelerated by an electric field and hit a low-work-function electrode, called a dynode, where each energetic electron causes the emission of two or more secondary electrons. By repeating this secondary electron emission process at successive, more positively charged dynodes, each initial electron can give rise to a large swarm of electrons at the last dynode from which the resulting charge pulse can be collected and further amplified, if desired. In general the

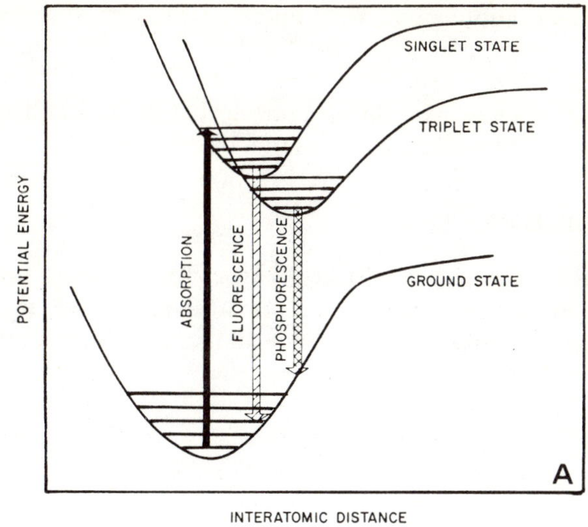

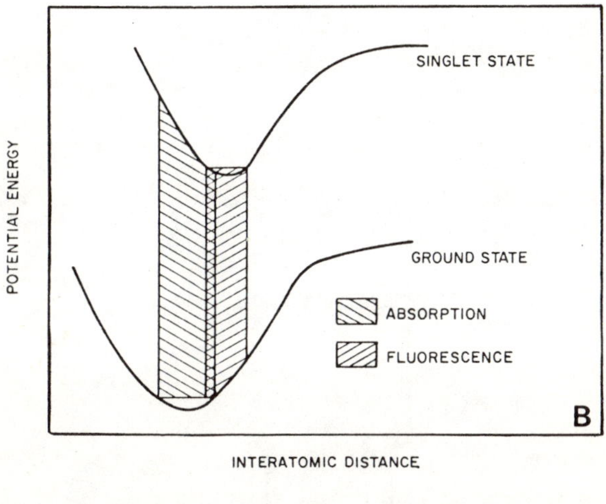

Figure 5.1 Simplified potential energy diagrams showing the relationship between light absorption, fluorescence and phosphorescence: (a) single line transitions; (b) total spectrum ranges.

magnitude of this pulse will be proportional to the number of photons reaching the photocathode and, hence, to the *energy* of the incident radiation. The *number* of pulses, representing the number of separate (in time) exciting events in the phosphor will be proportional to the *intensity* of the incident radiation. After amplification the pulse rate can be counted to obtain intensity information. With suitable treatment of the electronic pulse output from the multiplier, the distribution of pulse amplitudes can be used to obtain energy spectra, particularly for gamma radiation.

Figure 5.2 shows schematically the principal steps in scintillation counting. These will be discussed in more detail below.

PHOTOMULTIPLIERS

The purpose of the photomultiplier is to convert a rather low light intensity into an electrical signal of sufficient amplitude to facilitate further pulse treatment and pulse-height discrimination. Its sensitivity depends on

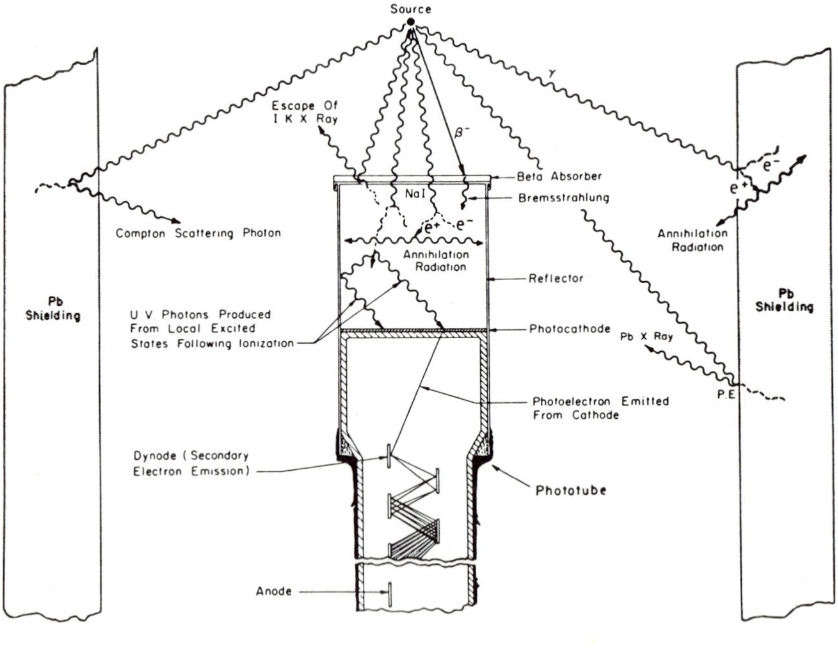

Figure 5.2 The scintillation detection process.

the efficiency of conversion of light photons to electrons at the photocathode, and on the degree of electron multiplication and its consistency in the multiplier section.

The photocathode consists of a material with a low work function, *i.e.,* its ionization energy is low enough that photon energies in the visible or near-ultraviolet region are sufficient to create a free electron and to give it enough kinetic energy to escape from the surface region of the solid film into the surrounding vacuum, yet high enough to minimize purely thermal emission at room temperature. Since this process is dependent on the incident energy, each photocathode has a frequency or wavelength range in which it will respond best. Since after emission the electrons must move in a vacuum, the photomultiplier must be placed in an evacuated envelope and the "window" material through which the incident light reaches the photocathode must be "matched" to the photon wavelength range, *i.e.,* be fully transparent to it. Table 5.1 presents the characteristics of some of the more common photocathode materials.

Since a shift towards the red end of the spectrum implies a longer wavelength and lower photon energy, photocathode materials in that region must have lower work functions. One of the problems associated with low work functions, that is, low ionization energies, can be that thermionic emission becomes possible at relatively low temperatures, such as room temperature. As a consequence, electrons may be emitted from the photocathode spontaneously due to thermal vibrations giving rise to spurious electron pulses that are referred to as dark current. This can be minimized only by cooling the photocathode which will also reduce somewhat the emission of photoelectrons.

As Table 5.1 indicates, different material formations result in photocathodes capable of responding over different regions of the spectrum. Figure 5.3 shows a plot of many of the commercial photocathode responses, some of which are employed by the television industry. Type S-4, S-11 and S-20 are encountered most often in scintillation detectors.

After emission the electrons from the photocathode must be accelerated by an applied voltage to reach energy sufficient to cause multiple emission of secondary electrons at the first dynode; subsequently the same conditions must be met between each successive pair of dynodes. Typical voltages between each multiplying stage are of the order of 50-250 V with each dynode being at a more positive voltage than the preceding one. This is accomplished by connecting each dynode to a tap on a resistive voltage divider chain as shown in Figure 5.4. To minimize dark-current contributions from the later stages, it is customary to make the first one or two voltages somewhat higher than the rest; also, to minimize instabilities from stray feedback, by-pass capacitors are often

Table 5.1 Designation and Properties of Photocathodes

Designation of Response	Cathode Surface	Window	λ'_s (mμ) (λ_s)	Max. Quantum Eff. (%)
S 1	As–Cs–O	Glass	800	0-42
S 4	Sb–Cs (opaque)	Glass	400	12-4
S 5	as S4	Corning 9741 glass	340	18-2
S 8	Bi–Cs (opaque)	Glass	420	0-68
S 9	Sb–Cs (semitransparent)	Glass	480	6-45
S 10	Ag–Bi–O–Cs (semitransparent)	Glass	540 (380)	3-6
S 11	Sb–Cs–O) (semitransparent)	Glass	440 (380-420)	11-3 (11-6) 15-8 (16-2)
S 13	as S11	Fused silica	440 (360-420)	13-2 (13-6)
S 17	Sb–Cs–O (thin layer on Al window)	Glass	490	21-5
S 20	Sb–K–Na–Cs (tri-alkali)	Glass	420 (410)	18-9 (19-3)
EMI S	Sb–Cs (semitransparent)	Glass	(380-420)	(10)

Properties of Photoemissive Surfaces

Cathode Surface	Max C_{pe} (%)	λ_s (m μ)	λ_o (m μ)	Max Sensitivity (μA lumen^{-1})	Thermionic Emission (A cm^{-2})
Ag–O–Cs	1	850	1,200	60	10^{-12}
$Cs_2Sb(O)$	20	400	620	90	10^{-15}
Bi–Ag–O–Cs	10	450	750	90	10^{-14}
Na_2KSb	25	370	650	60	10^{-16}
$(Cs)Na_2KSb$	35	400	850	230	10^{-16}

[a]Different entries represent different commercial tubes.

supplied on those stages. The overall voltage applied to a 10-stage photomultiplier may be of the order of 1,200 to 2,000 V. If the anode is grounded, this may result in a high potential difference between the photocathode and any metallic light shield surrounding it, leading to possible arcing damage. If the cathode is grounded, the output signal will be at a high potential and extreme care must be taken in running cables and in connecting any preamplifier. Since the preamplifier often is a field-effect transistor, it is usually safer to operate with the anode or collector near ground. In any case, the high voltage must always be shut off before any signal leads are removed or connected.

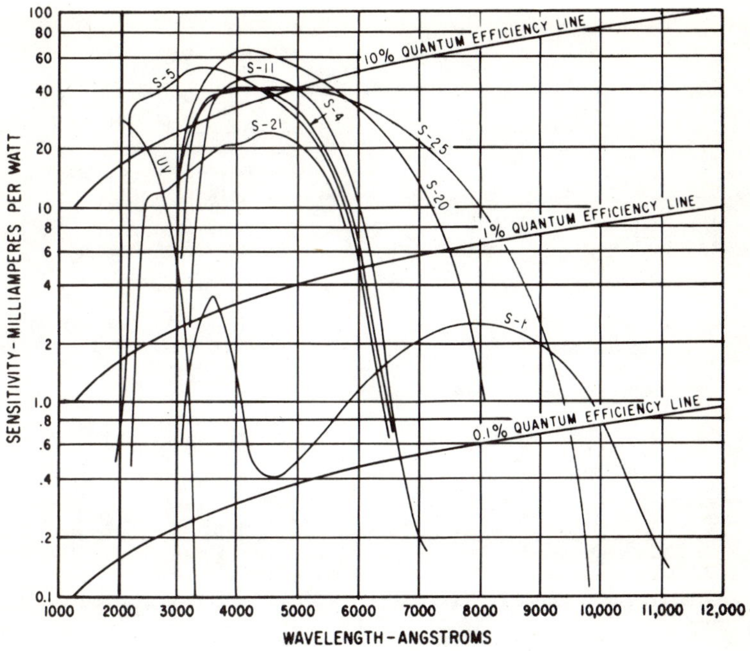

Figure 5.3 Photocathode response curves.

The voltage-divider arrangement should be located so that it will not affect tube operating temperature. Head-on type photomultipliers sometimes use zener diodes between cathode and dynode No. 1 to provide constant voltage when tube sensitivity is varied by adjustment of supply voltage. An important consideration is that the voltage-divider current should be maintained at a value of at least 10 times that of the expected average value of anode current. If this consideration is not observed, deviation from linearity and limitations on anode-current response to pulsed light may occur. The latter effect may be reduced by connecting capacitors between the tube socket terminals for the last 3 or 4 dynode stages and anode return. The values of the capacitors will depend upon the shape and the amplitude of the anode-current pulse, and the time duration of the pulse, or train of pulses.

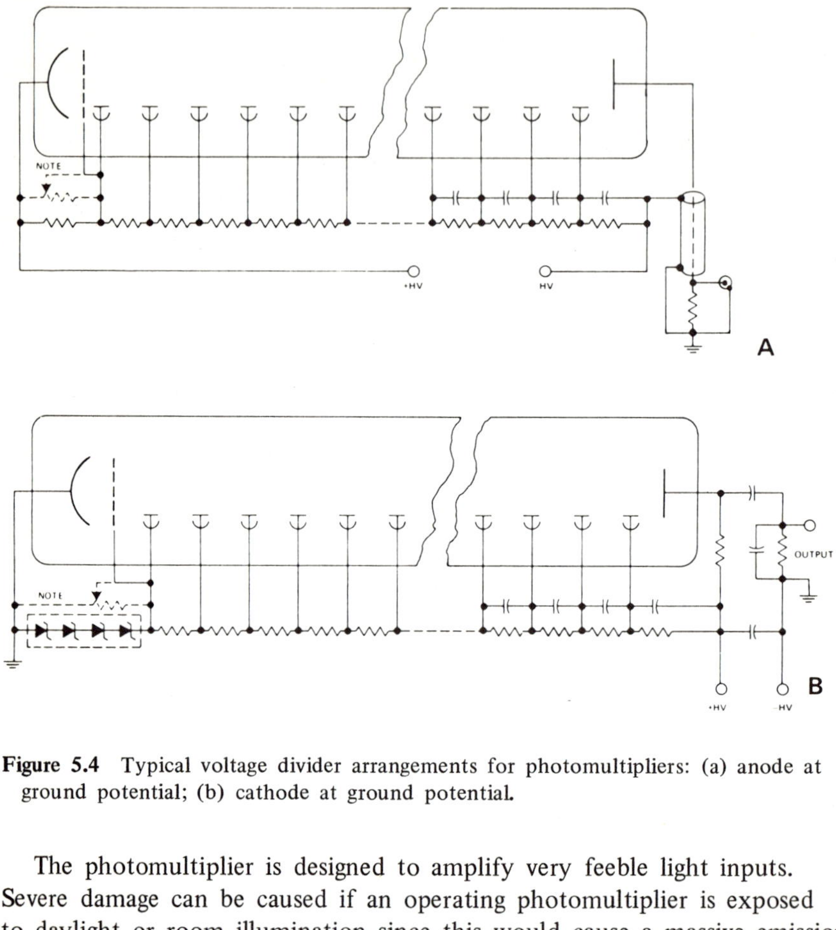

Figure 5.4 Typical voltage divider arrangements for photomultipliers: (a) anode at ground potential; (b) cathode at ground potential.

The photomultiplier is designed to amplify very feeble light inputs. Severe damage can be caused if an operating photomultiplier is exposed to daylight or room illumination since this would cause a massive emission of secondary electrons and could easily damage or even evaporate the photocathode film. The high voltage to the photomultiplier *must always be shut off* before the detector enclosure is opened for source or detector change or any other reason!

The voltage divider chain passes a steady current, parts of which must supply the charge to the dynode when secondary electrons are emitted. The rate of response of the dynodes and of the multiplier as a whole is, therefore, determined by the transit time of electrons between dynodes and the time constant of the voltage divider system consisting of R_1-R_{10} and the distributed capacitance of the detector, between the dynodes, the base pin connectors and any external capacitance C. Reducing the value of the resistors to speed up response may result in an excessively high standing current required from the power supply; the resistor chain typically consists of 0.5-1.2 megohm resistors.

The electron multiplication of the photomultiplier is described by the multiplication factor M.

$$M = A\eta r^n \tag{5-1}$$

where
r = secondary emission rate per dynode
n = number of dynodes
η = cathode conversion constant (electrons/incident photon)

The dynode emission depends on the energy of the impinging electrons and, hence, on the interdynode potential V_d. For Sb-Cs dynodes

$$r \propto V_d^{0.7}$$

typically

$$r = 0.2\ V_d^{0.7}$$

Hence,

$$M = A\ \eta\ \Sigma\ r^n = A'\ \eta\ \Sigma\ V_d^{0.7n} \tag{5-2}$$

For a 10-stage multiplier, M varies approximately as V^7.

It is evident that small changes in applied voltage can cause a severe change in M; for this reason a very well stabilized voltage supply is required. The electron emission varies with temperature according to Richardson's law

$$\eta = B\ T^{3/2} \tag{5-3}$$

Consequently the multiplication factor is also temperature-sensitive and detector performance may vary with ambient temperature changes, as does the dark current.

For very rapid response time, the stray capacitance can be reduced by using low-background, low-capacitance polystyrene tube bases. The other major factor controlling the distributed capacitance is the physical configuration of the dynodes. The dynodes must be so placed that the accelerating field steers all secondary electrons to the next dynode, there must be no diverted electrons that could cause oscillations, particularly if leakage occurred from a later dynode to an early one, the dynodes must be firmly mounted to avoid voltage fluctuations, their assembly must be compact to fit into a moderately small glass envelope and there must be no danger of arcing which would cause light emission and stray signals.

Various manufacturers have met these requirements by different configurations which are shown in Figure 5.5.

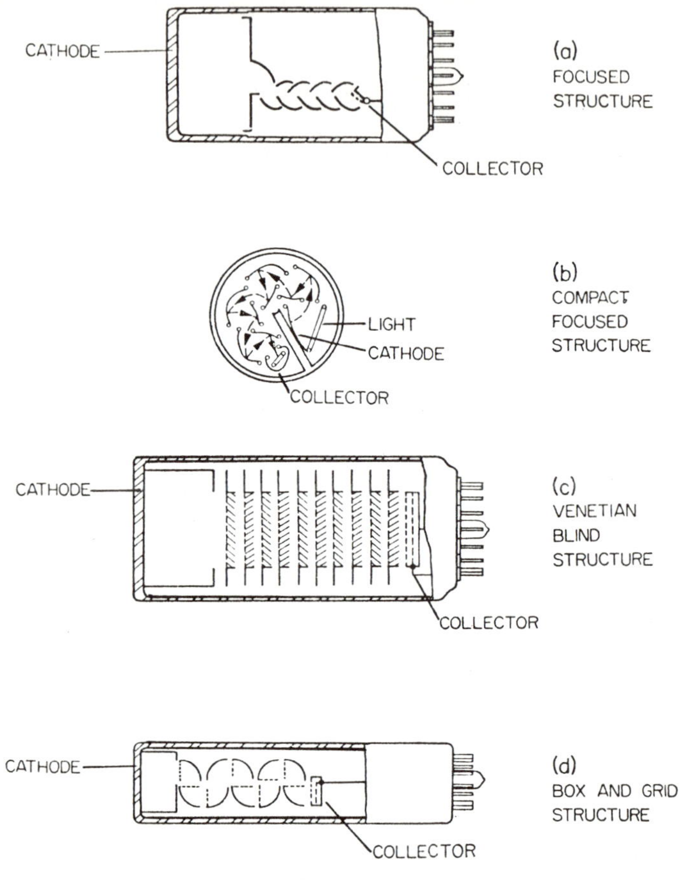

Figure 5.5 Photomultiplier dynode structures.

The commercial types shown are competitive in performance and in meeting most of the criteria mentioned above for major nuclear detection applications. In certain types of research, response time, low capacitance, low dark current, special temperature conditions or matched performance for multiple detectors may be of overriding concern and special versions of these photomultipliers may be obtained.

Since the flow of electrons between dynodes may be affected by stray electric and magnetic fields existing in the vicinity of the detector, it is usually necessary to contain the photomultiplier in a metal housing and preferably surround it with a cylindrical magnetic shield made of a high-permeability material.

Operating the multiplier at excessive voltage can lead to overloading, that is, larger electron emission at later dynodes than can be handled by the voltage divider network, and a buildup of space charge. This will result in nonlinearity in output voltage and can result in overheating of the tube.

In recent years some new developments of electron multipliers have appeared. One of these is the continuous resistive secondary emitter, two versions of which are shown in Figure 5.6. Secondary electrons are emitted along the continuous surface of a tubular channel whose resistive properties permit the maintenance of a continuous voltage gradient. Figure 5.6b shows a microchannel plate operating on this principle which can be used in image-forming scintillation cameras.

At present, such multipliers seem to find their main application as direct charged-particle detectors and in infrared detection, and less so in nuclear detection work because of their limited aperture.

SCINTILLATORS

The key component in scintillation counting is the scintillator, where conversion of incident radiation into light occurs. For efficient detection, the scintillator, or phosphor, should have high density, a high fluorescent light yield, be transparent to the emitted light with little internal scattering, have a refractive index compatible with the light guide and photomultiplier window, and have adequate physical and mechanical stability. The dimensions of the scintillator should be sufficient to stop most of the incident radiation and it should be readily contained in a light-tight shield. Not all of these criteria can be met by any one scintillator material and a variety of phosphors have been developed for different applications. Scintillator materials may be categorized as inorganic or organic and may be solid or liquid. Various gases, particularly the noble gases, will also exhibit fluorescence phenomena, but they offer no practical advantages over gas-filled ion collection systems and, thus, will not be considered further.

Inorganic Phosphors

The major parameters of interest are fluorescent yield and light output. Many solid inorganic substances will fluoresce readily under suitable excitation and various formulations have been marketed for fluorescent lamps and television picture screens. However, only a few materials have

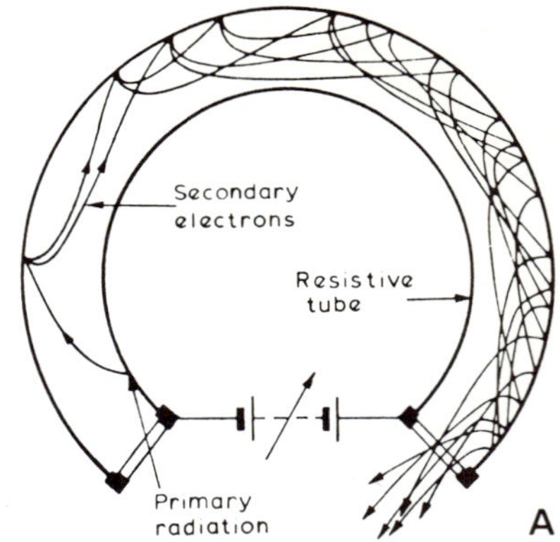

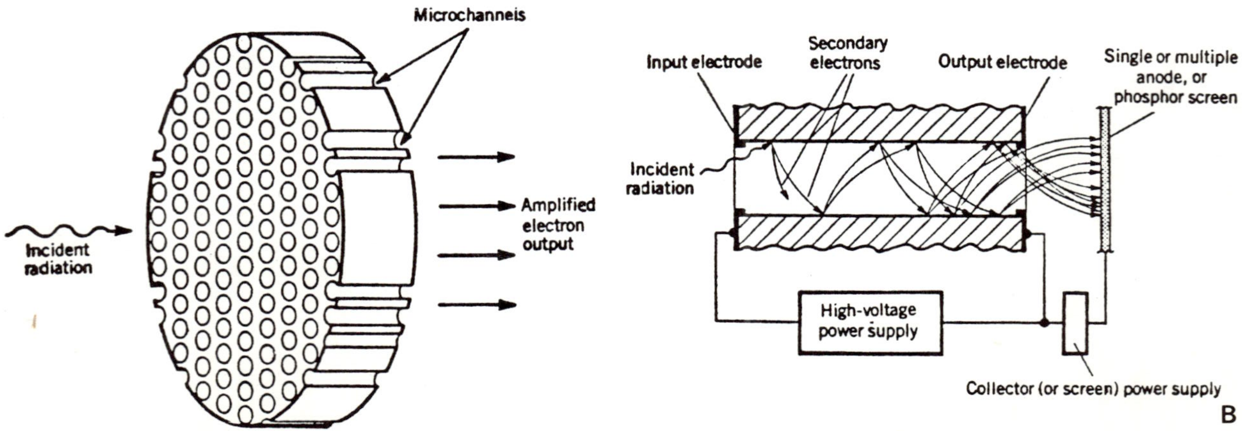

Figure 5.6 Channel electron multiplier arrangements: (a) single tube system; (b) microchannel plate.

found application in nuclear detection where radiation-induced events may be few in number and light yield and efficiency are critical.

One of the earliest and most widely used phosphors is zinc sulfide (ZnS), grown artificially or derived from the mineral compound *willemite.* It has been widely used on cathode-ray oscilloscopes and was employed in Rutherford's spinthariscope. Pure zinc sulfide emits light in the blue-green region; it is normally used with a silver or copper impurity which shifts emission to the yellow-green region. Its atomic number is high and it has a high fluorescent efficiency. Its drawbacks are that it is nearly impossible to grow in massive crystals and is not sufficiently transparent to its own emitted light.

Consequently, the use of zinc sulfide, as ZnS(Ag), is confined to the detection of heavy charged particles, such as protons or alpha particles, whose range is short enough that a thin layer of ZnS(Ag) is adequate to stop them. In practice, a thin granular layer of ZnS(Ag) is cemented by means of a dilute transparent binder on to the outer surface of the cathode window of the photomultiplier. To ensure that light from the outer exposed layer of the phosphor can reach the photocathode, the layer must be thin enough to be translucent, yet thick enough to stop the incident particles. For alpha particles this thickness is of the order of 20 μm; if it is much thicker or thinner, detection sensitivity falls off. Because of the fragility of the layer and the absorption effect of moisture or grease, the phosphor layer should never be touched with bare fingers! Despite these drawbacks, ZnS(Ag) is widely used for alpha particle detection and provides a comparison standard for the fluorescent yield of other detectors.

As anybody knows, who has watched the slow decay of light emission from a television screen after the set is turned off, there is a slow phosphorescent component in ZnS with decay times of the order of minutes to hours. This can be minimized by choosing a suitable dopant impurity; however, it also indicates the need to avoid any exposure of the scintillator to bright light, even with the photomultiplier voltage turned off, since the residual light emission may introduce an undesirable decaying background reading into subsequent measurements.

The limited thickness of zinc sulfide scintillators makes them unsuitable for the detection of radiations with longer range. For this one must look to high-Z materials that can be grown in moderately large transparent crystal blocks. After some initial work with tungstates, whose limited crystal size and high refractive index soon proved inconvenient, this field has been dominated almost exclusively by the alkali halides, NaI, KI, LiI and related compounds. These materials are ionic crystals, inherently poor electrical conductors, and crystallize in rather brittle, simple lattice structures. They are strongly hygroscopic and NaI, for instance, forms a brown-yellow iodide film on its surface when exposed to moisture. The crystals cleave readily and crack when subjected to thermal or mechanical shock. However, fairly large clear crystals of sodium iodide can be grown with care in a dry inert atmosphere, up to 30 cm (12 in.) in diameter, though the cost per crystal increases enormously past the "standard" 2-3 in. (5-7.5 cm) diameter size. Larger detector volumes have been assembled by means of "mosaics," tightly packed assemblies of individual crystals.

To obtain fluorescence emission with good efficiency and in a convenient frequency region, dopant impurities are added to the crystal melt before or during growth. For sodium iodide, thallium is preferred since it distributes itself uniformly through the crystal and gives rise to light at frequencies close to the maximum sensitivity of commercial photomultipliers. Note that thallium is a very toxic material and NaI(Tl) crystals that require cleaning, polishing or remounting should always be handled in a dry box and never with bare hands.

Table 5.2 compares the characteristics of the more important inorganic scintillators. Of the alkali halides, only NaI(Tl) and LiI are used to any significant extent as scintillators for gamma-rays and neutrons, respectively. LiF is also used, with different dopants, in thermoluminescent dosimetry and will be discussed in that connection later. It has also been used in curved-crystal X-ray spectrometry where the crystal diffraction patterns obtained from perfect, carefully bent single crystals are used to obtain precision energy determinations.

Because of its high efficiency and high atomic number, NaI(Tl) is employed as the conventional high-efficiency gamma-ray detector phosphor. It is usually supplied as a solid cylindrical crystal, mounted in a hermetically sealed casing or as a "well crystal" with a hole drilled partway into it to permit counting of small vials or cylindrical samples in a near-4π geometry. In removing material for the hole, the sensitivity of the detector is greatly reduced and this reduction may not always be offset by the improved detector geometry. Figure 5.7 shows cross sections through the two types of detector assemblies. One factor that also contributes significantly to overall sensitivity is the optical light collection efficiency of the detector assembly. Since the fluorescent light is emitted by the excited target atom in all directions, only part of it initially travels in the direction of the window to the photocathode. For this reason, the crystal must be coated or surrounded by an efficient, diffuse reflecting material that will redirect a high proportion of the light towards the window. Most NaI(Tl) crystals have their surfaces roughened and coated with Al_2O_3 or MgO which are good reflectors in the spectral region of interest. It is evident that good light collection from the upper regions of a

Table 5.2 Properties of Inorganic Scintillators

Scintillator	Density (g/cm^3)	Effective Atomic No. Z	Wave Length of Max. Emission (A)	Refr Index	Decay Time (μsec)	Rel. Efficiency (NaI(Tl) = 100)	Remarks
ZnS(Ag)	4.10	27	4,500	2.4	~ 5	100	Mostly crystalline powder
ZnS(Cu)	4.10	27	5,200	2.4	~ 15	100	Mostly crystalline powder
CdS(Ag)	4.8	44	7,600	2.5	~ 10	100	Yellow crystals
NaI(Tl)	3.67	50	4,100	1.85	0.25	100	Hygroscopic
KI(Tl)	3.13	49	4,100	1.68	> 1	25	Hygroscopic
CsI(Tl)	4.51	54	white	1.79	1.2	45	Not hygroscopic
LiI(Tl)	4.06	52	~5,000	1.95	1.2	10	Activation problems
LiI(Eu)	4.06	52	~4,400	1.95	1.4	35	Activation problems
CsF	4.11	52	3,900	1.48	0.005(β), 0.02(α)	5	Deliquescent
$CdWO_4$	7.9	63	5,200	2.25	~ 6	~65	Small rodlike crystals
$CaWO_4$	6.06	59	4,300	1.92	~ 4	~ 50	Small rodlike crystals

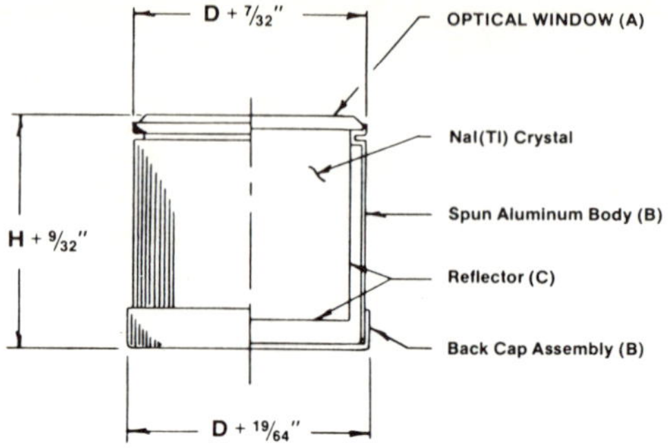

STANDARD SMALL CRYSTAL ASSEMBLY

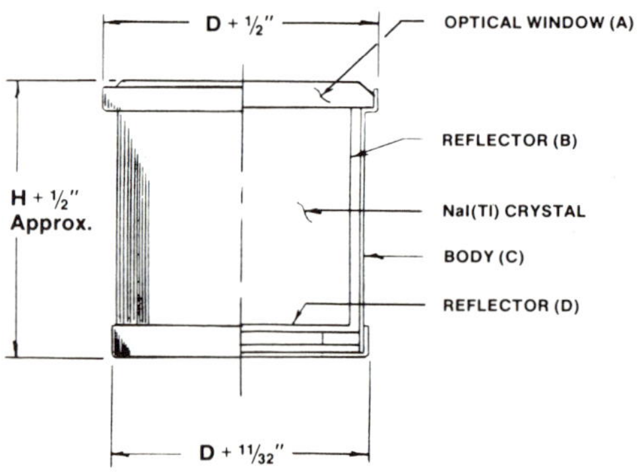

LOW MASS CRYSTAL ASSEMBLY (Up to 5" Dia.)

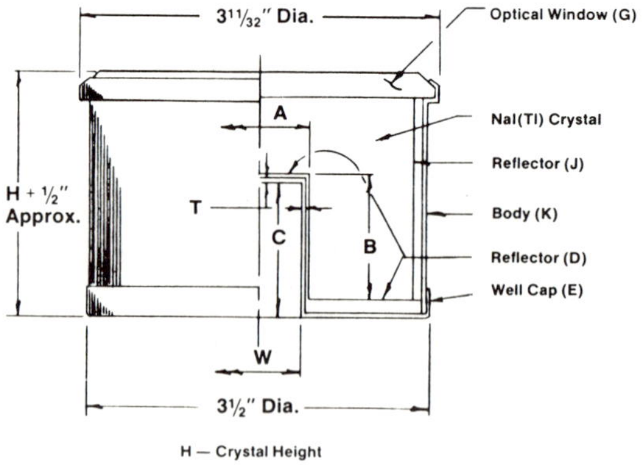

INTERMEDIATE SIZE WELL TYPE CRYSTAL
ASSEMBLY — 3" DIA.*

Figure 5.7 Typical scintillator mountings (Courtesy Harshaw Chemical Co.).

well crystal is relatively more difficult, and sets a practical limit to the maximum ratio of well diameter to crystal diameter.

To obtain good optical contact between the crystal and the window, a thin coating of mineral oil or silicone grease is applied and, in many cases, the crystal is pressed against the window by a spring.

Organic Scintillators

The size and volume of inorganic transparent crystals that can be obtained at reasonable cost and quality is clearly limited; any NaI(Tl) crystal above 12.5 cm (5 in.) in size involves great cost and considerable problems in handling and protection from thermal or mechanical shock. For applications requiring large areas or large volumes, such as wholebody counters or neutrino detectors, one has to look for other materials. Such requirements can be met to a large extent by organic materials in the form of solid crystals, plastics or liquids, though at some sacrifice in sensitivity and signal output amplitude.

Many organic substances have been known to fluoresce when excited by daylight or X-rays. It was shown in 1949 that anthracene and naphthalene could be grown into clear, thin crystal plates that can serve as satisfactory scintillators for beta radiation and soft X-rays. Anthracene is still widely used as a reference material for this purpose. Many other organic substances, particularly aromatic compounds related to the simple phenols, were found to scintillate in the near-ultraviolet and blue regions of the spectrum and their output can be wavelength-shifted by a suitable impurity. Among the interesting features of these materials are the relatively rapid exciton transfer and de-excitation time, which permit more rapid counting than is possible with inorganic phosphors, which typically exhibit relaxation times of the order of 3-10 μs. Table 5.3 compares some of the relevant parameters for various solid aromatic phosphors. It is seen that few materials rival anthracene for light-pulse amplitude, but several others, notably *trans*-stilbene, are faster.

With regard to detector volume, it proved fairly difficult to obtain clear stable organic crystals in dimensions of more than a few centimeters. On the other hand, it was found that the scintillating properties were retained by aromatic liquids and by certain polymers that may be regarded akin to the liquids. Many organic liquids will scintillate strongly if they are highly purified and can dissolve small quantities of other aromatics that serve as wavelength shifters. Here again it is important that the liquid be transparent to its fluorescent emission; however, it is obvious that with that provision, large-volume detectors of any shape can be devised provided that the liquid is not too volatile, flammable or attacks its container

Table 5.3 Properties of Important Organic Scintillation Crystals

Substance	Density (g/cm^2)	Wavelength of Maximum Emission (Å)	Decay Time (n sec)	Relative Efficiency[a]	Response for Heavy Particles (α/β)
Anthracene	1.25	4,450	23-38	100	0.1
trans-Stilbene	1.16	3,850	< 3-8.2	~60	0.1
p-Terphenyl	1.23	3,900; 4,100	4.5	~40	
p,p'-Quaterphenyl		4,350	4.2	~85	
Diphenylacetylene	1.18	4,000	~ 5	~45	
Naphthalene	1.145	3,450; 3,850	~ 6	~30	
Phenanthrene	1.03	4,100;4,300;4,500	~ 8	~80	

[a]Relative to anthracene = 100 for high-energy β-particle excitation ($E > 1$ MeV).

or any window material. Of the various materials tested, xylene, toluene and phenyl cyclohexane were found to be comparable in performance; biphenyl also is a suitable liquid in certain applications. All of these materials by themselves fluoresce in the near ultraviolet, and an effective fluorescent agent, such as *para*terphenyl, must be added. Even that mixture will not reach the optimum frequency region of the photomultiplier and a very small amount of an additional wavelength shifter must be included. Concentrations as low as 0.05 g/l of this secondary wavelength shifter are sufficient to ensure that de-excitation and light emission occur preferentially via its low-lying impurity levels. Table 5.4 shows the characteristics of a few of the scintillating liquids that have been prepared. Many of these are commercially available as "scintillator cocktails" and almost invariably consist of the matrix fluid, the principal scintillator and the wavelength shifter as shown in the table. Most liquid scintillators can provide pulse amplitudes up to 65% of that for anthracene crystals, for comparable volumes and geometries. By enabling one to use larger volume detectors, however, they permit the detection of gamma-rays and other high-energy radiations under conditions where solid detectors would have been impractical, too slow or too expensive.

The solvents usually applied are toluene and xylene, and less frequently phenylcyclohexane and triethylbenzene. The primary solute is in the most cases *p*-terphenyl (TP), but 2,5-diphenyloxazole (PPO), 2-phenyl-5(4-biphenylyl)-1,3,4-oxadiazole (PBD), and other compounds are also employed. For secondary solutes, 1,4-di[2-(5-phenyloxazolyl)]-benzene (POPOP), 2-(1-naphtyl)-5-phenyloxazole (α-NPO), 2,5-di(4-biphenylyl)-oxazole (BBO), or other compounds are used.

One of the major applications of liquid scintillators then is for large-volume detectors. Figure 5.8 illustrates such a detector that has been developed for whole-body counting. Although the scintillator fluid will

Table 5.4 Properties of Important Organic Liquid Phosphors

Solvent	Density (g/cm^2)	Primary Solute (g/l)	Secondary Solute (g/l)	Wavelength of Maximum Emission (Å)	Decay Time (nsec)	Relative Efficiency[a]
Toluene	0.87	TP(9)[b]	—	3,530	2.2 ± 0.3	48
		TP(5)	—	3,530	< 2.9	—
		TP(4)	POPOP(0.1)[b]	4,320	—	61
		TP(4)	α-NPO(0.1)[b]	4,150	< 3.2	54
		TP(4)	BBO(0.1)[b]	4,220	—	62
		TP(4)	DPH(0.1)[b]	—	—	40
		PBD(8)[b]	—	3,700	< 2.8	63
		PPO(4)	—	3,820	⩾ 3.0	54
Xylene	0.86	TP(5)	—	3,250-4,000[c]		48
		TP(5)	POPOP			44
		TP(5)	α-NPO			49
		TP(5)	BBO			57
		PPO(3)	—			
Phenylcyclohexane	0.94	TP(3)	—	3,200-4,100[e]	< 2.9	48
		TP(5)	POPOP			54
		TP(5)	α-NPO			46
		TP(5)	BBO			53
		TP(5)	PBD			48
		PPO(3)	—			55
Triethylbenzene	0.86	PPO(3)	—			52

Note: DPH = 1.6-diphenyl-1,3,5-hexatriene.

[a]Relative to anthracene = 100 for high-energy β-particle excitation.

[b]Saturated with argon.

[c]Spectral range.

Figure 5.8 Large-volume whole-body counter.

transmit its fluorescence light, it does not do so perfectly and, hence, for optimum sensitivity the scintillator tank is viewed by a multiplicity of photomultiplier tubes, cemented to the outside of the liquid cylinder.

Another advantage of liquid scintillators like toluene or xylene is their capacity as a solvent for a wide range of organic compounds. This makes it possible to introduce biological substances and, also, metal-organic compounds as converter materials into the phosphor for neutron or neutrino detection. Some of these materials are listed in Table 5.5.

One can also make deuterium analogues of many scintillator compounds. More importantly, the solvent properties of the phosphor matrix material

Table 5.5 Relative γ-Ray Scintillation Efficiencies of Solutions Containing Various Elements (after Kallmann)

Element	Solvent (Percent by mass)	Fluorescent Solute	Percent of Intensity of Same Solute in Xylene
Boron	Trimethylborate	PPO(3 g l^{-1})	5
	Trimethylborate 50% + xylene 50%		79
	Trimethylborate 47% + xylene 31% + naphthalene 22%		95
Silicon	Tetraethylorthosilicate	α-Naphthyl-phenyl-oxazole (3 g l^{-1})	5
	Tetraethylorthosilicate 75% + naphthalene 25%		74
Phosphorus	*n*-Butylphosphate	PPO (3 g l^{-1})	9
	n-Butylphosphate 70% + naphthalene 30%		72
Sulfur	Phenylcyclohexane 47% + thiophene 53%	9, 10 Diphenyl-anthracene (20 g l^{-1})	30
	Phenylcyclohexane 43% + thiophene 48% + naphthalene 9%		60
Chlorine	Xylene 85% + chloroform 15%	PPO (4·5 g l^{-1})	20
	Xylene 64% + chloroform 14% + naphthalene 22%		35
Arsenic	Xylene 96% + triphenylarsine 4%	9, 10 Diphenyl-anthracene (10 g l^{-1})	42
	Xylene 86.5% + triphenyl-arsine 3.5% + naphthalene 10%		83
	Xylene 81% + triphenylarsine 9.5% + naphthalene 9.5%		58
Antimony	Xylene + triphenylstibine 1%	9, 10 Diphenyl-anthracene (2 g l^{-1})	34
	Xylene 87% + naphthalene 12% + triphenylstibine 1%		81
	Xylene 81% + naphthalene 16% + triphenylstibine 3%		57
Mercury	Xylene 90% + di-*n*-butyl-mercury 10%	α-Naphthylphenyl oxazole (5 g l^{-1})	9
	Xylene 66% + naphthalene 23% + di-*n*-butyl-mercury 11%		15

enable one to introduce organic materials for "internal" or windowless beta counting. In this procedure the material to be counted must be soluble in toluene or other phosphor matrix, or be capable of being converted to a soluble form. It is then dissolved and its beta/gamma emission detected in an essentially 4π geometry. In practice, the amount of sample material that can be thus introduced is limited, because the sample acts as an impurity introducing its own energy levels that may divert excitons from the normal fluorescent process; also, its presence may interfere with light collection by causing discoloration or turbidity. This reduction in detection efficiency attendant on introducing a sample of a given type into the scintillator is called *quenching.* The quenching follows a trend of the type shown in Figure 5.9, which indicates the limit of sample material that may be introduced before serious reduction in detection sensitivity occurs. For any sample material introduced, a quenching curve of that type must be obtained to allow correction for the change in detection sensitivity.

Three types of quenching can be defined and any one substance may quench by more than one type. They are as follows:

a. Impurity ("chemical") Quenching

This arises due to the reduction of fluorescence light yield by the presence of nonfluorescent molecules which absorb some of the excitons

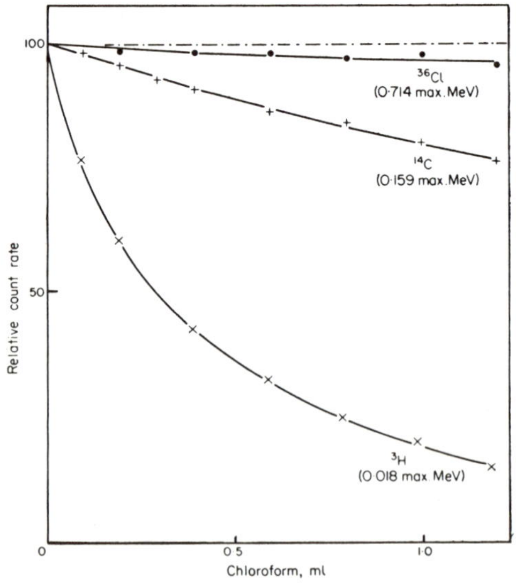

Figure 5.9 Typical quenching curve for liquid scintillators.

from the solvent molecules and dissipate the energy by radiationless transfer, *i.e.*, conversion to heat. They can also combine with solvent or solute molecules to form encounter complexes which consist either of unexcited molecules (static quenching) or excited molecules that undergo deactivation reactions (dynamic quenching). The magnitude of the quenching by any particular impurity or dissolved sample matter depends on the solvent excitation lifetime, the solute fluorescence lifetime and the chemical nature and molar concentration of the quencher.

b. Color Quenching

If the labeled material causes chemical or photochemical reactions that result in changes in the absorption characteristics of the scintillator solution or if an inherently colored material, *e.g.*, urine, is introduced, the light output from the phosphor may be diminished and a lower detection signal arises than appropriate to the sample activity.

c. Photon Quenching

The incorporation of incompatible substances that are not soluble in the scintillator solution may result in a heterogeneous counting mixture. In this case the detection efficiency of the scintillator is reduced and light collection, also, may be incomplete.

Quench effects can be corrected by obtaining a quench correction curve. In this procedure various known amounts of inactive quenching material are added to the scintillator solution which is then excited by an external gamma source. A plot of count rate versus concentrations yields relative efficiency.

Another method, the channel shift method, is based on the fact that the addition of a quenching agent and the resultant reduction in light output have the effect of shifting the apparent beta spectrum to a position of lower energy. A calibration curve can again be established by counting a series of samples containing a known constant amount of radioactivity but varying concentration of quencher and comparing the count rate in adjoining pulse-height analyzer channels.

Quenched standards, usually based on ^{14}C- or ^{3}H-labeled hexadecanes, are available commercially for calibration. These are dissolved in toluene and quenched by CCl_4 or chloroform; the solutes are PPO or POPOP. They are provided in argon-flushed vials. The preferred method of withdrawing aliquots is by chemical syringe injection into a capped vial.

The use of internal counting by means of liquid scintillators is particularly convenient for carbon-14 and hydrogen-3, which emit such low-energy beta particles that most other detection methods are rarely practical. For organic samples labeled with C-14 or H-3, such as may arise in medical or biochemical research, liquid scintillation counting finds widespread use.

Not all C-14- or H-3-containing samples are soluble in aromatic solvents or are even organic in nature. This applies particularly to water samples and other materials of environmental origin. In that case, 1,4 dioxane is often used as the scintillator matrix, as it is able to dissolve aqueous samples subject to the same quenching limitations as the other scintillators. Dioxane will freeze at 11.8°C and care must be taken if this detector is used in conjunction with a cooler system, intended to reduce phototube dark current.

Another device used to overcome quenching problems or solubility limitations are "phosphor spaghetti" or scintillation gels. In these cases the radioactive material is introduced into plastic fluorescent rods or dispersed in suspended gels that are placed inside the scintillator solution without actually diluting it. The energy transfer and the light collection may be less efficient in these cases but quenching is avoided and sometimes the samples remain reusable.

Most liquid scintillation detectors employ a single sample vial, in which the sample to be counted has been dissolved or dispersed in a liquid scintillator mixture. This vial is then viewed by two photomultipliers, whose outputs are connected in coincidence, so that only fluorescent light from the vial that is seen by both phototubes is accepted and recorded, whereas dark current and other spurious events originating in one or other phototube alone will not be counted. To further reduce chance coincidences from dark current, the whole system is usually mounted in a cooler operating at about 8°C. This also stabilizes system performance; much lower temperatures become undesirable as they would reduce photocathode conversion efficiency too much and would affect the solubility of many compounds in the scintillator matrix. Figure 5.10 illustrates such a liquid scintillator system.

Care must be taken in the choice of sample vials in such a system to ensure the absence of background activity in the glass from K-40 or uranium daughters, adequate transparency, and resistance to solvent attack if plastic vials are used.

Among the practical problems of liquid scintillators, when used as large-volume detectors, is the need for a resistant and strong container vessel, their volatility and flammability, and the difficulties in handling large volumes of these materials safely. One attractive way to overcome these problems is by the use of plastic scintillators. These materials are typically based on polystyrene which acts as a matrix in which the principal scintillator and the wavelength shifter are dissolved, much as in a liquid. The ingredients can be added to the styrene monomer and the

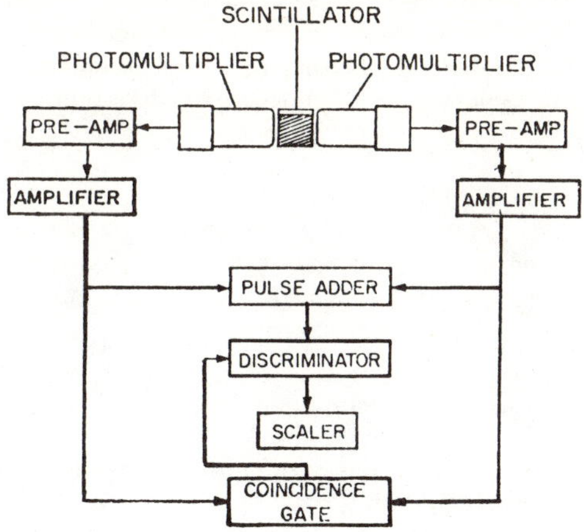

Figure 5.10 Two-tube coincidence system for liquid scintillation.

mixture then polymerized in a mold, or they can be added to a powdered polymer and recast. The size of a solid plastic scintillator is limited by the mold available, by the need to remove substantial amounts of heat during polymerization, and by the cooling characteristics of the scintillator in the mold. The scintillator material can be readily cut, machined, coated and polished, and for large volumes can be combined into multiblock mosaics. Table 5.6 summarizes some properties of typical plastic scintillators. Since the material is not hygroscopic, moisture sensitivity like that for other phosphors is no problem. Plastic scintillators are used widely where the use of liquid organics would be inconvenient or unsafe, where their higher cost is offset by their greater ease of mounting or installation, and for dE/dx detectors where they can be used in the form of thin films or foils.

LIGHT GUIDES

Efficient light collection must be supplemented by adequate optical coupling between the scintillator and the photocathode. This implies matching of the optical characteristics of all intervening materials and avoidance of reflections at interfaces. If the scintillator is contained in a light shield or container, then the window in this container must be coupled to the photomultiplier by a transparent glue or light pipe. For direct connections, clear silicone greases are often used whose transmission characteristics and refractive index are close to that of glass and whose viscosity should be high enough to ensure permanence of contact, even if

Table 5.6 Properties of Important Organic Plastic Phosphors

Solvent	Density (g/cm^2)	Primary Solute (% by wt)	Secondary Solute (% by wt)	Wavelength of Maximum Emission (Å)	Decay Time (nsec)	Relative Efficiency[a]
Polystyrene	1.054	TPB(1.7)[b]	—	4,450	4.6	36
		TP(2.5)[b]	—	3,550	⩽3.0	28
		PPO(1.0)[b]	—	3,800	—	24
		TP(0.9)	α-NPO(0.05)	4,150	2.2±0.3	36
		TP(3.4)	TPB(0.02)	4,450	4.0	39
Polyvinyltoluene	1.032	TPB(1.4)[b]	—	4,450	4.6	37
		TP(4.0)[b]	—	3,550	⩽3.0	32
		TP(4.0)	TPB(0.02)[b]	4,600	4.0	45
		TP(3.4)	DPS(0.09)[b]	~3,800	⩽3.0	52
		TP(3.4)	POPOP(0.1)	~4,300	—	59
		PBD(1.3)	POPOP	~4,300	—	47

Note: DPS = *p,p'*-diphenylstilbene.
[a]Relative to anthracene = 100 for high-energy β-particle excitation.
[b]Dissolved gases removed.

the detector assembly is lying on its side or the ambient temperature rises. For experiments around accelerators whose magnetic field or residual radiation field may affect the performance of the photomultiplier, a longer light pipe may be employed so that the photomultiplier can be located in a field-free region. The light pipe must be internally reflecting to minimize transmission losses; quartz or lucite rods or fiber optics have been employed for this purpose. In all cases great attention must be paid to the optical connections at each end of the light pipe. Low-attenuation fiber-optics light guides are now available that can guide the scintillation light signal over appreciable distances. This is particularly valuable when a scintillator mosaic is viewed, such as in certain "cameras" used in scintigraphy, a technique employed in nuclear medicine to map distributions of radioactive tracers in patients (see page 163).

DETECTOR SYSTEMS

It should be evident from the foregoing discussion that no single scintillation detector will meet the needs of all types of measurements. However, certain general requirements have emerged that are characteristic of most systems. The system usually must preserve the most valuable characteristics of scintillation detectors, their high sensitivity, linear energy response and fast response time. Table 5.7 summarizes the major characteristics of the scintillation detectors most commonly employed.

Table 5.7 Characteristics of Principal Scintillators

Radiation Detected	Preferred Scintillator	State	Form	Range of Available Dimensions	Special Applications
Alpha	ZnS(Ag)	powder	thin layer	up to 20 μm thick any area	
Beta	Anthracene	solid	crystal block	3-6 mm thick up to 5 cm long	
	Plastic	solid	thin disk	up to 15 cm diam	
	Toluene + p terphenyl, Diphenyl	liquid	solution	—	internal β counting (low energy)
X-Rays	NaI(Tl)	solid	thin cylinder	3 mm x 5 cm	X-ray fluorescence Dosimetry
	Plastic	solid	sheet	3 mm x 5 cm	
Gamma	NaI(Tl)	solid	cylinder	2 cm-15 cm diam	Portable detectors Spectrometry
			well	2 cm-15 cm diam	Small sample, 4π geometry
	Plastic	solid	cylinder	1 cm-40 cm diam	Mosaic detectors high-energy
	Tolueme + p terphenyl	liquid	large volume, annular	—	whole-body counting
Neutrons	LiI(Eu)	solid	cylinder	2-5 cm diam	neutron spectrometry
	ZnS(Ag) in Lucite	solid	disk	~5-10 cm diam	fast neutrons (Hornyak detector)
Neutrinos	Loaded liquid	liquid	tank	—	

From the operational point of view one may distinguish between:

- single-detector counters
- single-detector spectrometry systems
- dual-detector coincidence systems
- multiphototube systems, either operated in summation or anticoincidence

All systems require very stable power supplies, preferably a constant-temperature environment, absolute light-tightness around the detector-phototube assembly and protection from thermal and mechanical shocks. Figure 5.11 illustrates schematically the main elements of such systems. The preamplifier usually is a field-effect transistor; care must be taken

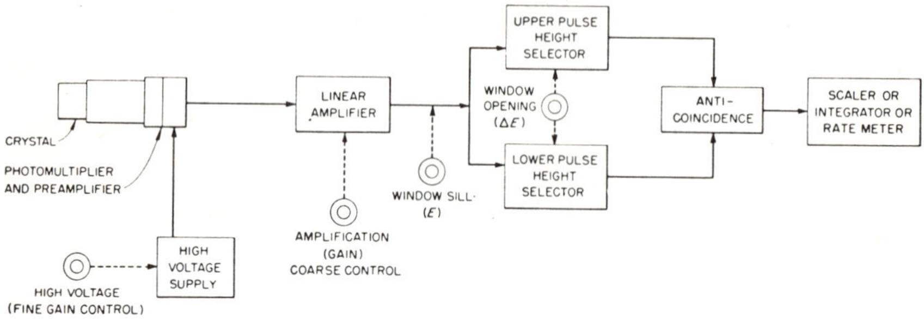

Figure 5.11 Block diagram of scintillation detector system.

not to subject it to excessive overvoltages due to poor grounding or by changing connections with the voltage turned on.

The use of NaI(Tl) scintillators in gamma-ray spectrometer systems has received the greatest attention, because for many years, before the advent of lithium-drifted semiconductors, they provided the most convenient and widespread means of analyzing gamma-ray emitters. Even now they are widely used for this purpose owing to their greater sensitivity and convenience, though their resolution is inferior.

Gamma-Ray Spectrometers

Sodium iodide crystals can be used in gamma-ray spectrometers because they are linear over a wide range of energies. That means that the fluorescent light intensity produced by photoelectric interaction is directly proportional to the incident gamma energy. Consequently, the electron pulse amplitude from the photomultiplier due to a given gamma interaction should also be proportional to this energy. Hence, by sorting these pulse amplitudes in a pulse-height discriminator, which assigns pulses to different counting "channels" according to their peak amplitude, one can obtain a number distribution of pulse amplitudes and hence of the energy of successive photon-induced events. How detailed this distribution can be defined depends on the stability and step width (window width) that the discriminator can separate and assign to successive adjoining channels. The wider the windows the less well the energy distribution is defined; however, too narrow a window width may involve difficulties in clearly distinguishing adjoining channels and may result in insufficient counts in any one channel to be statistically significant. Figure 5.12 illustrates the discrimination process.

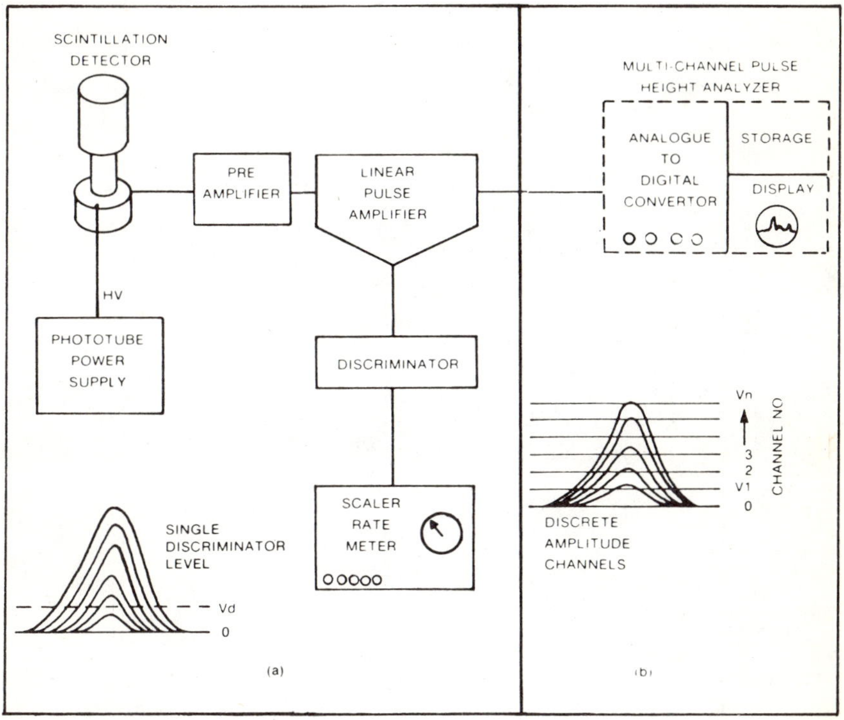

Figure 5.12 Principles of pulse height discrimination.

By plotting the count rate against the channel number, one obtains an energy spectrum since each channel, by representing a certain pulse height, counts the number of events produced by photons of corresponding energies. Figure 5.13 illustrates such a spectrum (after Heath).

By calibrating the detector system with sources of well-defined transition levels, such as cobalt-60 (1.17 and 1.33 MeV) or cesium-137 (0.662 MeV), one can identify the channels corresponding to these energies for a given voltage and amplifier setting and plot a calibration line relating channel number to photon energy; the slope of the line gives the number of keV per channel width.

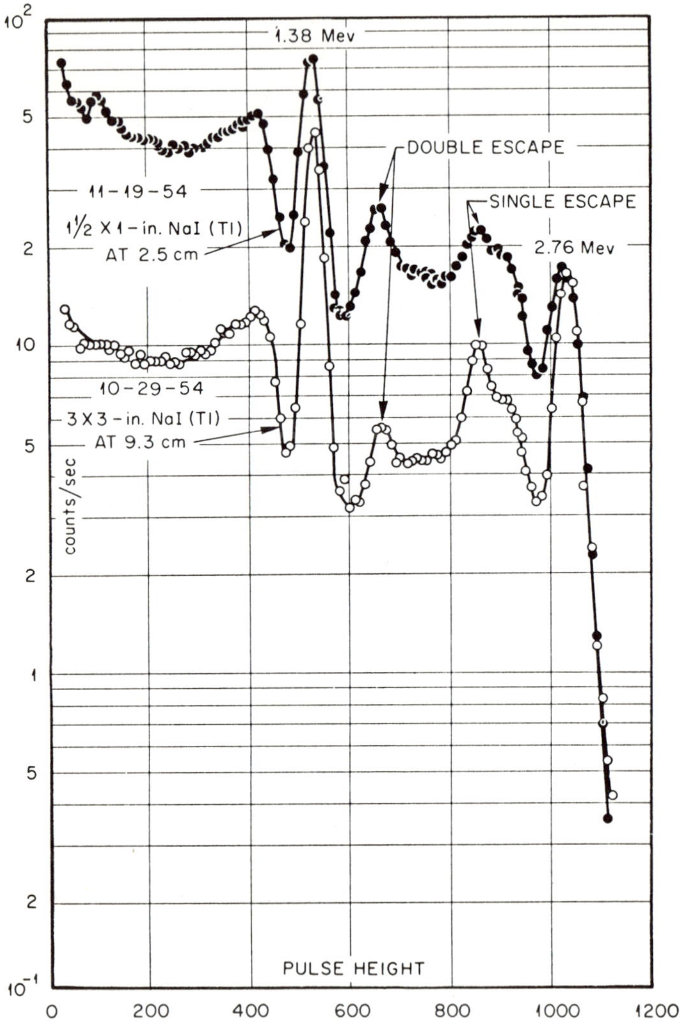

Figure 5.13 Typical scintillation γ-ray spectrum-Na-24 (from Heath, 1964).

On inspecting a spectrum, such as that in Figure 5.13, two features become evident. One is the so-called Compton distribution: the broad, unarticulated shelf of scintillation events, due to Compton interactions which lead to fairly arbitrary energy deposition in the detectors, on top of which the photoelectric peaks seem to ride, and which seems to represent a certain background to them that must be subtracted from the channel counts if a quantitative count rate for just the photoelectric peaks is needed. The other is the fact that the photoelectric peaks invariably have a finite width, even though the originating photon energy is sharply defined. By assigning each peak the characteristics of a Gaussian curve, we can describe this finite width of each peak in terms of its *resolution*, by analogy with optical line width. The resolution is usually defined in terms of the full width of the peak at half maximum amplitude (net) [FWHM] expressed in energy units. For scintillation detectors, for comparison purposes, conventionally the Cs-137 line at 0.662 MeV is used; the best attainable resolution for 3-5 cm diameter NaI(Tl) crystals is about 7.4%, *i.e.*, the FWHM at 662 keV is about 50 keV. Any two peaks closer than 45-50 keV in that energy range could not be fully distinguished from each other.

A great deal of effort has been devoted to the identification of the causes for this lack of good resolution in scintillation detectors, which far exceeds any source-related random fluctuations. The following factors have been shown to be the principal contributors to this limit in resolution:

1. location dependence of the interacting event in the phosphor crystal (increases with crystal size)
2. variations in optical path length
3. fluctuations in photocathode emission
4. variations in secondary emission from dynodes
5. dark-current contribution
6. voltage fluctuation on photomultiplier
7. gain variations in amplifiers
8. discriminator bias fluctuations
9. fluctuations in Compton background
10. fluctuations in background counts

To obtain the best resolution, all of these factors should be controlled as well as possible, the temperature should be stabilized and the voltage supply ideally should have operated long enough to reach internal temperature equilibrium. Most scintillation spectrometers are mounted inside a heavy lead shield to reduce background; to minimize bremsstrahlung, the inside of the shield should have a graded layer of copper and aluminum, to provide diminishing atomic numbers towards the detector.

The availability of transistorized spectrometer systems has led to an increasing sophistication in data handling and presentation. Multichannel analyzer systems employing 400, 1,000 or 1,024, or 4,096 channels are widely available with print-out or plotter output. Background and Compton contribution can be directly subtracted and identification of peaks by nuclide can be programmed into the system. However, for proper use and appreciation of such systems, the basic principles of the detection process must be fully understood.

Coincidence Systems

Reference has already been made to the use of coincidence circuits in liquid scintillation counters to reject events originating in one photomultiplier tube only and to accept events seen by both tubes. Figure 5.14 illustrates the operation of such circuits in simplest terms.

The pulse from one detector is used to "trigger" the gate, allowing a pulse from the other detector arriving within a set time interval to pass through the gate, carrying any energy information with it. Such circuits also find use to select events of common origin in high-energy physics research, in neutron time-of-flight measurements and for cosmic-ray suppression in low-background counting (see also Chapter 11). Another use of coincidence circuits is for the reduction of the Compton continuum in gamma-ray spectra. The ambient background can be reduced by surrounding the detector with heavy shielding such as iron or lead. Compton events are then suppressed by surrounding the primary detector by an annular secondary NaI(Tl) detector (see Figure 5.15a). This "anticompton" detector detects the Compton-scattered gamma-rays, which are not fully absorbed in the primary detector, and causes them to be rejected electronically in an anticoincidence circuit.

To obtain high-efficiency detection of low-level radiation in the presence of an ambient background, one can also employ a phosphor sandwich or "Phoswich"™, illustrated in Figure 5.15b. It consists of a thin NaI(Tl) disk optically coupled to a larger scintillator of a different material, *e.g.*, CsI(Tl). Both are coupled to a common photomultiplier tube which now views pulses of both detectors with different characteristics. The different signals are separated in an electronic gating system on the basis of the different phosphor decay times. In this way, one can discriminate between photon interactions, from, say, X-rays occurring only in the thin front crystals from those occurring in both crystals. Hence, Compton background from high-energy gamma-rays can be suppressed by proper anticoincidence gating. Figure 5.15c shows an adaptation of such a system for environmental monitoring.

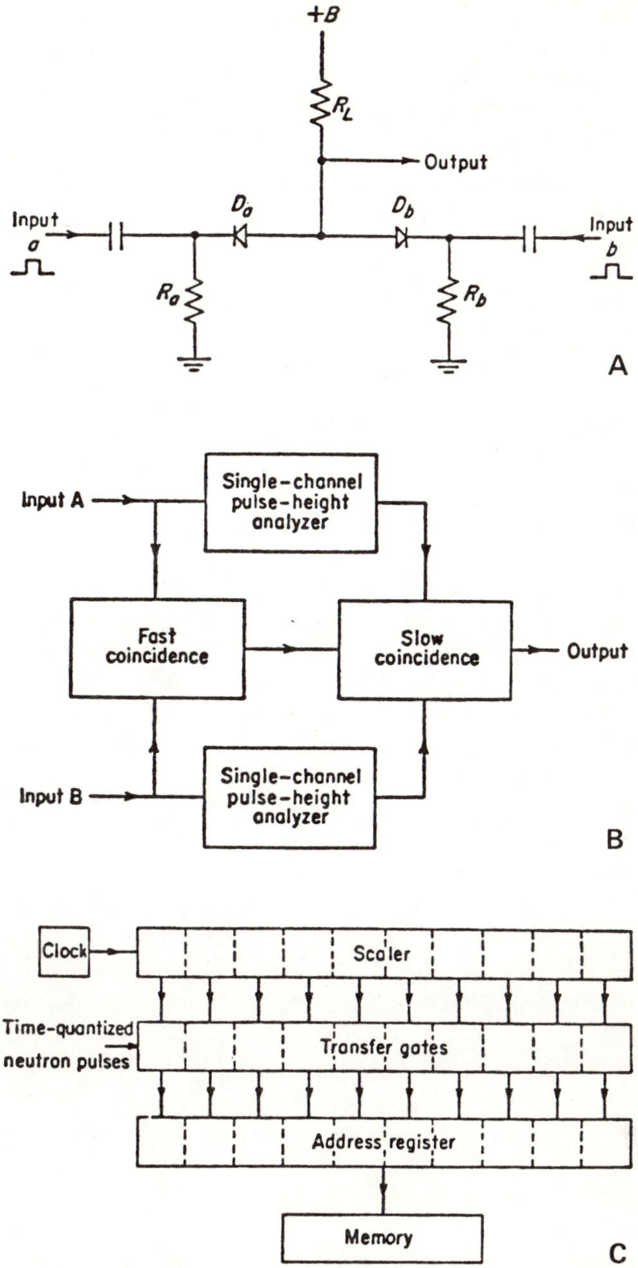

Figure 5.14 Examples of coincidence circuits: (a) simple diode circuit; (b) "fast-slow" coincidence system; (c) time-of-flight spectrometer.

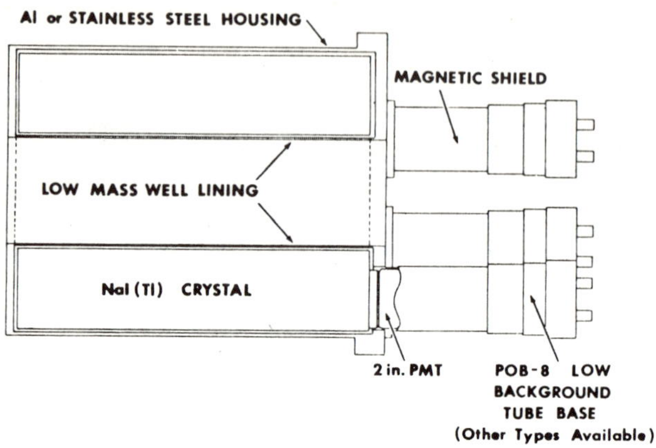

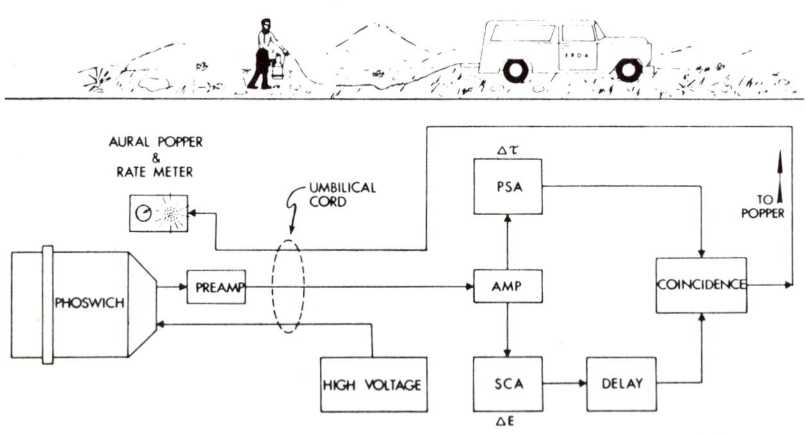

Figure 5.15 Special scintillator arrangements: (a) anti-Compton annulus; (b) "Phoswich" detector: matched window type for X-ray and low-energy gamma detection (courtesy Harshaw Chemical Co.); (c) schematic diagram of mobile Phoswich detector system (after Ahlquist, 1978).

Scintillation Camera Systems

In nuclear medicine, maps of the distribution of radioactive tracers in organs that are suspected to be morbid are obtained by "scanning" the organ with a well-collimated detector. NaI(Tl) scintillation detectors are usually employed by this purpose because of their high gamma sensitivity. With one or more detector assemblies moving in steps across the patient, a point-by-point intensity diagram is obtained, as shown in Figure 5.16. To obtain adequate spatial resolution and sufficient counts in marginal regions, the detector must have a narrow collimator aperture. As a result, total counting times tend to be rather long or possibly unacceptably high levels of radioactivity would have to be employed.

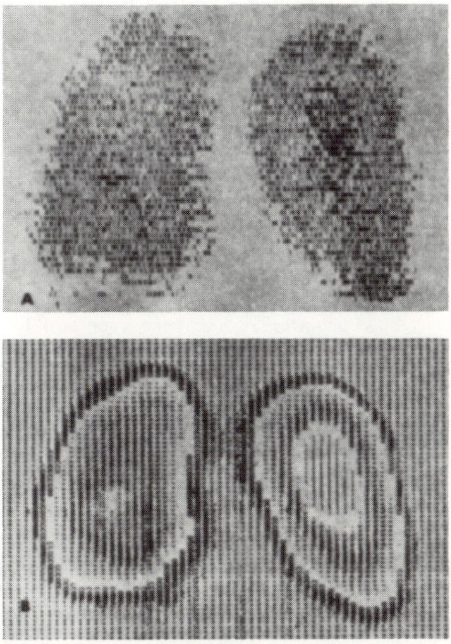

Figure 5.16. Lung scintigram. (A) Stylus scan; (B) Computer output. Activity distribution left/right: (A) 53.6 : 46.4 (B) 52.4 : 47.6.

One way of overcoming this problem is to greatly increase the number of detectors and to treat the outputs by means of a computer to enhance image contrast. Figure 5.17a shows a cross section through a 10-detector head with individual photomultiplier outputs. An alternative approach is to use a large single flat crystal scintillator which is viewed by a number of photomultipliers simultaneously; each area seen is illuminated by a different portion of the source volume as a result of the collimator

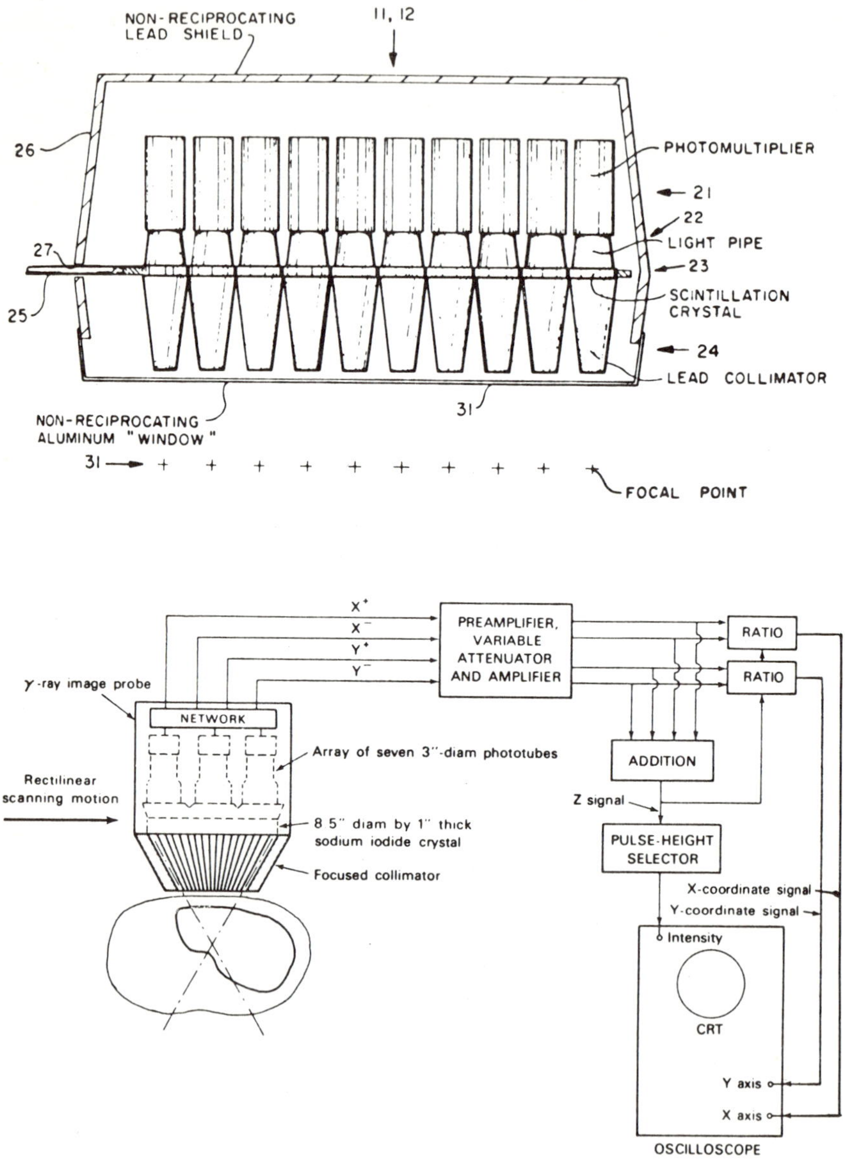

Figure 5.17 Scintillator cameras: (a) scintillator camera ("Cleon scanner") showing the 10 detectors and 10 collimators; (b) sectional drawing of Anger camera.

configuration shown in Figure 5.17b. In such an *Anger camera,* stray light from other portions of the crystal is suppressed by pulse-height discrimation. The spatial resolution is again determined by collimator aperture. Both multi-image systems appreciably reduce patient exposure and scanning time. Nevertheless, contrast and resolution are inherently limited by the irregular nature of the source volume and overlying tissue and by the necessary compromise between collimator aperture and counting statistics. This is illustrated in Figure 5.18.

CHERENKOV DETECTORS

One of the most impressive views in a swimming pool nuclear reactor, and one often pictured, is the bright blue glow in the water surrounding the reactor core when operating. A similar glow is seen surrounding a strong gamma-ray source, such as cobalt-60, when stored under water. In both cases this light is due to the so-called *Cherenkov Effect,* which may be considered an electromagnetic bow wave occurring when a nuclear particle traverses a material with a speed greater than the velocity of light in that medium. The direction of this bow wave and, hence, the direction in which light is emitted with respect to the path of the moving particle is given by

$$\cos\theta = \frac{1}{n\beta} = \frac{c}{nv} \tag{5-4}$$

where c = velocity of light
v = velocity of particle
n = refractive index of the medium in the appropriate frequency range

The condition for the Cherenkov Effect to occur is

$$n\beta > 1 \quad \text{or} \quad v \geqslant \frac{c}{n} \tag{5-5}$$

and hence the effect can be used only for very energetic radiations. Table 5.8 presents the minimum, threshold energies ($v_{min} = c/n$) for various particles for some media of interest. It is evident that the Cherenkov Effect can be used mainly in high-energy physics research, where the directional correlation with particle energy implied in Equation 5-4 makes this effect uniquely capable of "filtering" out specific energy regions by observing light emitted into particular directions.

Cherenkov detectors are usually categorized as focusing and nonfocusing types. In a nonfocusing detector, the total light produced by the energetic

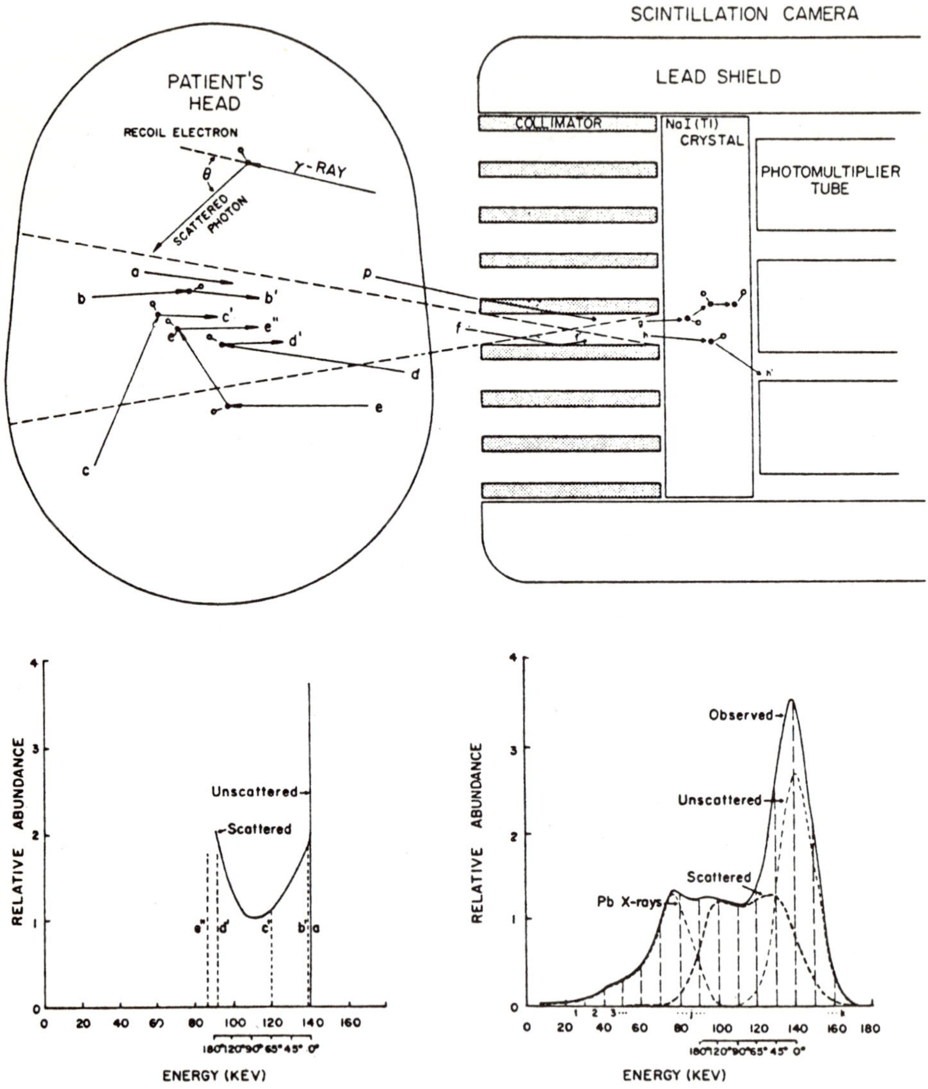

Figure 5.18 Imaging process with collimated camera. NaI(Tl) detector and Tc-99m tracer (after Zimmerman).

Table 5.8 Properties of Optical Media and their Cherenkov Thresholds (after Moyer)

Material	n	Threshold $\beta = v/c$	Threshold Kinetic Energies		
			Protons	Pions	Electrons
Liq. N_2	1.205	0.830	760 MeV	112 MeV 4	410 keV
Liq. O_2	1.221	0.819	710 MeV	104 MeV	380 keV
FC-75[a]	1.277	0.782	580 MeV	86 MeV	315 keV
Water	1.332	0.751	500 MeV	73 MeV	265 keV

[a]A fluorochemical available from Minnesota Mining & Mfg. Co. (Composition $C_8F_{16}O$).

radiation is observed much the way a liquid scintillator is observed, by a single one or a coincident pair of photomultiplier tubes. Figure 5.19 shows a particularly simple type of nonfocusing Cherenkov detector that can be used to measure radiation intensity in the energy range above threshold. The number of photons emitted per centimeter path length over the visible frequency interval (400-700 nm)

$$\frac{dN}{dx} = 490\ Z^2\ \sin^2\theta \tag{5-6}$$

Depending on photocathode and electron multiplier characteristics, one can detect (25-50) $\sin^2\theta$ photoelectrons per cm path length observed.

For pulsed particle beams the count rate is essentially limited by photomultiplier tube parameters. By combining a plastic scintillator and a lead-glass Cherenkov detector in series, connected in a coincidence gate system, high-energy γ-radiation in the 100-1,000 MeV range can be detected.

Most applications utilize focused detector systems that select the energy range "window" to be detected by observing only light emitted into a specific angular nimbus. Figure 5.20 illustrates two of the many configurations that have been devised for the purpose, the Frisch counter (Figure 5.20a) and the Kinsey counter (Figure 5.20b). The energy window that the detectors will accept can be quite narrow with relatively low background. The principal difficulty arises from the need to define sharply the flight path observed and to protect the detector system from damage from the surrounding high-energy radiation fields.

To illustrate the improvements in such detectors obtained in recent years, Figure 5.21 shows the differential focusing Cherenkov counter developed by Fitch and associates (1978). This detector utilized water as the detector medium, contained in a Lucite meniscus 1.5 in. (3.75 cm) thick. The counter was designed to measure $\beta = v/c$ in the range 0.95 - 100 with a $\Delta\beta$ resolution of about 0.015 corresponding to a measurement

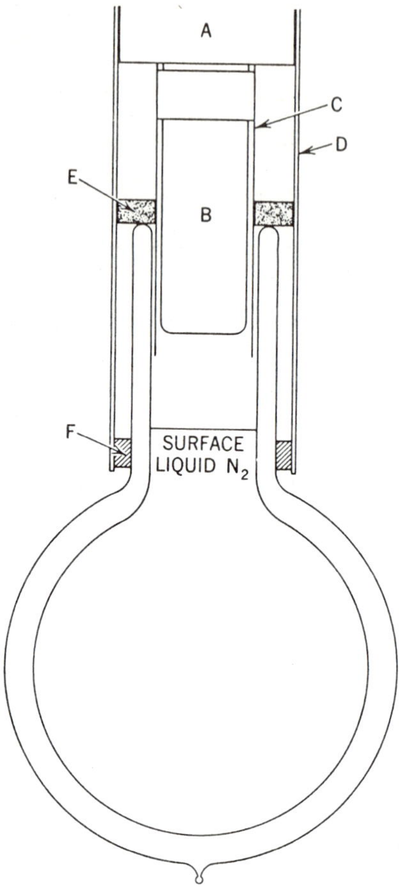

Figure 5.19 Nonfocusing Cherenkov counters: liquid N_2 or O_2 Detector in Dewar flask.

angle in Cherenkov angle θ of ±0.015 radian with an angular acceptance of ±0.070 radian. Such a detector can observe pions, kaons and protons in the momentum range above 3.2 GeV/c.

The advantage of Cherenkov detectors lies in providing a highly selective means of measuring very-high-energy particle spectra in a range that is accessible only with difficulty to other observational methods, such as bubble chambers. In a less dramatic application, Cherenkov radiations are also used to measure β activities in aqueous samples, provided the mean beta energy is above the 260-keV threshold, *i.e.*, $E_{\beta max}$ is above 500 keV.

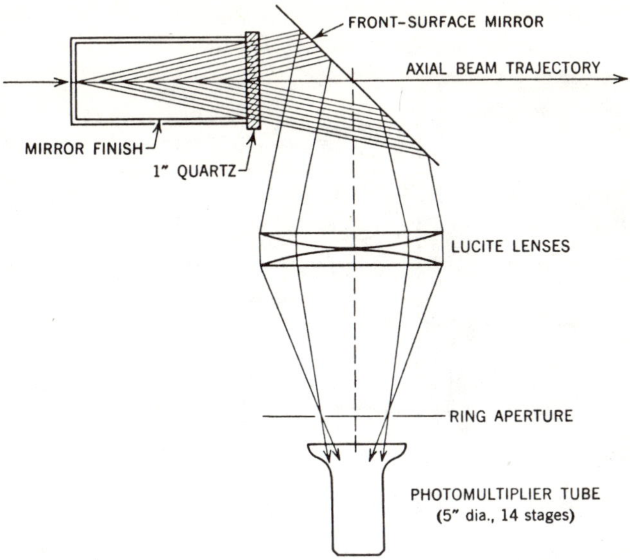

A

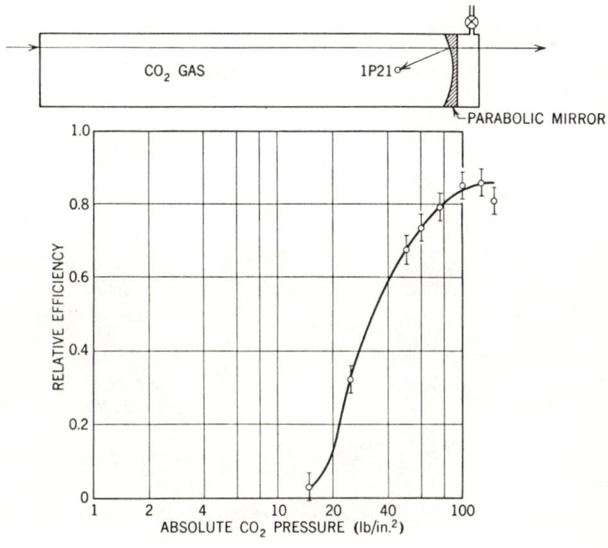

B

Figure 5.20 Focused Cherenkov detectors: (a) optical principle of Frisch counter; (b) diagram of Kinsey counter and threshold response curve of CO_2 gas counter to 3.2 GeV pions.

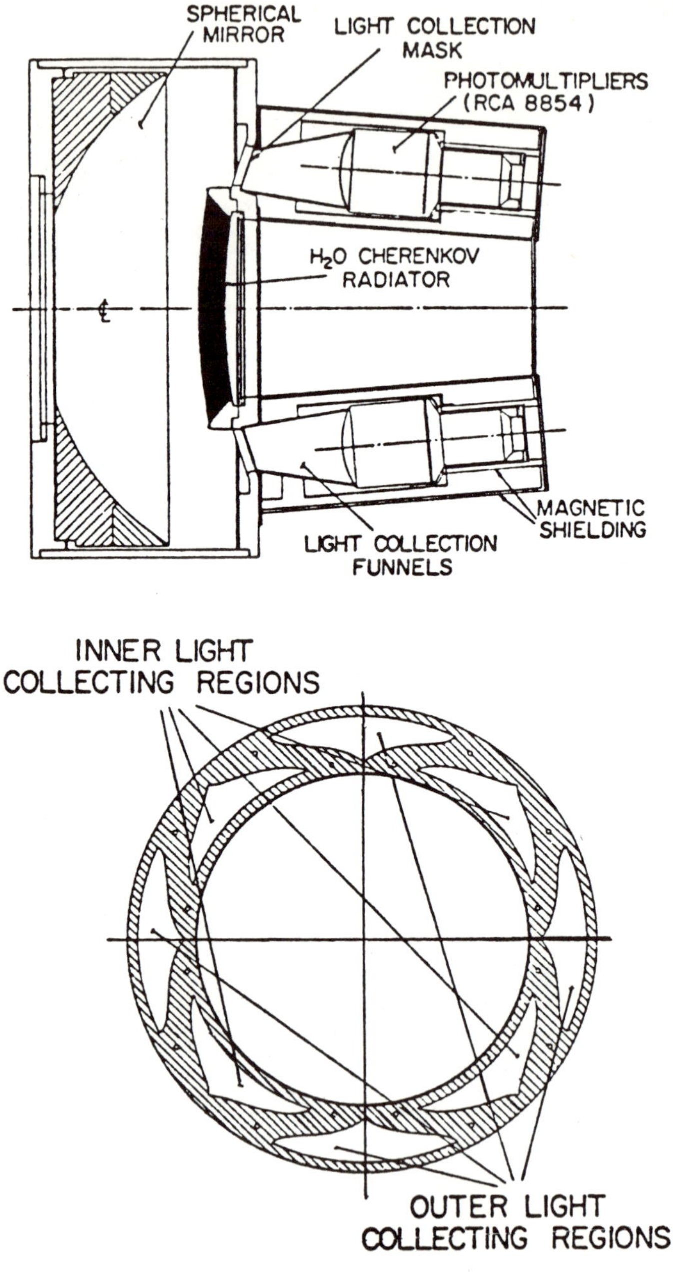

Figure 5.21 Differential Cherenkov counter: (a) assembly (b) light collecting mask (after Cester *et al.*).

SUMMARY

As Table 5.7 has shown, scintillation detectors can be used to detect most nuclear radiations, provided the appropriate scintillator is employed. Their advantages are high sensitivity, linearity with energy, and great flexibility in use and in possible configurations. Disadvantages include relative fragility, high cost, sensitivity to temperature and voltage variations and, for some types, problems in source preparation and source introduction.

REFERENCES FOR FURTHER READING

Birks, J. B. *The Theory and Practice of Scintillation Counting* (New York: Macmillan Publishing Co., 1964).

Bransome, E. D., Ed. *The Current Status of Liquid Scintillation Counting* (New York: Grune & Stratton, 1970).

Cester, R., V. L. Fitch, R. W. Kadel, R. C. Webb, J. D. Whittaker, M. S. Witterell and M. May. *IEEE Trans.* NS25:525 (1978).

International Atomic Energy Agency. *Medical Radionuclide Imaging,* Proceedings of the Los Angeles Symposium (Vienna: IAEA, 1976).

Leskovar, B. "Microchannel Plates," *Physics Today* 30(11):42-49 (1977).

Neame, K.D., and C. A. Homewood. *Liquid Scintillation Counting* (New York: Halstead Press, 1974).

Ramm, W. J. "Scintillation Detectors," in *Radiation Dosimetry,* Vol. II, F. H. Attix and W. C. Roesch, Eds. (New York: Academic Press, 1968).

Schram, E. *Organic Scintillation Detectors* (Amsterdam: Elsevier, 1963).

Siegbahn, K., Ed. *α-β-γ Ray Spectrometry* (Amsterdam: North-Holland Publishing Co., 1965).

PROBLEMS

5.1 Using the data in Table 5.4 estimate the increase in signal amplitude that would result from a change from a toluene-PBD to a toluene-TP-POPOP scintillator. What photocathode response would be appropriate in each case and what increase in sensitivity would arise from the shift to longer wave lengths?

5.2 A 10-stage photomultiplier is operated with an interstage voltage of 120 V. Calculate the change in multiplication if the overall voltage is raised to 1,300 V, or if the first two stages only are operated at 160 V between dynodes.

What would be the advantage of the latter arrangement?

5.3 A photomultiplier tube has an effective collector capacitance of 40 pF, an interdynode capacitance of 15 pF; it is supplied with 1,000 V applied across a resistor chain of ten 1 MΩ, and a collector load of 150 kΩ. Estimate the maximum output signal that can be obtained, and the rise and decay times of the output pulses.

5.4 In the previous case, pulses of 1,000 light photons per sec reach the photocathode where they are converted with a 6% efficiency. Estimate the output pulse amplitude and charge per pulse making reasonable assumptions.

5.5 A cylindrical scintillation detector is used to measure the activity from a planchet containing a cerium-144 gamma-ray source. Assuming a well-shielded counter, reasonable geometry and efficiency, calculate the minimum source activity that may be detected for a net count rate above five times background.

How long would that sample have to be counted to determine the activity within a 95% confidence limit?

5.6 Compare the detection efficiency, background count, geometry, overall sensitivity and mental stress to the patient for the two whole-body counting configurations in Figure 5-8, making reasonable assumptions.

5.7 Estimate the resolution of the detector employed to obtain Figure 5-13. Explain the particular channel location of the various peaks.

5.8 Making reasonable assumptions regarding the collimator design employed and the distance from the patient, estimate the approximate iodine-131 activity in the thyroid required to provide a scan like that of Figure 5-16.

5.9 Calculate the number of Cherenkov photons produced by a particle traveling at $v/c = 0.95$ in water ($n = 1.33$) in the S-11 response range (3,500-5,500 A). Assuming 100% light collection and 6% photoelectrode conversion efficiency, what photomultiplier output signal would be produced?

5.10 An anthracene crystal and a 12-stage photomultiplier tube are to be used as a scintillation spectrometer for β-rays. The phototube output circuit has a combined capacitance of 45 pF. If an 8-mV output pulse is desired whenever a 55-keV beta particle is incident on the crystal, calculate the electron multiplication required per stage. Assume perfect light collection and a photocathode efficiency of 5%. (Assume 550 photons per beta particle.)

5.11 A 3.75-cm cylindrical scintillation detector is used to measure the radiation emitted from a zirconium powder sample filling a 100-ml beaker. Assuming a nearest approach of 4 cm between the detector and the surface of the beaker, and adequate shielding, calculate the minimum amount of Zr-95 that would have to be present in the sample to give a net count rate of 10 times background.

5.12 Calculate the threshold energy for the production of Cherenkov radiation in Lucite for the following particle types: alpha, protons, μ mesons and electrons. The index of refraction of Lucite is 1.50.

CHAPTER 6

SEMICONDUCTOR DETECTORS

PRINCIPLES OF OPERATION

In Chapter 2 it was explained how irradiation of a semiconductor material can result in the creation of a hole-electron pair, provided enough energy is given to the electron to raise it into the conduction band. Since the material starts out with very few mobile electrons, *i.e.,* as a poor conductor, the absorption of even a few quanta can cause a disproportionately large change in conduction properties. The nature of these changes depends on the type of semiconductor involved, whether conduction occurs predominantly by electrons (n-type material) or by holes (p-type material) and how rapidly either type of charge carrier is removed by annihilation, that is, a return of the excited electron to its vacant place or an equivalent one in the valence band. This return can be delayed by the presence of impurities which produce traps for either holes or electrons, where they may stay for long or short periods depending on the general state of the material and the presence of any applied potentials. Impurities that favor conduction of electrons, by trapping holes, are called donors; those that trap electrons and promote charge conduction by holes, *i.e.,* positive charges, are called acceptors. By adding impurities of one type or another to an intrinsic semiconductor, such as very pure silicon, one can produce n-type or p-type silicon. Similar effects can be obtained with germanium; semiconducting Si and Ge form the basic ingredients of modern semiconductor electronics and microcircuitry. Both silicon and germanium belong to the central Group IV in the Periodic Table with four electrons and four vacancies to fill an electronic shell. Elements in Groups V and VI with more electrons can serve as donor impurities, those in Groups II or III as acceptor impurities. Binary compounds composed of near-equal proportions of donor and acceptor elements by themselves, such as GaAs, InSb, CdTe, HgI_2, BiTe and InP, also exhibit semiconductor properties. Because most of these materials show significant changes in conductivity while irradiated, they all could serve, to varying extent, as radiation detectors. Table 6.1 lists the major properties of some semiconductors of interest.

Table 6.1 Properties of Semiconductors

	Si	Ge	Ga As	Cd Te	In Sb
Atomic Number (Z)	14	32	31,33	48,34	49,51
Density (ρ) (300°K) (g/cm^3)	2.33	5.33	5.32	6.20	5.78
Band Gap (Eg) (eV)	1.08	0.75	1.43	1.5	0.17
Mean Energy per Carrier Pair (W) (eV)	3.65	2.95	4.2	4.43	0.6
Intrinsic Resistivity (300°K) (Ωcm)	2.3×10^5	47			
Electron Mobility (300°K) (μ_n) (cm^2/Vsec)	1,350	3,800	8,500	~650	78,000
Hole Mobility (300°K) (μ_p) (cm^2/Vsec)	480	1,800	420	~45	750
Carrier Lifetime (τ) (μsec)	100-3,000	1,000	0.1	40	~0.1

Very early in the development of semiconductors it was found that a particular radiation sensitivity was obtained if the exposed material started out essentially free of any mobile charge carriers. This could be obtained by applying an electric potential to draw off any charge carriers or, better, by applying it to a paired junction of p-type and n-type material in close contact, Figure 6.1. By applying reverse bias as shown, holes and electrons are drawn away from the junction region leaving a *depleted* layer which is now particularly sensitive to the injection of new charge carriers by external irradiation.

Because of the bias voltage, any current flowing will be unidirectional, that is, the junction acts as a rectifier or diode. Some backward flow is possible up to a maximum voltage when the forbidden gap is overwhelmed and a large breakdown current occurs. This voltage is called the *Zener Voltage,* and the abrupt nature of the effect, at a fixed potential, makes the use of *Zener diodes* convenient to provide reference voltages. Figure 6.2 shows the current voltage characteristics of a semiconductor diode. By adding additional n-type or p-type electrodes, a diode can be converted into the equivalent of a vacuum triode and exhibits amplifying properties, *i.e.,* application of a small voltage variation between two electrodes results in a large change in current flow between two other electrodes. Since semiconductors can be made extremely small, capacitances can be minimal and response times of semiconductor devices can be made enormously faster than for the corresponding

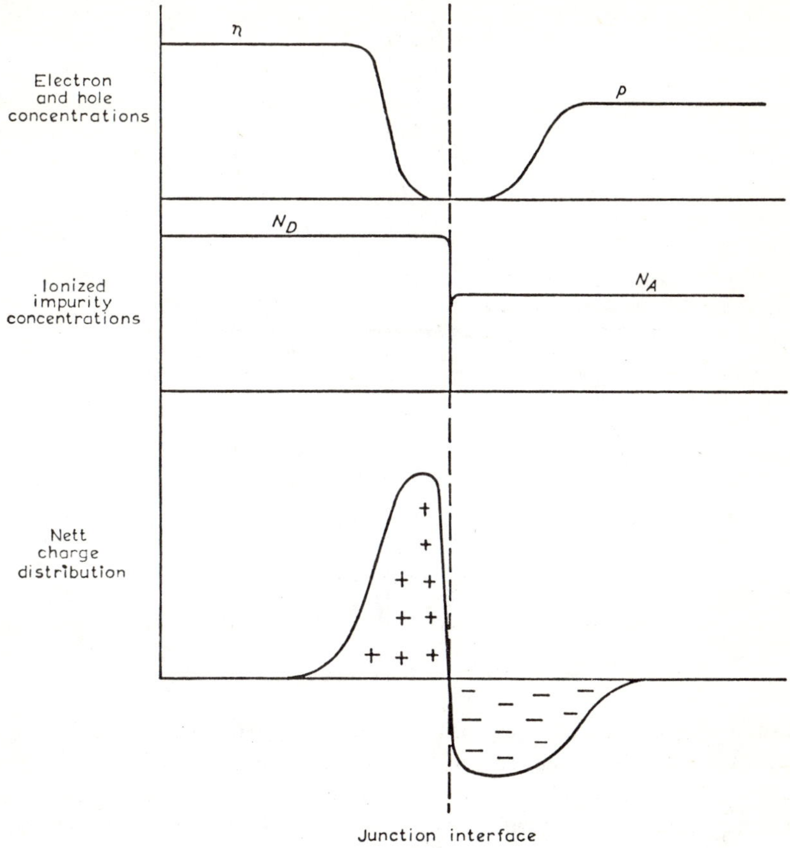

Figure 6.1 Charge distribution in a p-n junction.

vacuum tube devices. As a result, the latter have been largely displaced in most counting and data handling systems. Only where high voltages are required for many gas-filled and scintillation detectors are vacuum tubes found as rectifiers and stabilizers, and even there they are gradually replaced by solid-state devices.

For purposes of radiation detection one requires a semiconductor of suitable atomic composition, a depleted volume of a size comparable to the range of the radiation to be detected, and associated circuits that permit utilization of the fast response time of which the semiconductor is capable. This depends, of course, on the carrier lifetime, that is, the time until carriers are annihilated, and the quantum efficiency, *i.e.*, the rate at which excitons are produced per unit flux traversing the depleted region. Silicon ($Z = 14$) and germanium ($Z = 32$), the most efficient semiconductors, have relatively low atomic numbers,

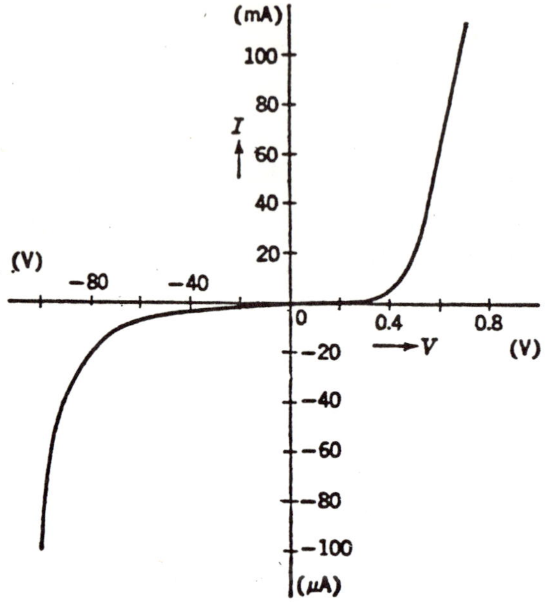

Figure 6.2 Characteristic of a junction diode (Note scale changes).

but for photon detection Ge, with its higher atomic number, is the more efficient. On the other hand, silicon has a large forbidden gap (Figure 2.1) and, hence, is less susceptible to higher temperatures than germanium. In both cases extremely high purity, much higher than for ordinary transistor circuits, is required for efficient detector application. Compound semiconductors can be "dirtier," but most of them make less satisfactory radiation detectors because of their narrow forbidden gap, which makes them vulnerable to thermal-induced noise due to thermal excitation of electrons, transferring them to the conduction band. Cadmium telluride is a notable exception as a compound semiconductor that is finding use as a high-Z radiation detector. Because of the low-bias voltages required to operate semiconductors, they are more convenient than other types of detectors. Their small size makes it possible to insert them into phantoms and to use them in confined spaces. However, large-volume detectors may be more difficult to make and often are made up of "mosaics" of individual detectors. Because of their sensitivity to radiation, they are very susceptible to radiation damage and are not suitable for use in intense radiation fields. Many semiconductors are sensitive to intense light and most of them deteriorate if exposed to moisture.

The generation of a hole-electron pair may be considered to be analogous to the production of an ion pair in a gas-filled detector. It then immediately becomes obvious why semiconductors offer greater sensitivity and compactness: with their greater density they have a greater stopping power than gases and they may completely absorb some energetic particles. On the other hand, detectors can be made so thin that they absorb only a small part of the energy of the incident particle and that energy loss, dE/dx, over a known path length is a good means of identification. Hence, a thin dE/dx detector coupled with a totally absorbing (E) detector provides a convenient means of energy determination and particle discrimination.

The accuracy of any energy determination is high because the average energy, W, to produce a charge-carrier pair is 3.65 eV in silicon, compared with about 30 eV in gases and about 300 eV in a scintillator; hence, the available signal charge is about eight times as big as in gases and statistical fluctuations are reduced by a factor $\sqrt{8}$.

The movement of the electrons and holes is limited by the rate of recombination which depends on the carrier concentrations and the temperature. The concentrations n_e and n_p are given by

$$n_e = n_p = 10^{19} \exp\left(-\frac{E_g}{2\,kT}\right) \qquad (6\text{-}1)$$

At room temperature, kT $\sim$ 0.025 eV and for a gap voltage $E_g \sim 1$ eV, $n_e \approx n_p \mp 2 \times 10^{10}$ per cm^3. The energy of the highest electron energy level filled at absolute zero is called the thermodynamic potential or the Fermi energy, E_f. It is a convenient way of describing the energy state of a material. In an intrinsic semiconductor

$$E_f = -\frac{E_g}{2} - \frac{3}{4}\ kT \log_e \frac{m_e}{m_h} \qquad (6\text{-}2)$$

where m_e and m_h are the effective masses of electrons and holes. If m_e and m_h are the same, which is not always the case,

$$E_f = -\frac{E_g}{2}$$

i.e., the Fermi level lies in the middle of the energy gap.

If an electric field E' is applied, net momentum changes occur in the valence and conduction band and the Fermi energy shifts. The total current density in a field E' is

$$J = E'e\,(n_e\,\mu_e + n_p\,\mu_p) \qquad (6\text{-}3)$$

where μ is the mobility, *i.e.*, the average drift velocity in the field direction. The mobility is affected by lattice imperfections which disturb the potential distribution around the atoms among which the carriers drift. We can define

$$\mu = \frac{e\tau}{m} \tag{6-4}$$

where τ = mean free time between scattering events, and the drift velocity

$$v = \mu E' \tag{6-5}$$

According to Ohm's law

$$J = \sigma E' = nev \tag{6-6}$$

Hence, the conductivity

$$\sigma = ne\mu \tag{6-7}$$

In general $m_e \neq m_h$; therefore, $\mu_e \neq \mu_h$ and the conductivity is determined only by the density and mobility of the majority carrier. In n-type material the effective carrier density depends on the number of traps for the holes, *i.e.*, the donor concentration N_d, and $\sigma = N_d \mu_h e$. Similarly, in p-type materials

$$\sigma = N_a \mu_h e \tag{6-8}$$

In silicon, $\mu_e \sim 1{,}500\ cm^2/V\ sec$ in low electric fields, decreasing in higher fields; $N_d \sim 10^{13}\ cm^{-3}$, leading to a resistivity $1/\sigma$ of about 100 ohm-cm. When p-type and n-type materials are joined to form a p-n junction, the positions of the valence and conduction bands will not be the same at both sides of the junction and the Fermi energy will be displaced as shown in Figure 6.3. When a reverse-bias voltage V is applied, the energy levels are further shifted and the charge carrier will move away from the junction region to lower-energy positions. If the conductivity σ is the same on both sides, by Poisson's equation

$$\frac{d^2V}{dx^2} = -\frac{4\pi q}{\epsilon} \tag{6-9}$$

where q = charge density ne and $\epsilon = \kappa\epsilon_o$; κ = dielectric constant.

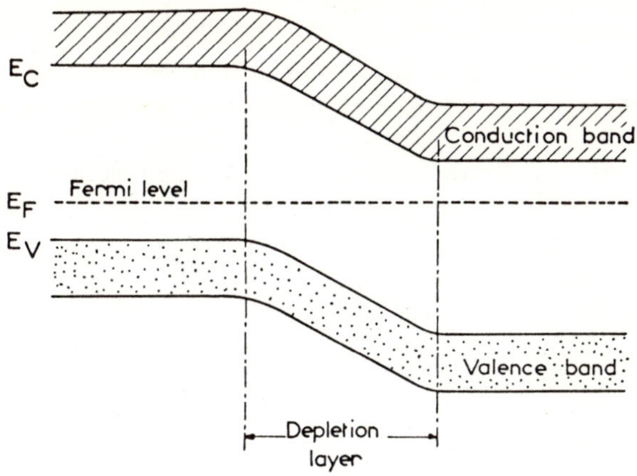

Figure 6.3 Energy band diagram for an n-p junction.

Hence
$$\frac{d^2V}{dx^2} = -\frac{4\pi e}{\epsilon}(N_c + n_e - n_p)$$

where N_c is the donor or acceptor concentration

$$N_c \gg n_e \text{ or } n_p$$

The depth of the depleted layer

$$d = \sqrt{\frac{\epsilon V \rho' \mu}{2\pi}} \tag{6-10}$$

where $\rho' = 1/\sigma$ = resistivity. The carrier collection time for a parallel electrode system of width d to which a field E′ has been applied can be calculated from the equation of motion of the electron

$$\frac{dx}{dt} = \mu_e E'$$

$$E' = -\frac{dV}{dx} = -\frac{4\pi e N_d}{\kappa}(x - d) \tag{6-11}$$

where x is the distance into the n layer.

Hence,

$$\frac{dx}{(d - x)} = \frac{4\pi e N_d}{\kappa} \mu_e \, dt = \frac{dt}{\tau_o}$$

where $\tau_o = \kappa/4\pi\sigma$ = dielectric relaxation time of the junction material. The corresponding equation of motion for holes is

$$\frac{dx}{(d - x)} = -\frac{dt}{\beta\tau_o} \tag{6-12}$$

where $\beta = \mu_e/\mu_h$

If a hole is located at $x = x_1$ at $t = 0$

$$x(t) = d - (d - x_1) \exp(t/\tau_o) \tag{6-13}$$

Holes have a finite collection time τ_h, found by putting $x = 0$ in Equation 6-13

$$\tau_h = \beta\tau_o \log_e \frac{d}{(d - x_1)} \tag{6-14}$$

The charge collection time for electrons equals τ_o, the dielectric relaxation time for the initial material; collection goes to completion exponentially.

CHARGED-PARTICLE DETECTORS

Because of their compactness and rapid response, semiconductors have found particular applications in two areas: as alpha-particle detectors and as gamma-ray spectrometers. We have seen that zinc sulfide phosphors in conjunction with a photomultiplier tube provide an efficient, but rather cumbersome, means of alpha-particle detection; they also require avoidance of exposure of the phosphor to light and, hence, made this detector difficult to use for contamination monitoring unless a thin metal window protects the phosphor, which then reduces the detector sensitivity substantially.

Some of these difficulties can be overcome by the use of semiconductors. The problem here is to create a depleted region in a semiconductor close enough to the surface to be within the range of any alpha particle falling on that surface and still to apply the necessary bias voltage to maintain the detector in a sensitive condition. Two types of detectors have been developed for this purpose: the surface-barrier detector and the diffused-junction detector.

Surface-Barrier Detectors

The construction of surface-barrier detectors is illustrated in Figure 6.4. A slice of n-type silicon is etched with HF-HNO_3, then exposed to air to produce a thin film of silicon oxide, a p-type semiconductor. A thin film of gold is deposited on the oxide layer to produce a conducting "ohmic" layer to which the bias potential can be applied. The other bias voltage is applied to the ohmic indium contact in the base of the detector. On applying reverse bias, a depleted region is created just below the oxide film, up to 50 μm wide, well within the reach of incident alpha particles, provided the gold layer and oxide film are thin enough compared with the range of the α-particles. By encapsulating the detector, it can be protected from moisture which would short-circuit the semiconductor. The gold layer is also separated by a ring-shaped cut from an outer ring which serves as a guard ring and is at capsule potential. Because of the tenuous nature of the oxide film and the guard ring arrangements, the voltage that can be applied is limited, thus limiting the depth of the depleted layer that can be obtained. Although surface-barrier detectors are convenient for alpha-particles, for more energetic charged particles, such as

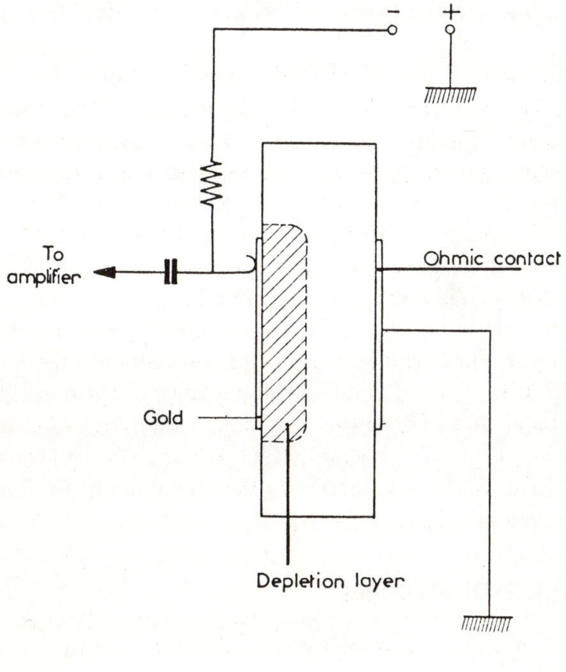

Figure 6.4 Schematic representation of a surface barrier detector (generally gold-n-type silicon).

high-energy protons, a wide depletion volume is required such as that supplied in a diffused-junction detector. Energy resolutions obtainable for surface-barrier detectors are of the order of 20-40 keV for alpha-particles and 15-35 keV for beta-particles, depending on detector size and capacitance.

Diffused-Junction Detectors

In a surface-barrier junction, the depth of the depleted region is limited by the extent to which the presence of oxide can induce a space-charge effect into the underlying n-type material. To obtain a deeper depletion region, the junction proper has to be moved further into the substrate. One way of doing this is by *ion implantation.* In that process a particle accelerator, such as a Van de Graaff generator, is used to embed doping ions, *e.g.,* boron ions, in the surface of an n-type crystal. The range of the bombarding particles determines the depth of implantation. At least 5×10^{11} atoms per cm^2 are required and the target must be cooled to liquid-nitrogen temperatures to avoid heating it, which would result in diffusion of the dopant ions. One problem arises from radiation-damage effects which, by causing atomic displacements into interstitial lattice positions, introduce traps and reduce carrier lifetime, mobility and population.

An alternative way of doping the material to substantial depth is the diffused junction. This type of detector is usually produced by diffusing a high concentration of donor impurities into a p-type material, usually high-resistance silicon. One method is to coat one side of a Si wafer with P_2O_5 dissolved in Glycol and to heat the wafer at 800°C in dry N_2 for 30 min. The phosphorus diffuses into the base material; after ohmic connections are made, the arrangement is as shown in Figure 6.5. Again the detector is hermetically encapsulated, with a guard-ring system in most cases (Figure 6.5b).

Such detectors cannot be operated at high temperatures as the dopant may become remobilized. Furthermore, the dopant concentration tends to fall off exponentially from the surface, limiting the thickness of the sensitive layer to $\sim$ 1-2 millimeters at most. While this may match the range of α- and some β-particles, it is not sufficient for γ-rays. Much wider sensitive layers with a leveled electric field can be obtained with the technique of *lithium drifting* in either silicon or germanium.

LITHIUM-DRIFTED DETECTORS

Lithium is a donor atom. It is relatively small with a diffusion coefficient in Si about 10^7 higher than that for phosphorus. It enters interstitial sites and is suitable for the preparation of deep-diffused junctions. Highly pure silicon usually retains traces of boron and hence is slightly p-type. If lithium donor

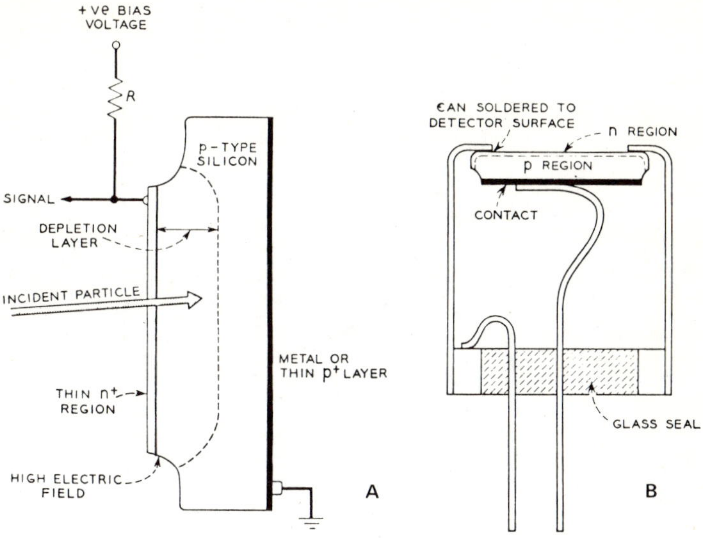

Figure 6.5 Diffused junction detectors; (a) principle of operation; (b) encapsulation method.

ions are introduced with a concentration almost equal to the ionized acceptor (B^-) concentration, then a volume of essentially intrinsic semiconductor is created with a low, but balanced, hole and electron concentration.

Lithium is coated onto one side of a single crystal slice into which it is diffused by heating. If N_o is the surface concentration, D the diffusion coefficient of Li at the appropriate temperature, and N_a the ionized acceptor concentration, then the donor concentration at depth x is

$$N_D = N_o \operatorname{erfc}\left[\frac{x}{2(Dt)^{1/2}}\right] \tag{6-15}$$

The junction interface at time t is at x = c, where $N_D = N_A$ (Figure 6.6). Then a strong reverse-bias voltage V is applied, while the crystal is held at a temperature at which the ion drift in the depletion region is appreciable. Under these conditions the Li ions drift from the donor-rich region $x < c$ to the region $x > c$. In both regions $N_A = N_D$ ultimately, since an excess of opposite charge would lead to a reverse force on the ions. Thus, an accurately compensated region is built up at both sides of the junction line c. The maximum width w is given by

$$w^2 = 8L\int E'\mu\, dt \tag{6-16}$$

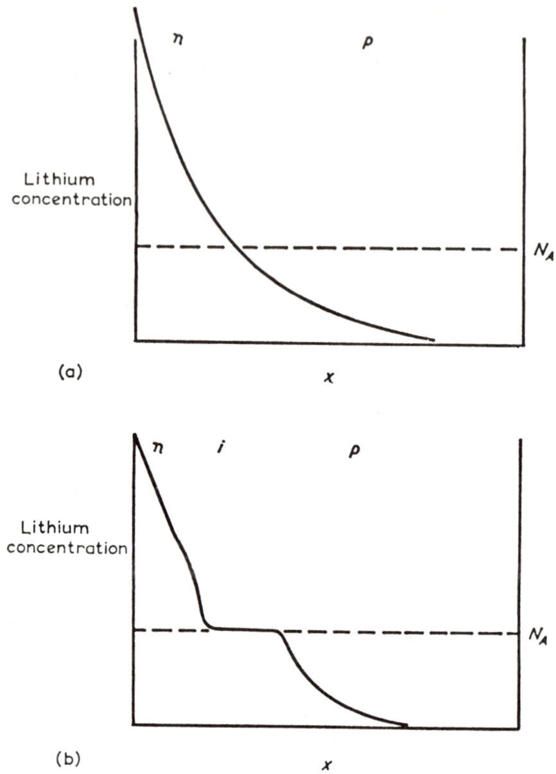

Figure 6.6 The lithium drift process: (a) after diffusion; (b) after drifting.

where $L = \frac{2Dt}{c}$ and $E' = \frac{V}{w}$

In the case of long drift times

$$w = (2\mu Vt)^{1/2}$$

and is proportional to $\sqrt{V}$. The intrinsic region constitutes the sensitive volume of the detector and such detectors are also known as p-i-n devices. Active volumes of more than 80 cm^3 have been achieved.

In germanium, which has a smaller energy gap, a lower drift temperature (~50°C) is used than in silicon and must be followed by a "clean-up" at a lower temperature (−10-0°C) to improve compensation.

After completion of the lithium drift, the detector must be kept at a low temperature, as otherwise Li^{3+} ions tend to diffuse even without an electric field and to fill lattice vacancies, resulting in a significant loss in sensitivity and resolution. It is customary to mount lithium-drifted diodes on a "cold finger," that is a metal base which is constantly cooled by liquid nitrogen. One such arrangement is shown in Figure 6.7. Both Ge(Li) and Si(Li) are normally supplied mounted above or below a Dewar flask, *i.e.*, a double-walled cryogenic reservoir that must be kept filled with liquid nitrogen. Failure to keep the detector cooled leads to deterioration of detector performance, which can be repaired only partially by a renewed clean-up drift. This dependence on a perpetual liquid nitrogen supply is the most serious drawback of lithium-drifted detectors.

Because of its higher density and atomic number, Ge(Li) is the preferred material for gamma-ray spectrometry. Si(Li) detectors are frequently used for low-energy X-rays where their lower Z and lesser temperature sensitivity is an advantage.

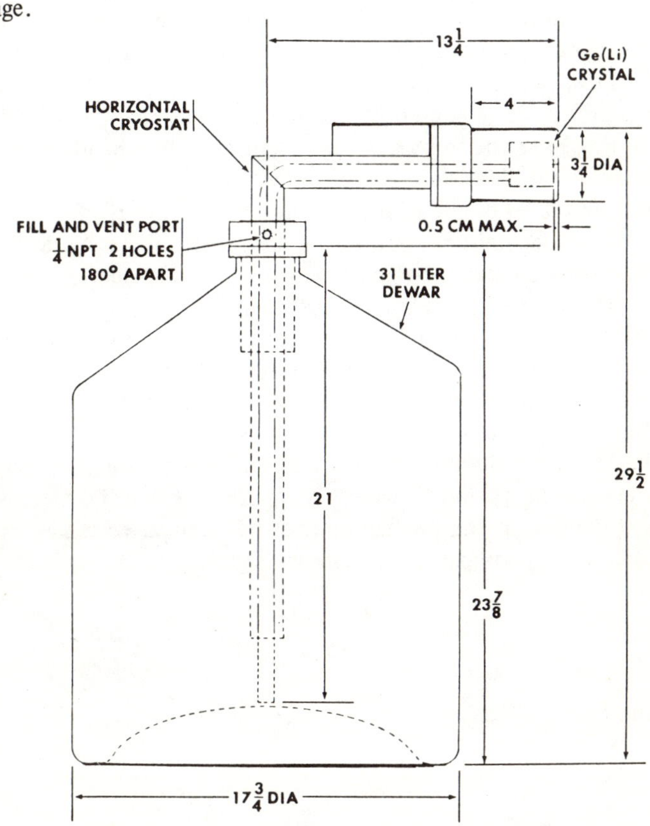

Figure 6.7 Ge(Li) detector mounted on Cryostat Dewar.

Ge(Li) detectors have almost completely superseded other detector systems for γ-ray spectrometry, because of the high-energy resolution obtainable and the relative convenience of use, despite the cooling problem. Figure 6.8 compares the resolution obtainable for the cobalt-60 peaks for a Ge(Li) detector and a NaI(Tl) scintillator. Although the scintillator is more sensitive for equal volumes, the sharp photoelectric peaks still give superior peak definition if used with a multichannel pulse-height analyser, by rejecting Compton and background events outside a very narrow channel.

Figure 6.9 illustrates the main features of a gamma-ray spectrogram as obtained with a Ge(Li) detector. The photoelectric peaks ride on a wide Compton distribution which can be canceled out electronically in routine situations. The energy resolution conventionally is measured at 0.662 MeV, the Cs-137 photoelectric peak, in terms of the full width at half maximum. The detector is fairly linear over the energy range from 50 to 1,500 keV, so that it is permissible to interpolate discriminator channel numbers in terms of energies from a calibration with 3-4 known peaks. For spectra including high-energy peaks, one may observe *summation peaks*, representing accidental coincidence of major peaks, and *escape* and *double escape* peaks, 0.511 and 1.02 MeV below major peaks (Figure 6.9). Resolution is better for small detectors, as collection times rise with sensitive volume; on the other hand, large detectors are more sensitive for high-energy gamma-rays.

The basic limitation of the energy resolution of a Ge(Li) detector, aside from the preamplifier noise limitations, is the statistical fluctuation in the number of ion pairs created for a given energy E. This fluctuation involves the *Fano factor.* Consider a gamma-ray whose energy E is completely absorbed within the intrinsic region of a Ge(Li) detector, with part of the energy going into the formation of ion pairs and part of it going into heating the lattice crystal structure (phonon or thermal energy). The ratio of the variance to the yield is defined as the Fano factor.

The yield, Y, is the number of electron-hole pairs produced for a given amount of energy deposition E, while the variance, σ^2, is the mean square variation in the yield. If the average number of electron volts that result in the production of any ion pair in germanium is given by ϵ, then

$$Y = \frac{E}{\epsilon} \tag{6-17}$$

and the Fano factor is

$$F = \frac{\sigma^2}{E/\epsilon} \tag{6-18}$$

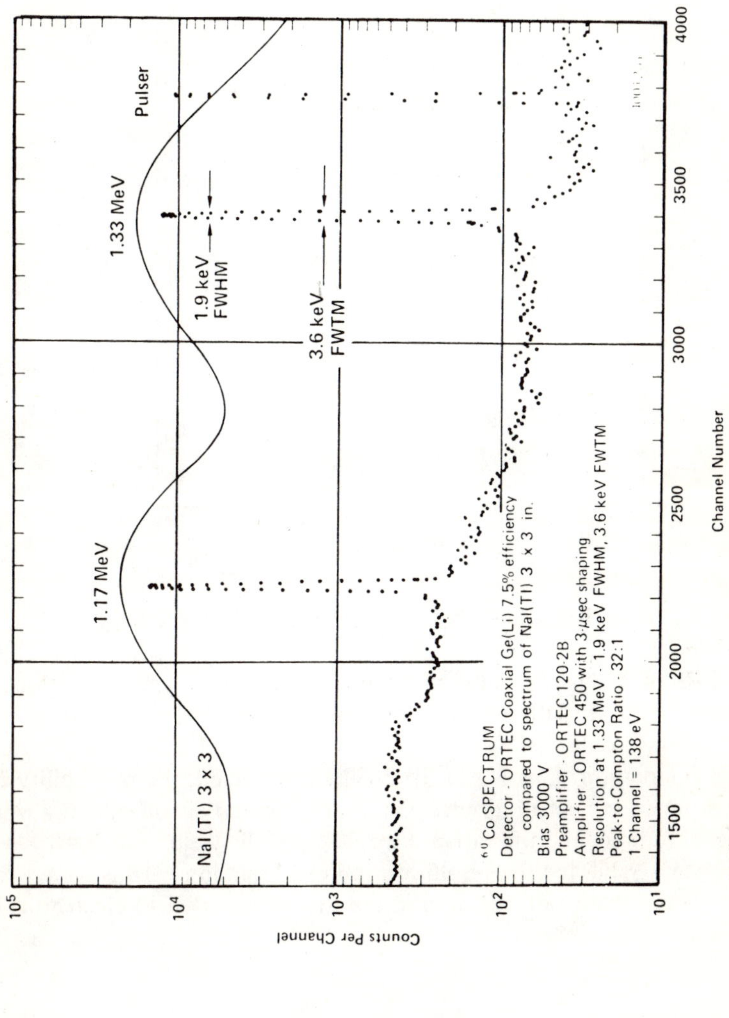

Figure 6.8 Cobalt-60 γ-ray spectrum from Ge(Li) and NaI(Tl) detectors.

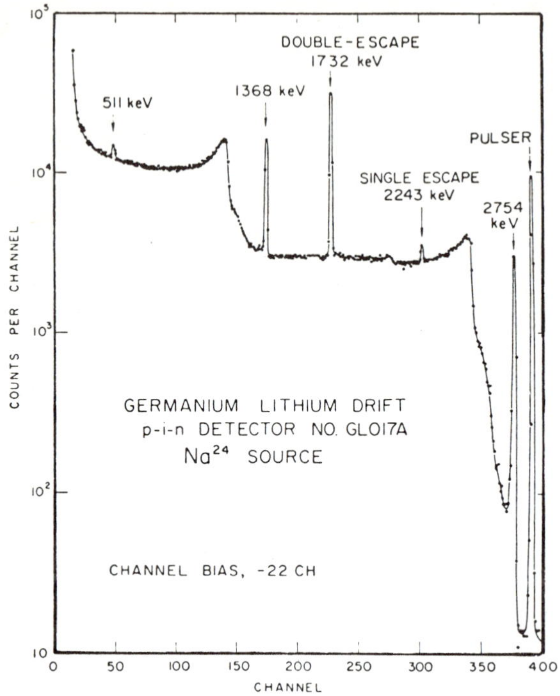

Figure 6.9 Sodium-24 γ-ray spectrum from Ge(Li) detector.

The division of energy between energy-loss processes and ionization is essentially statistical. Without the competing processes of energy loss, all the incident energy would result in ion-pair production and there would be no statistical fluctuation; thus F = 0. If, on the other hand, the probability of ion-pair production is small, resulting in a very large ϵ, Poisson statistics would be expected to hold, and the variance would equal the yield; thus F = 1.

To obtain a value for the resolution in keV, Equation 6-18 is solved for the root mean square variance, σ, and the result converted to electron volts when σ is multiplied by ϵ

$$\epsilon\sigma = \epsilon\sqrt{\frac{FE}{\epsilon}} = \sqrt{\epsilon EF} \tag{6-19}$$

The full width at half maximum (FWHM) is obtained by multiplying the result by 2.35. Hence, if ϵ is 2.9×10^{-3} keV/ion pair, then ΔE, the FWHM in keV for radiation of energy E(keV), becomes

$$\Delta E = 2.35\sqrt{0.0029EF} = 0.1286\sqrt{EF}$$

for Ge at 77°K.

For an example of the 1.332-MeV gamma-ray of ^{60}Co the ΔE(FWHM) would be, using F = 0.13 for germanium:

$$E = 2.35\sqrt{(0.0029)\,(1332)\,(0.13)}$$

$$= 2.35\sqrt{0.502164} = (2.35)\,(\sim 0.7)$$

$$= 1.645 \text{ keV (FWHM)}$$

Figure 6.10 shows the three major designs of coaxial Ge(Li) detectors in common use. The center conductor forms one biasing electrode and serves as the output terminal to the preamplifier. The outside of the cylinder is grounded.

The pulse amplitudes from the lithium-drifted detectors are small and it is necessary to mount a sensitive preamplifier, usually a field-effect transistor, close to the detector. Pulse rise times are of the order of nanoseconds and to take advantage of fast rise times and high count rates possible, both the amplifier and the discriminator circuits must include high-quality fast circuit elements. For this reason, also, and to obtain full benefit of the 400-4,000 discriminator channels available, the system must be electrically very stable and be operated ideally in a stable temperature and humidity environment.

In a typical system for energy spectroscopy, such as that shown in Figure 6.11, the amplifier output is fed directly into a multichannel pulse-height analyzer (MCA). Most MCA inputs are designed for a 0 to 10 V span of signals to be measured; amplifiers are usually designed to furnish other requirements of a few of the multichannel analyzers, such as lower signal amplitudes. The multichannel analyzer input must not contain any further pulse shaping and must deliver the input pulse directly into its ADC (Analog-to-Digital Converter). It is especially important to avoid connection to low-level negative inputs that are provided in many analyzers and are intended for direct connection from the photomultiplier tube anode in a scintillation counter. These inputs are followed by a shaping amplifier in the MCA and would destroy the desired overall pulse shaping for an output from the external amplifier. When used as shown in Figure 6.11, the analyzer must include provisions for gating and timing.

The region of interest in a spectrum frequently is only a small portion of the total spectrum obtained at the main amplifier output, and the total number of available channels in the MCA may not provide sufficient definition of this region. Spectrum expansion can be obtained by using a biased amplifier after the main amplifier.

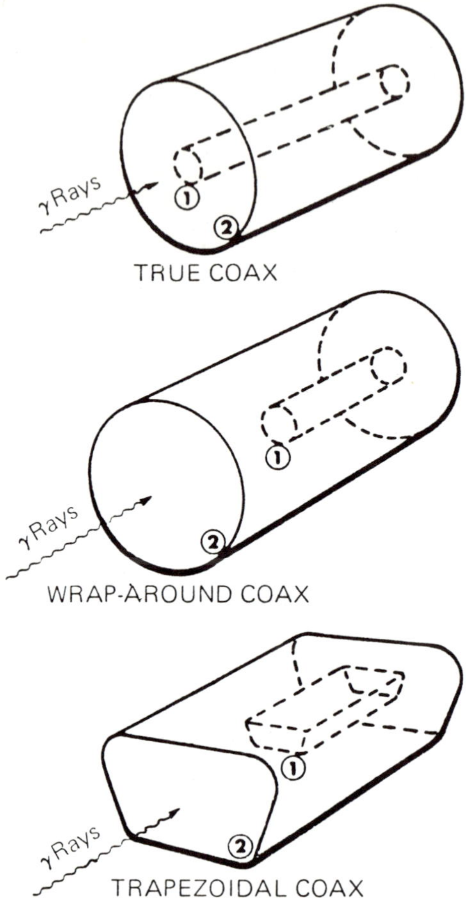

Figure 6.10 Three types of coaxial detectors.

When a biased amplifier is used, a pulse stretcher should also be included to provide equal pulse widths into the analyzer. The pulse stretcher can be located between the biased amplifier and the analyzer, but a location between the main amplifier and the biased amplifier is preferred. This location improves the response of the biased amplifier to amplitudes that just exceed its adjusted threshold level.

Many other systems are available commercially to provide automatic Compton suppression, direct readout in energy and, after appropriate calibration, direct identification of elements and their concentration in a standard sample volume.

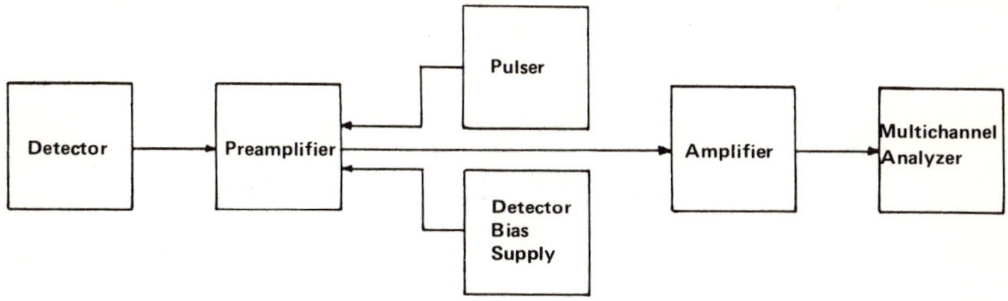

Figure 6.11 Simple multichannel spectrometry system.

In summary, Table 6.2 compares the properties and applications of the three major semiconductor detectors. Figure 6.12 compares the resolutions obtained with various gamma-ray spectrometers.

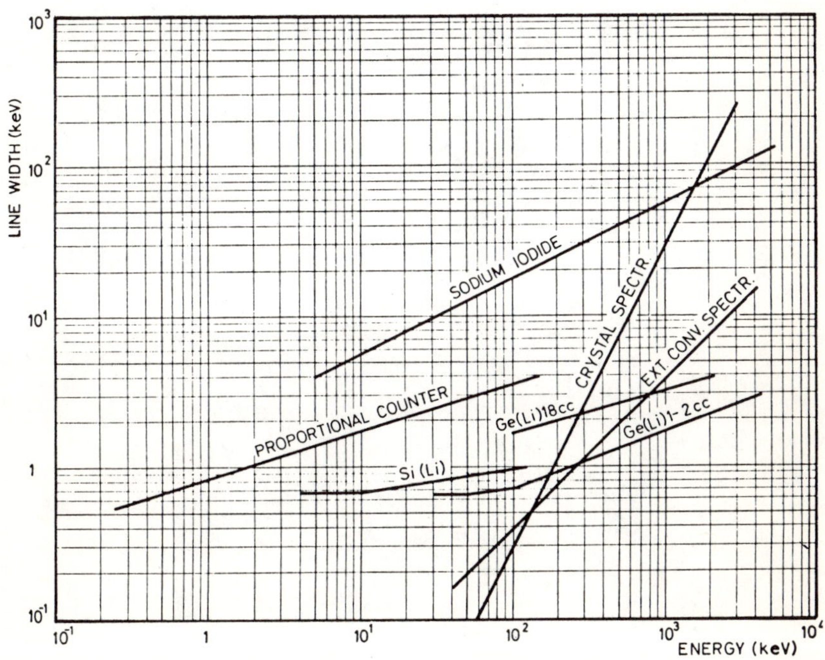

Figure 6.12 Resolution vs photon energy for different gamma-ray spectrometers.

Table 6.2 Summary of Semiconductor Detector Characteristics

Material		Sensitive Thickness	Best Known Resolution	Limitations	Most Suitable Applications	Comments
Diffused Junctions	Ge or Si. High resistivity. Long lifetime.	up to 4 mm	20 keV for 5 MeV alphas	Thickness limited by resistivity. Resolution limited by current noise in small area devices and inhomogeneity in large ones	Monitoring Health Physics. Medical work etc. where robustness is more important than resolution.	More robust and more readily encapsulated than surface barriers.
Surface Barriers	Ge or Si. High resistivity. Long lifetime.	up to 2 mm	3.8 keV for Cs^{137} betas; 11 keV for 6 MeV alphas	Thickness limited by resistivity. Resolution limited by current noise in small area devices and inhomogeneity in large ones.	High resolution charged particle spectrometry and applications involving special geometry.	Main problems are obtaining a low noise, stable back contact and the selection of good material. Encapsulation is possible only with epoxy resins.
Lithium Ion-Drift Devices	Ge or Si. Medium resistivity. Long lifetime.	up to 30 mm	Silicon 3.8 keV for Cs^{137} betas; 15 keV for 6 MeV alphas. Germanium 3.5 keV for Cs^{137} betas; 2.8 keV for Co^{57} betas	Thickness limited by experimental technique. Resolution limited by current noise for betas and gammas, by recombination for alphas and heavy ions.	Silicon High resolution charged particle spectrometry at high energies; also X-ray spectrometry. Germanium Gamma-ray spectrometry.	Main problems are in control of drift process and, for germanium, the stability of the finished device.

Note: All germanium counters require cooling to liquid air temperature. Resolution figures include amplifier contribution.

THERMOLUMINESCENT DOSIMETERS

Introduction

Thermoluminescence has been observed for centuries, whenever certain fluorites and limestones have been heated. Sir Robert Boyle and his colleagues reported on thermoluminescence as early as 1663. However, it was not until 1950 that Daniels proposed that thermoluminescence could be used as a radiation detector and more specifically, as a radiation dosimeter.

When a thermoluminescent phosphor is exposed to ionizing radiation at a low temperature, *e.g.*, room temperature, many of the freed electrons become trapped in lattice imperfections in the crystalline solid. These electrons may remain trapped for long periods when the crystals are stored at room temperature. If the temperature is raised, the probability of release is increased and electrons may be released from the traps and return to stable energy states with the emission of light.

In thermoluminescence dosimetry (TLD) this property is used to measure the radiation dose to which the phosphor has been exposed. This is done by means of a "TLD reader," consisting of a controlled heating element and a photomultiplier system which determines the light fluence emitted during the heating of the dosimeter material. In most TLD readers the integrated light intensity is measured as a function of the heating temperature cycle. The resulting graph is called a *glow curve*. Thus, thermoluminescence dosimetry consists of two steps: (1) radiation exposure, which leaves some excited electrons in metastable traps; and (2) read-out, which consists of controlled heating of the exposed TLD and measurement of the integrated light intensity emitted.

The Thermoluminescence Process

The basic phenomenon can be explained by use of a hypothetical energy diagram of an insulating crystal exhibiting thermoluminescence due to ionizing radiation, to illustrate the fundamental processes that occur in the phosphor (see Figure 6.13).

When a crystal is exposed to ionizing radiation, electrons are released from the valence band to the conduction band. This leaves a hole (a positively charged site) in the valence band. The electron and the hole may move through the crystal until they recombine or until they are trapped in metastable states. These metastable states are thought to be associated with defects in the crystal such as impurity sites. At this point there are two possible ways in which a thermoluminescence photon may be emitted. First, as the crystal is heated, sufficient energy may be given

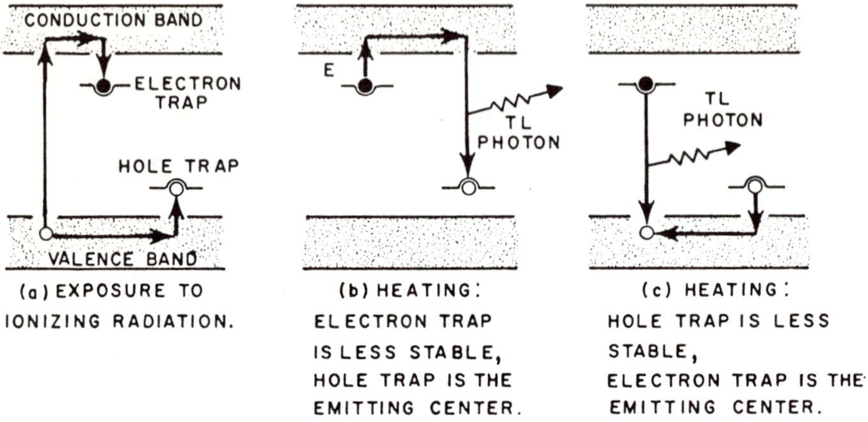

Figure 6.13 Schematic energy level diagram of an insulating crystal exhibiting thermoluminescence.

to the electron to raise it to the conduction band, removing it from the trapping site. The electron may move around in the crystal until it recombines with a trapped hole and a thermoluminescence photon is emitted. On the other hand, the hole trap may be less stable than the electron trap and, when the crystal is heated, the hole receives sufficient energy to wander until it recombines with a trapped electron and a thermoluminescence photon is emitted. Since the two processes are similar, usually only the first possibility is considered in discussions of thermoluminescence.

The energy gap between the valence and conduction bands determines the temperature required to release the electron and produce the thermoluminescence, and is characteristic of the particular material employed. In the practical situation, many trapped electrons and holes are produced. As the temperature of the crystal is increased, the probability of releasing any electron from a trap is increased; thus, the emitted light will be weak at low temperature, pass through one or more maxima at a higher temperature, and decrease again to zero as no filled traps remain.

The Glow Curve

A graph of the light emitted as a function of time or temperature is called a *glow curve.* The usual procedure is to plot the light emitted by the phosphor versus the temperature of the bulk material. The curve is obtained by recording on an electronic plotter the light signals from a photomultiplier tube viewing the material and the heat from a thermocouple in close contact with the thermoluminescence material.

A typical glow curve would show one or more peaks (maxima) as traps as various energy levels are emptied. The relative amplitudes of the peaks indicate approximately the relative populations of the electrons in the various traps. Either the total light emitted during part or all of the glow curve or the height of one or more peaks may be used as a measure of the absorbed dose in the phosphor or the exposure in air. When the peak height is used to measure absorbed dose or exposure, the heating cycle must be closely reproducible to avoid peak-height fluctuations.

To prepare the dosimeter material for reuse, it must be heated again or *annealed.* After the traps have been emptied by heating at a high temperature for a sufficient length of time, and the phosphor has been allowed to cool, it is ready to be reused. The exact prescription for the heat treatment of the crystal depends on the material itself and on the intended use of the crystal, *e.g.,* the exposure-level range. The reusability of TLD is one major advantage over other dosimeters. On the other hand, the annealing process causes a loss of information stored in the dosimeter and thus the loss of any record the exposed dosimeter may represent.

There are a large number of thermoluminescent materials; in fact, most materials thermoluminesce to some extent. To be more useful for a normal dosimetric application, a thermoluminescence phosphor must have a relatively strong light output and be able to retain trapped electrons for reasonable periods of time at temperatures encountered in the application. This requirement limits useful phosphors to those with trap energy levels (kT) at or above 80°C. Table 6.3 lists some of the most popular phosphors and some important characteristics. More detailed information will be given in the sections discussing individual phosphors.

Thermoluminescence Readers

The basic components of a device to evaluate (or, as it is commonly expressed, "to read") thermoluminescence dosimeters or the thermoluminescence of any material are simple. They consist of a system to heat the material, a detector sensitive to the light emitted by the phosphor, and a measuring or recording instrument. A typical arrangement is shown schematically in Figure 6.14. In research it is common to employ both a digital system to record pulses from the photomultiplier tube and an x-y recorder to produce glow curves.

Table 6.3 Characteristics of Some Commonly Used TLD Phosphors

Phosphor	Low Z	Utility at 10 mR	Low Fading	Thermal Neutron Response	
				Low	High
LiF (TLD-700)[a]	X	X	X	X	
LiF (TLD-100,600)[a]	X	X	X		X
CaF_2 (Fluorite)		X	X	X	
CaF_2:Mn		X	X	X	
CaF_2:Dy		X		X	
$Li_2B_4O_7$:Mn	X	X	X		X
$CaSO_4$:Mn		X		X	
BeO	X	X	X	X	
2MgO·SiO_2:Tb		X	X	X	

[a]Designations of the Harshaw Chemical Company which are used widely in the literature to indicate LiF.

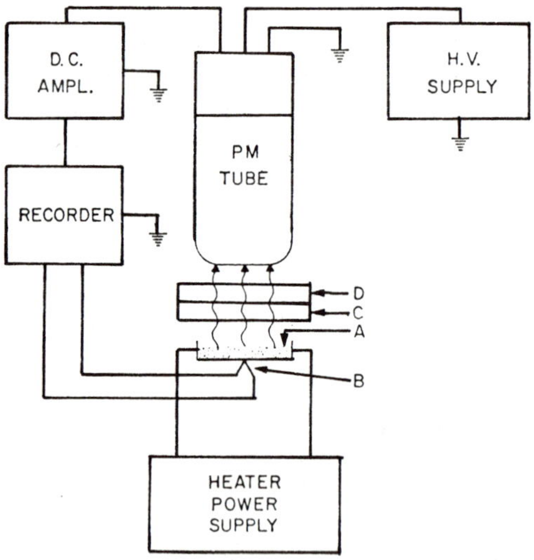

Figure 6.14 Schematic diagram of TLD reader.

The characteristics of commercially available TLD readers vary widely and the choice of a particular model and the complexity of the reader depend on the application of the TLD system.

As stated previously, the heating rate is an important parameter which should be controlled carefully. Figure 6.15 shows typical heating cycles for CaF_2 and LiF TLD phosphors. The constant heating rate above 100°C was about 8°C per second; LiF phosphors were evaluated at a constant temperature of 240°C, whereas 400°C was used for CaF_2. Some readers have an automatic anneal step included in the LiF reading cycle. In this case, rather than allowing the temperature to cool after reading, the temperature of the phosphor was raised rapidly to 340°C and held there for about 37 sec. This treatment allows reuse of the LiF phosphor immediately provided the maximum total exposure of the TLD has not exceeded 400 R. The LiF-anneal cycle is recommended for all new dosimeters or for those dosimeters that have not been read within one week.

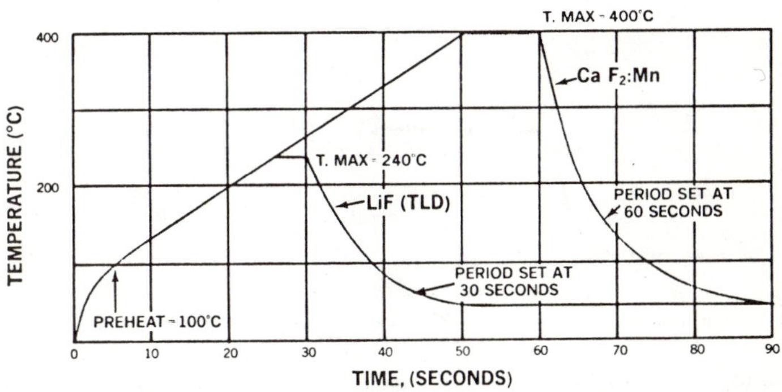

Figure 6.15 Suggested glow curve profiles for evaluation of LiF and CaF_2:Mn TLDs.

Figure 6.16 is an example of an alternative heating cycle for LiF recommended by another supplier. The figure illustrates the relation between heating temperature, heating time and light emission from the phosphor. Note that the low-temperature cycle is intended to empty all the low-energy (and less-stable) traps. The evaluation portion of the readout does not begin until this has been accomplished and the heating rate has again been increased.

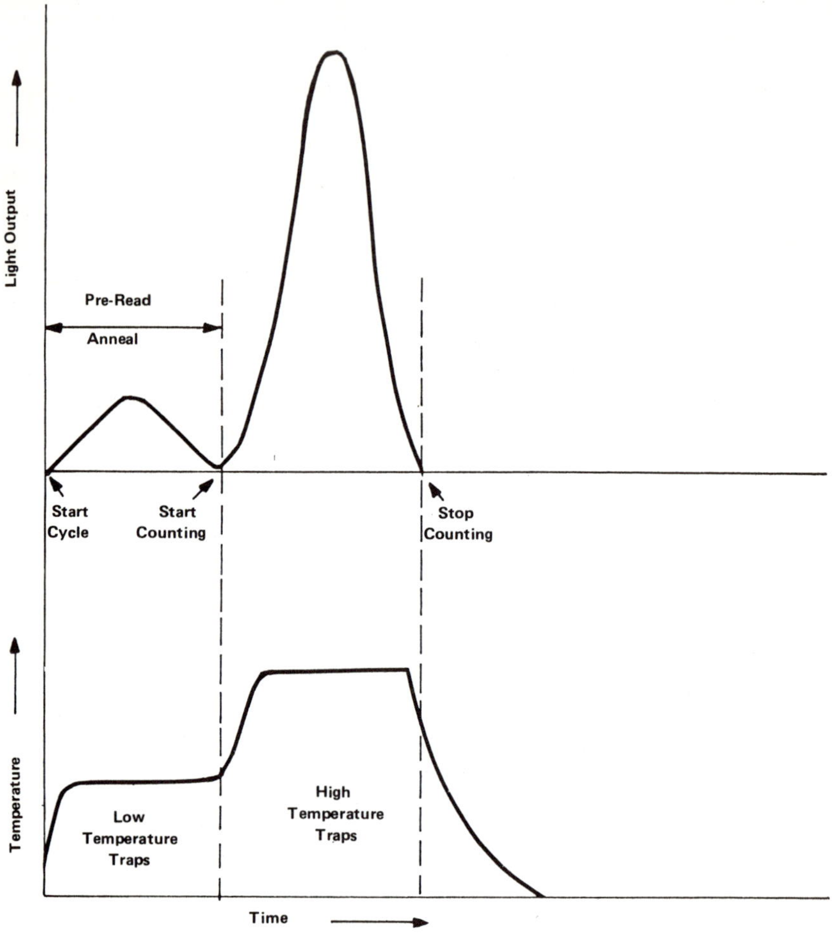

Figure 6.16 Typical read-out cycle for LiF TLD.

There are a number of factors which may affect the shape of the glow-curve. These include the heating rate and its uniformity; the size, shape and thermal conductivity of the sample; the radiation and annealing history of the sample; the recording instrument selected for use; and some spurious effects which may appear.

One of the most important parameters is the heating rate of the sample. If all other factors mentioned above are held constant, then doubling the heat rate will double the height of the glow peak. This effect is illustrated in Figure 6.17. For measurements of thermoluminescence by integration of the peak or by peak height, the heating rate need not be uniform, the heating rate need only reproducible.

Another parameter of importance is *fading*. Fading is the apparent loss of TL signal between exposure and evaluation. This is especially important when using the phosphor for personnel or environmental monitoring.

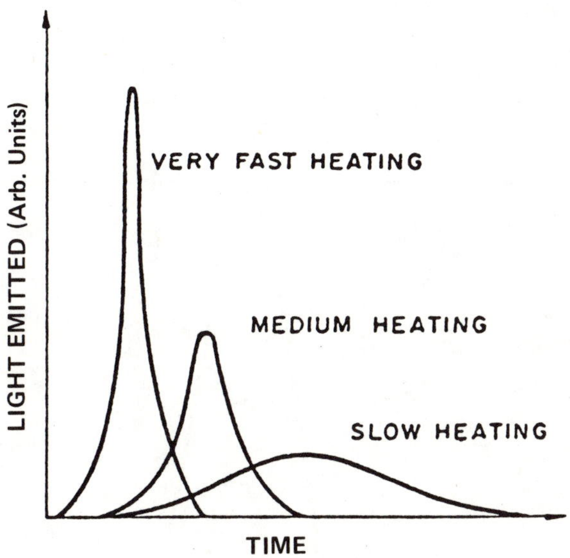

Figure 6.17 Effect of heating rate on the glow curve (after Schulman, 1967).

Dosimeter Materials

Calcium Sulfate

Manganese-activated calcium sulfate ($CaSO_4$:Mn) has a long history as a thermoluminescence dosimeter for ionizing radiation. The phosphor has a glow curve with a single peak occurring in the range of 80° to 120°C. There have been a wide variety of fading rates reported for $CaSO_4$:Mn and this represents one of the major problems or disadvantages in the use of this phosphor. However, this disadvantage is offset by its high sensitivity. In the past, the lowest detectable exposures were in the range of 1 to 10 mR, but a measurement of 20 μR, using large phosphor samples, has been reported. However, the standard deviation of these measurements was ±50%. Using Sm instead of Mn as a dopant, less fading and a higher light output are obtained.

The response of the phosphor as a function of exposure is linear, but there has been some disagreement as to the upper limit at which saturation commences with values from 400 to 10,000 R being reported. In addition, it has been demonstrated that visible light can cause marked fading in the phosphor and that the glow curve was changed drastically if the gas atmosphere in which the phosphor was heated was changed.

The $CaSO_4$:Dy phosphor has shown a single glow peak at about 200°C (see Figure 6.18). Except for the nontissue-equivalence of the material, the characteristics of this phosphor are all good. These include:

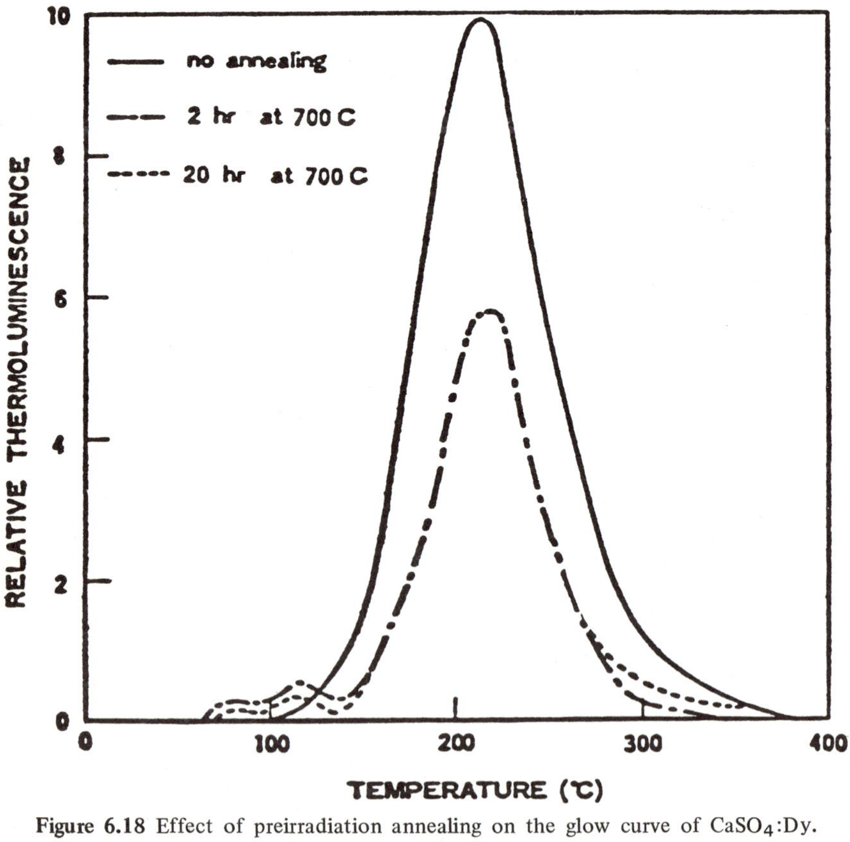

Figure 6.18 Effect of preirradiation annealing on the glow curve of $CaSO_4$:Dy.

1. little fading (1-2% per month);
2. a minimum detectable limit of 500 μR with a standard deviation of ±20%;
3. a linear dose response up to 3 x 10^3 R, superlinear in the range up to 10^4 R, and saturation at about 10^5 R;
4. tribothermoluminescence* is very low;
5. it is highly resistant to spurious effects such as sunlight, artificial light and ambient gases; and
6. the phosphor shows a constant thermoluminescence efficiency for more than 100 irradiations after repeated heatings to erase the previous dose.

The characteristics of $CaSO_4$ activated with Mn and Dy are shown in Table 6.4

Calcium Fluoride

Calcium fluoride (CaF_2) exists in nature as the mineral fluorite. Fluorite exhibits a strong radiation-induced thermoluminescence and, after special treatment, can be used satisfactorily for radiation dosimetry purposes. The first reported use of the radiothermoluminesce of natural calcium fluoride occurred in 1903.

This phosphor exhibits three principal peaks in the glow curve which occur at 70-100°C, 150-190°C and 250-300°C. The lowest glow peak has shown serious fading characteristics as a function of storage time. The response of calcium fluoride as a function of gamma-ray exposure is linear from a few mR to about 500 R with a standard deviation of ±2%. The response of the natural phosphor to fast neutrons is negligible. However, the response to thermal neutrons is about the same as for gamma-rays per rem in tissue. By enclosing the material in a metal filter (*e.g.*, lead), the gamma-ray response can be made constant, within ±20 to 30%, over the energy range of 80 keV to 1.2 MeV.

A synthetic calcium fluoride material is also available. This phosphor is activated with manganese and shows only a single glow peak located at about 260°C. The thermoluminescent emission spectrum shows a maximum at about 5,000 Å. The response of the commercial calcium fluoride dosimeters is about a factor of 10 higher at 40 keV than the response measured at ^{60}Co energies (1.25 MeV). Thermal neutron response has been measured and found to be such that 4 x 10^9 n/cm^2 produced a thermoluminescence signal equivalent to that produced by rad of ^{60}Co gamma-rays. The sensitivity to neutrons of CaF_2:Mn has been reported to be negligible. However, by coating the phosphor with ^6Li it has been

*Tribothermoluminescence is a signal induced by mechanical rubbing or grinding of the phosphor crystals against each other.

Table 6.4 Summary of Performance of TLD Materials

Property	$CaSO_4$:Mn	$CaSO_4$:Dy	CaF_2:Mn (TLD-400)[b]	CaF_2:Dy (TLD-200)[b]	LiF (TLD-100)[b]	$Li_2B_4O_7$:Mn (TLD-800)[b]
Density (g/cm^3)	2.6	2.6	3.18	3.18	2.64	2.4
Effective Atomic Number	15.3	15.5	16.3	16.3	8.2	7.4
TL Emission Spectrum	4,500-6,000 Å (5,000 Å max)	4,800 Å, 5,700 Å	4,400-6,000 Å	4,835, 5,765 Å	3,500-6,000 Å	5,300-6,300 Å
Temperature of Main Peak	110°C	220°C	260°C	180°C	195°C	200°C
Efficiency to ^{60}Co Relative LiF	70	20	10	30	1.0	0.15
Energy Response[a]	~10	~12.5	~13	~12.5	1.25	0.9
Useful Range	μR - 10^4 R	μR - -10^5 R	mR - 3 x 10^5 R	10 μR - 10^6 R	mR - 3 x 10^5 R	50 mR - 10^6 + R
Fading	50% in 24 hr	2% in 1 mo 8% in 6 mo	10%, 16 hr; 15%, 2 wk	10%, 24 hr; 16%, 2 wk	5%/yr (20°C)	<5% in 3 mo
Physical Forms	Powder, Teflon®[c]	Powder, pressed ribbons	Powder, Teflon, pressed, glass encapsulated	Powder, crystals, glass bulbs	Powder, chips, glass (any)	Powder, chips, Teflon

[a]Ratio of response at 30 keV to response at ^{60}Co energies.
[b]Commercial designation (Harshaw Chemical Company).
[c]Registered trademark E. I. duPont de Nemours and Company, Inc., Wilmington, DE.

possible to detect 5 x 10^3 neutrons/cm^2. Table 6.4 summarizes the TLD properties of calcium fluoride.

Figure 6.19 shows the TL emission spectrum of CaF_2:Dy after an exposure of 5 x 10^4 R of ^{137}Cs gamma-rays. Also sketched in the figure is the S-11 response of a photocathode commonly used in photomultiplier tubes. The light emitted by the TLD appears white to the eye; actually the peaks at 4,600, 4,835 and 5,765 Å lie in the blue and yellow regions of the visible spectrum, and it is readily detected by a photomultiplier with an S-11 response.

Linearity of response with exposure for CaF_2:Dy has been observed over the range of exposures from 0.5 mR to 100 R. CaF_2:Dy (TLD-200) has been shown to exhibit "supralinearity" above about 600 R. This is due to the fact that there are several peaks in the glow curve for calcium fluoride and these peaks grow at different rates based on the exposure. The supralinearity can be destroyed by preirradiation annealing; that is, by heating the dosimeters, before irradiation, to about 600°C for at least one hour. After this treatment the glow curve structure stays the same at all exposure levels.

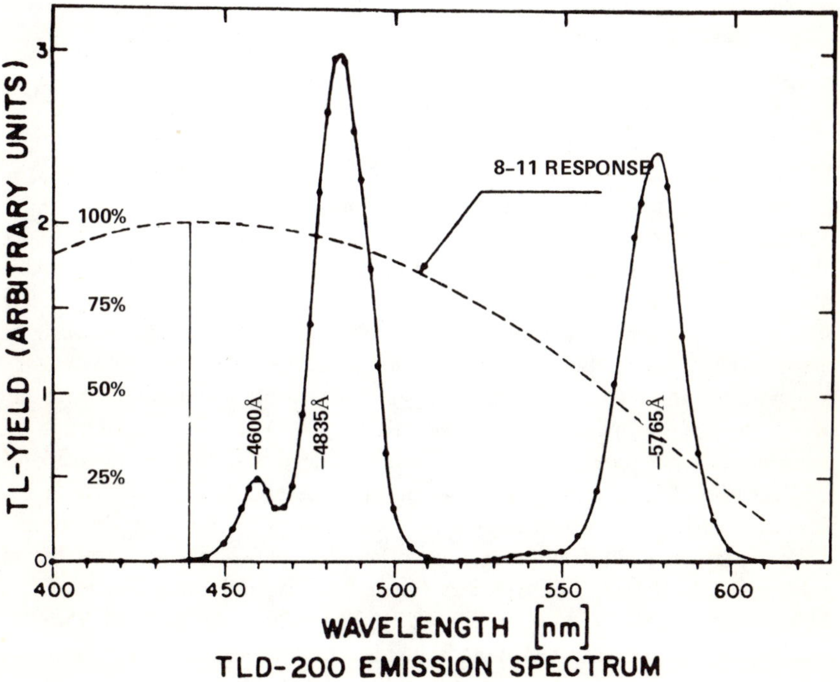

Figure 6.19 The emission spectrum of CaF_2:Dy (TLD-200) after an exposure of 5 x 10^4 R of Cs-137 gamma-rays.

Lithium Fluoride

Lithium fluoride is used as a TLD material in the form of pellets of pressed powder of LiF. The use of pellets, instead of powder, makes handling of the material easier.

Cameron, working closely with the Harshaw Chemical Company, developed a thermoluminescence-grade lithium fluoride called TLD-100, consisting of a loose powder, which was annealed at 400°C before being pressed into pellets. LiF (TLD-100) has received a great deal of attention due to its advantageous characteristics. These are:

1. a nearly flat response per roentgen over a wide range of energies. In the previous sections we have indicated some significant over-responses for photons in the 30-40 keV energy range. For LiF the response at 30 keV is only ~25% larger than that at 1.25 MeV. This difference in response can be overcome by use of a suitable energy compensation screen;
2. the gamma-ray sensitivity is only a factor of 2-3 smaller than the sensitivity of synthetic CaF_2:Mn; and
3. the light emitted in the 210°C glow peak shows little fading with storage time at room temperature; of the order of 10 to 15% within three weeks after irradiation, if the area under the glow peak was measured.

In addition, the phosphor is usable over a wide range of gamma-ray exposures. The response versus exposure relation is linear up to about 700 R and then the response becomes proportional to the exposure to the 1.2 power, *i.e.*, $R = E^{1.2}$, up to about 3×10^4 R. Above this exposure level, saturation begins and is quite pronounced at 10^5 R.

The lower limit of exposure for LiF is set by spurious effects caused either by oxidation in air or by tribothermoluminescence or both. Cameron reported that these spurious signals could be on the order of the equivalent of a 1 R exposure. Filling the reading volume of the TLD reader with dry nitrogen will reduce the effects; permanent encapsulation of the powder in glass also minimizes the effects of gas and tribothermoluminescence and allows the measurement down to 10 mR. However, there is a sacrifice of tissue equivalence due to the glass capsule. The dose-rate dependence of LiF has been found to be negligible up to at least 2×10^8 rad/sec, as measured with 15-MeV electron beams with pulses of 1.3 μsec duration.

For measuring neutron fields, two other LiF phosphors are available in addition to TLD-100. TLD-100 has the natural isotopic abundance of lithium isotopes (7.5% Li-6, 92.5% Li-7); TLD-700 is made from 95.62% pure lithium-7, whereas TLD-600 is highly enriched (99.993%) lithium-6. The thermal neutron cross section of ^{7}Li is negligible when compared with the 945-barn cross section for the (n, α) reaction in ^{6}Li. Thus, TLD-700 has essentially no response to thermal neutrons. Either of the other two LiF phosphors may be used as a pair with TLD-700 to differentiate between gamma-rays and thermal neutrons in a mixed radiation field. TLD-700 has been reported to have less than 0.5% of the thermal neutron response of TLD-100; thus, TLD-700 measures only the gamma-radiation exposure. The other materials, TLD-100 or TLD-600, measure the radiation exposure due to both gamma-radiation and thermal neutrons. The amount of light from a TLD-100 that has been exposed to 2.7 x 10^7 n/cm^2 thermal neutrons equals the light from a gamma-ray exposure of 1 R. Fast neutrons at 10^8 neutrons/cm^2 cause less response in either TLD-100 or TLD-700 than does 1 R of gamma-radiation.

Lithium Borate

Another phosphor that has been studied is lithium borate. Usually manganese activator is added to this material; this appears to be the most promising of the lithium-based phosphors. Like $CaSO_4$:Mn, $Li_2B_4O_7$:Mn can be prepared from readily available raw material. The method of preparation is simple and can be carried out in a well-equipped laboratory. The phosphor has been studied extensively because of its similarities to LiF (TLD-100). As can be seen from Table 6.4, lithium borate is almost exactly tissue-equivalent and has very favorable characteristics relative to LiF.

Figure 6.20 compares the glow curves from samples of LiF and $Li_2B_4O_7$ exposed to 1 R of ^{60}Co radiation. The low temperature peak seen in $Li_2B_4O_7$:Mn decays rapidly; however, the peak at about 200°C is relatively stable. Actually the peak consists of more than one component and the shape and temperature of maximum change are dependent on the exposure. The maximum of the peak appears at about 180°C for exposures $< 10^3$ R, at about 210°C for exposures in the range 10^4 R, and again at about 180°C for exposures around 10^6 R. The emission from the main glow peak is in the yellow-orange region of the spectrum.

Figure 6.21 compares the energy response of several concentrations of manganese. Also included in the figure is an energy-response curve for LiF. This figure also illustrates the possibilities of using the tandem technique to determine the average or effective energy of a photon beam. As can be seen, the ratio of the response of LiF to $Li_2B_4O_7$:Mn has a unique value at essentially all photon energies. Thus, the ratio of the responses, after proper calibration, should give the energy of the radiation field.

The TL response as a function of exposure for $Li_2B_4O_7$:Mn is linear to about 10^4 R; there is a region of supralinear response in the exposure range 10^4 to 10^6 R and saturation occurs at about 5 x 10^6 R, well above the dose range of practical significance.

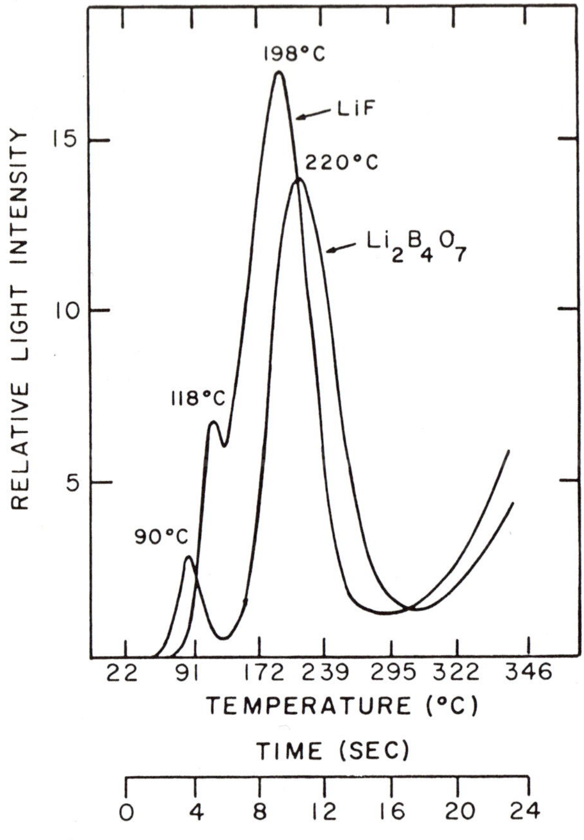

Figure 6.20 Glow curves from 100-mg samples of LiF and $Li_2B4\ O_7$:Mn (0.1% Mn) exposed to 1 R of Co-60 gamma-rays.

Summary

The use of thermoluminescence dosimetry in radiation detection has spread rapidly throughout the world. In many cases it has replaced more

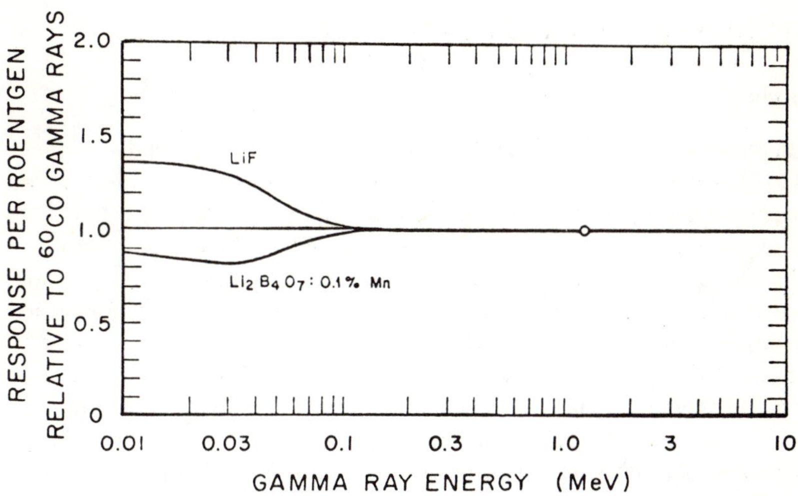

Figure 6.21 Calculated energy-dependence curves of LiF and $Li_2B_4O_7$:Mn.

conventional techniques in personnel monitoring (*e.g.*, film badge, pocket ion chambers) and it has been used extensively as an environmental monitor, in medical applications and in radiobiology. There are a number of TLD materials available commercially, and readers for evaluating the materials are also in a high state of development. The simplicity, sensitivity and reusability of the phosphors are the primary attractions for the user. Through careful calibration, TLD materials can serve as a reliable radiation detector in many situations.

There is no clear-cut choice of a dosimeter material to be selected for a specific purpose. For example, for routine personnel monitoring there are several choices: CaF_2 doped with either Mn or Dy, LiF and $Li_2B_4O_7$. LiF and $Li_2B_4O_7$ are both nearly tissue-equivalent and are, therefore, normally recommended. For monitoring in neutron-radiation fields the LiF materials (LiF normal, ^{6}LiF and ^{7}LiF) are recommended. Either CaF_2 or $CaSO_4$ may be used for environmental monitoring. These phosphors exhibit a high sensitivity relative to LiF. However, $CaSO_4$:Mn as well as CaF_2:Mn and CaF_2:Dy show a high fading rate which influences the use of these dosimeters over extended periods. For environmental monitoring $CaSO_4$:Dy with a fading rate of only about 8% in six months is recommended.

RADIOPHOTOLUMINESCENT DETECTORS

Introduction

Radiophotoluminescence (RPL) was first described by Goldstein in 1912 when he discussed the ultraviolet-excited luminescence spectrum of inorganic compounds after exposure to β- and γ-radiation. The term radiophotoluminescence was actually introduced by Przibram in about 1922 and the term was redefined in 1925. At that time radiophotoluminescence was described as a property of "a material that is originally non-luminescent under visible or ultraviolet light and is made responsive to such excitation by pretreatment with ionizing radiation." The mode of operation is very similar to thermoluminescence with the substitution of the illumination step. Essentially deexcitation, *i.e.,* emptying of traps, is effected by illumination with light of the appropriate frequency.

Radiophotoluminescence and Dosimetry

There are a number of phosphate glasses available for use in an RPL system. This section will concentrate on only three of these glasses. The compositions of these glasses are listed in Table 6.5.

Table 6.5 Compositions of Glass Suitable for RPL[a]

Dopant Element	High-Z	Relatively Low-Z	Low-Z
Ag	4.3%	4.3%	4.2%
Al	4.7%	4.7%	4.6%
K	7.7%	—	—
Ba	10.8%	—	—
P	28.4%	33.7%	33.3%
O	44.1%	52.3%	53.5%
Li	—	1.9%	3.6%
Mg	—	3.1%	—
B	—	—	0.8%

[a]Glasses are available as 1.6 x 1.6 x 0.5 cm blocks or 6 mm long x 1 mm diameter rods.

A schematic diagram of a RPL reader is shown in Figure 6.22. This reader is very simple and in many ways is similar to the TLD reader discussed previously. Usually a mercury vapor ultraviolet lamp is used to excite the dosimeter. The fluorescent light from the dosimeter, in the orange region, is detected by a photomultiplier tube in the reader. There are several techniques that can be used to collect and/or evaluate the signal from the photomultiplier tube; the usual technique is to use a milliammeter calibrated in the appropriate radiation units (*i.e.*, rads, grays or roentgens).

Several problems have been encountered in the use of glass dosimeters for RPL. First, the glass will fluoresce when exposed to UV light; thus, a predose reading must be made on each piece of glass. This is similar to measuring the background in a more conventional radiation detector.

Second, fluorescence due to surface contamination can interfere with the readings; hence, cleanliness and care in the use of the glass are extremely important. For example, the oil from one's fingers is sufficient to invalidate a reading. Thus, in handling the dosimeters it is recommended that forceps be used. In addition, to assure cleanliness the dosimeters are

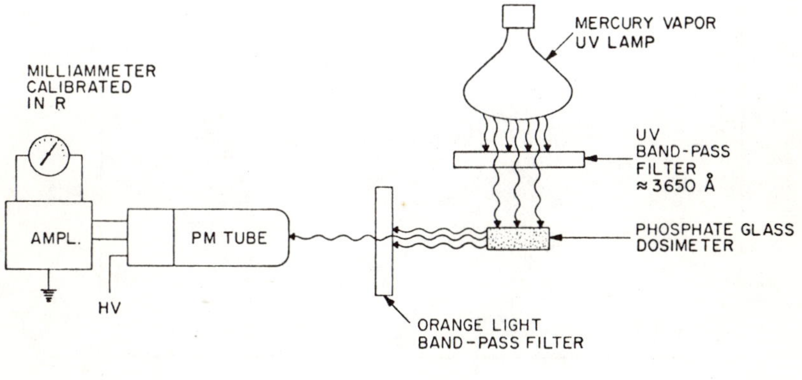

Figure 6.22 Schematic diagram of glass-dosimeter reader.

usually washed before reading. A typical washing regime would include swishing in acetone, washing in distilled water and then absolute alcohol to remove the water. At the end of this procedure the dosimeters are air-dried before being placed in the reader. Ultrasonic cleaners have been employed in recent years but the above procedure is so simple that it is difficult to justify such an expenditure. Chips on the glass will also affect the reading by interfering with the light path. Careful handling with soft-tipped forceps can reduce this problem. Chipping is more common in the small glass rods than in the much larger glass blocks.

The usable dose range depends on the size of the dosimeter, on the orientation of the glass with respect to the UV beam in the reader, the treatment of the glass, and the reader employed. The minimum detectable limit is in the range of 10 rads, dependent on the treatment of the dosimeter and also on the reader. This point is illustrated in Figure 6.23 which shows the instrument response as a function of exposure. Before any heat treatment, the two readers yielded different results above 10^4

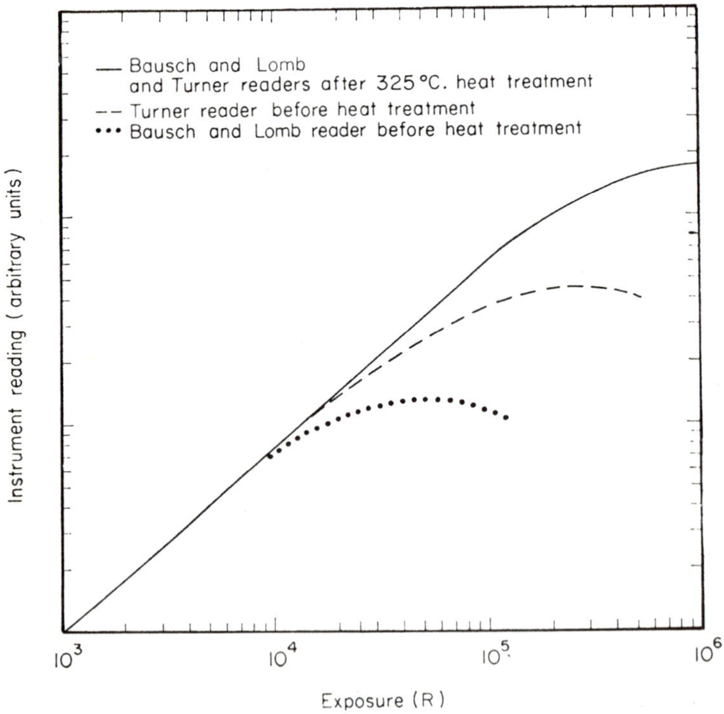

Figure 6.23 RPL response of Ag-activated phosphate glass rods as a function of gamma-ray exposure (Lee *et al.,* 1961).

rads. After heat treatment, the range of linearity of response was extended more than an order of magnitude and the readers yielded similar results.

Most glasses show a rapid buildup of fluorescence with time after exposure. This buildup reaches a maximum then decreases over a long period of time. The buildup period may be quite long (1-2 months) but can be reduced by heat treatment. Storage temperature can have a significant effect of the luminescence intensity. For example, glasses stored after exposure at 400°C (an extreme temperature) lost essentially all their signal in about 10 min. Over the same time period (~10 min) glass stored at 100°C showed a buildup of nearly a factor of 2 and remained relatively constant for nearly 12 hr.

The fluorescence response of all three glasses has been found to be independent of dose rate (for X-ray or electron pulses) at dose rates up to at least 10^8 rads/sec. The high-Z glass rods are about 30 times more responsive per roentgen at 70 keV than for ^{60}Co gamma-rays. The low-Z glass shows about one-third as much energy dependence.

As with TLD, the energy dependence of the glasses can be used to determine the effective energy of an unknown gamma-ray field. The ratio of the readings of two or more pieces of glass gives a measure of the effective energy. In contrast, screens have been used to reduce the energy-dependence. The exact procedure (shielded-unshielded, etc.) depends on the particular application.

Response to Neutrons

The high-Z glass is sensitive to thermal neutrons due to the activation and subsequent decay of the silver component. The thin glass rods appear to be less sensitive because the β-particles from the activated silver have a greater probability of escape from them. Some gamma-radiation is also produced and the energy absorption is less in the rods than in the blocks.

A lithium shield can be used to reduce and, sometimes, eliminate the thermal neutron sensitivity. However, lithium-bearing glasses are available and have a much higher sensitivity to thermal neutrons than does the high-Z glass.

The sensitivity for fast neutrons is quite small. The response is due chiefly to elastic collisions. The response per rad relative to ^{60}Co is 0.5 to 0.7% for neutrons in the energy range 0.5 to 1.5 MeV and is only about 3.5% for 14-MeV neutrons. Fast neutron response can be enhanced by surrounding the glass with a hydrogenous material (*e.g.*, polyethylene) which serves as a proton radiator. A dosimeter suitable for measuring neutron dose from thermal to 18 MeV has been designed. This dosimeter simply surrounds the glass rod in a 20-cm-diameter "Pressdwood" sphere.

The response per rad in most glasses is dependent on the LET of the radiation to which it is exposed. The response for α-particles and protons is less than that for gamma-radiation.

Applications to Routine Monitoring

Radiophotoluminescence can be applied to routine monitoring situations if the glass is suitably enclosed. To be effective, an encapsulation device should make the response of the glass independent of energy and the direction of the incidence of radiation. In most cases this requirement cannot be met, but it should always be kept as a goal. The capsule should reduce the neutron sensitivity to a negligible level or alter the response of the dosimeter such that it can be related easily to a dose-equivalent response.

The encapsulation should protect the glass from the effects of ultraviolet light, scratches and other mechanical damage, dirt and possible radioactive contamination. It should allow for easy removal of the glass, *e.g.*, for washing before reading or heating before reuse. And, lastly, the glass dosimeter and its capsule should be identifiable unequivocally.

As an example, consider the system introduced by Maushart and Piesch. Their system of encapsulation was designed specifically to extend the energy response of the glass and to reduce its directional dependence. The dosimeter (Figure 6.24) consisted of an 8 x 8 x 4.7 mm glass block in a plastic sphere. The sphere was covered with two hemispherical shells of tin in which conical holes had been drilled to reduce the area of the tin by about 15%. The plastic sphere provided protection of the glass from ultraviolet light, contamination, dirt and mechanical damage. The energy response for optimum incidence was independent of energy within ±8% over the range 40 keV to 1.2 MeV. For adverse incidence, *i.e.*, around the area of the attachment device at the top of the sphere, the response was independent of energy within ±18%. The dosimeter yielded an approximately rem response for thermal neutrons as well as for gamma-rays. This was accomplished by the addition of 19 w/o boron to the plastic.

Also shown in Figure 6.24 is a soft-beam dosimeter; it consists of two glass blocks with filters of 1 mm of plastic and 0.06 mm of copper. This dosimeter allows the measurement of the dose fraction in the energy range 15 keV to 40 keV. The energy response of this detector was independent of energy, within ±15%, over this narrow range of energies. The directional response was independent of angle, within ±25%, for gamma-rays incident within a 60° cone from the perpendicular.

The soft-beam dosimeter and the spherical dosimeter may be worn together to cover the energy range 15 keV to 1.25 MeV. In addition, by

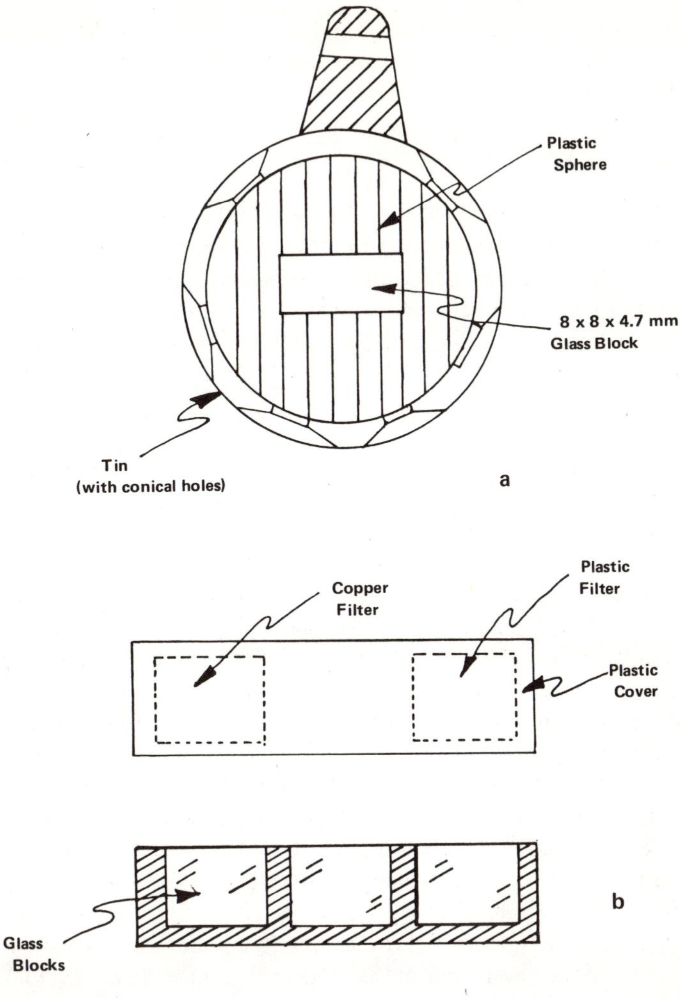

Figure 6.24 Sketch of RPL personnel dosimeter: (a) spherical dosimeter; (b) soft-beam dosimeter.

considering the ratios of the RPL intensity in plastic-covered glass to that in the copper-covered glass, and the ratio of the RPL intensity in the plastic-covered glass to that in the sphere, it is possible to estimate the effective energy of the incident radiation.

Several multiglass dosimetry systems have been described and most have properties similar to the system discussed above. In addition, there are at least two automatic systems available commercially to process and analyze the dosimeters. In some cases the glass does not have to be removed from its holder in order to be read. In addition, small computers have now been added to these systems to facilitate record-keeping requirements.

Measurement of Induced Activity

In addition to their normal use, the phosphate glasses may be used as activation detectors. The primary reactions are (n, γ) and (n, p) reactions with certain elements which constitute the glass. Using the spherical dosimeter arrangement discussed above as an example, there are basically three categories of neutron-activation reactions considered. These are:

1. *Activation of phosphorus.* The reactions are $^{31}P(n, p)^{31}Si$, which is a threshold reaction for neutrons above about 2.5 MeV, and $^{31}P(n, \gamma)^{32}P$ to give an indication of the thermal and epithermal neutron fluence. The usable dose range for these reactions is 0.5 to 1000 rads.
2. *Activation of silver.* The primary reaction is the $^{109}Ag(n, \gamma)^{110}Ag$ reaction for thermal and epithermal neutrons. The usable dose range is above 10 rads.
3. *Activation of the dosimeter capsule.* There are a number of reactions possible, mostly for thermal and epithermal neutrons. Three of these are $^{112}Sn(n, \gamma)^{113}Sn$, $^{116}Sn(n, \gamma)^{117}Sn$, and $^{114}Cd(n, \gamma)$ ^{115}Cd. The usable dose range is above 10 rads.

The evaluation of the dosimeter using activation methods is relatively straightforward. The system uses the difference in half-life between ^{31}Si (2.6 hr) and ^{32}P (14.3 days) to separate the activation fractions due to fast neutrons from the activation fraction due to thermal and epithermal neutrons. This requires two measurements of the induced activity. The first is made within the first 15 hours after exposure to determine the ^{31}Si activity. The second is made 24 hours after exposure to determine the activity of ^{32}P. It is usually necessary to correct the measured induced activity of the ^{31}Si (measured at 15 hours) for the activity of the ^{32}P.

Both ^{31}Si and ^{32}P are beta-emitters and the recommended method of evaluation is to observe the Cherenkov light produced in the glass by the high-energy beta particles. Figure 6.25 shows the results of such an evaluation. Figure 6.25a presents a composite decay curve for glass exposed to about $10^{10} n/cm^2$ of either thermal or fission-spectrum neutrons. Figure 6.25b shows the activity in the glass measured after exposure to 3 MeV, fission or thermal neutrons. Note that the response, as might be expected, is linear over a range of neutron doses of three orders of magnitude. Thus, the phosphate glass dosimeters employed in routine personnel monitoring can be used as an emergency neutron dosimeter in the dose range 0.5 to more than 1,000 rads of fast or thermal and epithermal neutrons.

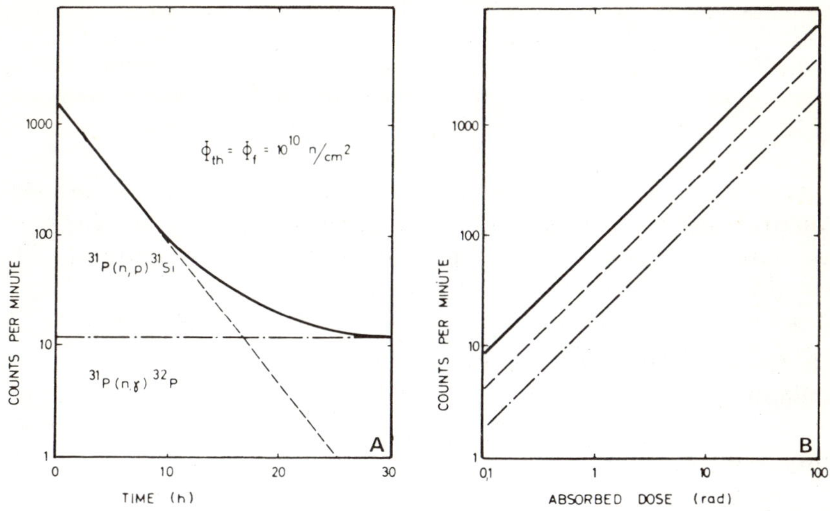

Figure 6.25 Count rate of Cherenkov light from Toshiba FD-1 glass: (a) decay curve; (b) dose response.

A special arsenic phosphate activation glass has been developed. This glass was designed to allow the determination of the total neutron fluence and the total neutron dose due to thermal, intermediate and fast neutrons (*i.e.*, > 2.5 MeV). However, the measurements of induced activity must be made within 6 hours of the exposure. In this glass, the neutron fluence is determined by observing the count rate of ^{31}S and of ^{76}As ($\tau_{1/2}$ = 26 hr) measured at known times after exposure. The accuracy of the measurement is reported to be ±15% in the time interval 4 to 8 hours after exposure.

Other Applications of Glass Dosimetry

Phosphate glass has been used in nuclear track-counting applications. In this application the glass is placed near a foil of fissile material: uranium, thorium or neptunium. After exposure to neutrons, the glass is placed in a chemical etching solution and tracks due to fission fragments from the fissile radiator are made visible in the glass. Usually these tracks are counted undeı a microscope and the neutron dose is determined based on proper calibration techniques. However, this method is not as satisfactory as the plastic foil track-etch detectors described in Chapter 7.

Summary

Radiophotoluminescence has a number of potential applications in radiation detection, especially in personnel monitoring. However, because of its low sensitivity compared to thermoluminescence dosimetry, it is not in wide use in the United States at this time. Nevertheless, the discussion illustrates the application of an observed phenomenon to radiation dosimetry and the ingenuity that can be employed to alter the energy response, usable exposure range and sensitivity to various radiations of a material.

THERMALLY STIMULATED EXOELECTRON EMISSION (TSEE)

Introduction

In 1927 it was reported that the emission of low-energy electrons may occur from the surface of certain inorganic solids undergoing specific chemical reactions. Later, in 1931, it was found that some inorganic crystals that had been exposed to ultraviolet light emitted electrons when heated. Kramer assumed that the electron emission involved an exothermic process and thus he introduced the term "exoelectrons." The process also was called, by others, the "Kramer Effect." Many other names have been suggested and used to describe the same phenomenon, including "tribo-thermostimulated-coelectron emission," "thermoemission," "electron exo-emission," "exoelectron emission" and "cold electron emission."

Kramer studied $CaSO_4$ extensively and found a linear relationship between radiation dose and TSEE. He found a minimum detectable dose of about 0.1 mR but there was evidence of a 55% fading in about 3 days and 65% in about 8 days. Even with these discouraging results reported by Kramer, the research on TSEE continued because of its apparent advantages. These advantages, compiled by Becker, are summarized below:

Apparent Advantages of TSEE

- Wide Variety of Substances
- Low Costs
- No Light Sensitivity Prior to Exposure
- No Complicated Processing
- Wide Dynamic Range
- Reusability
- Dosimetry for Low-Energy Radiation

Disadvantages

- Sensitivity Changes Due to Surface Alterations
- Not a Permanent Document of Exposure (Radiation Effect Destroyed During Reading)

Of the various materials studied, BeO exhibited several attractive characteristics for TSEE. These characteristics included a high sensitivity, chemical and physical stability, insolubility in water, and low Z; most practical applications use this material.

Exoelectron emission is observed in many processes. Exoelectron emission is seen during the solidification of metals and during some chemical reactions. Of most interest, at least in the present context, is the exoelectron emission noted in organic crystals, semiconductors and other organic materials after exposure to ionizing radiation. Even though only the radiation-induced effects are of interest here, other effects (some of which were noted above) may interfere with the successful use of the technique in radiation detection.

The exoelectrons appear to originate from a very thin surface layer probably not thicker than 10 to 100 Å. In addition, the exoelectron energy is small; an average electron energy of 0.65 eV with a maximum energy of 3 eV has been measured. Further, the similarities between TSEE and TLD indicate a similar process. In TSEE surface imperfections of the crystal are assumed to serve as traps for electrons liberated during irradiation. These electrons, or perhaps only a small percentage of the electrons, are liberated during thermal or optical stimulation. If these electrons can be detected, for example, in a large volume Geiger-Müller (GM) counter, the emission rate can be related directly to the radiation exposure.

Most of the more recent work has used BeO as the dosimeter. In a manner similar to that during the early development of TLD, there was a wide variation in the absolute TSEE intensity. In addition, there were variations in peak heights* and locations depending on the source of the BeO material, the amount of impurities present in the material, the thermal pretreatment of the material, and the grain size of the crystals.

Criteria for Selecting a TSEE Material

As stated in the introduction, many materials exhibit TSEE under certain conditions. In order to be usable as a radiation detector the TSEE materials should satisfy certain general requirements.

The exoelectron emission should not occur below about 120 to 150°C in order to keep fading to a minimum. On the other hand, the emission should not occur at a very high temperature since the GM counter cannot discriminate between thermionic electrons emitted from the heater element

*Following the techniques used in TLD, glow curves were obtained from materials exhibiting TSEE. In this case, the TSEE intensity (or more usually the count rate of the G M counter) is plotted against the temperature of the sample.

and those emitted from the sample. The material should be stable chemically against the normal atmospheric constituents. It should not be affected by humidity nor should it absorb gases. The material should be thermally stable in air and in the GM counting gas through all prereading, reading and postreading heat treatments.

As a dosimeter, the material should be as sensitive as possible to radiation exposure. Thus, it is necessary that a large number of electrons be emitted from the surface per unit exposure. Other optimum properties include linearity of response and LET dependence.

In addition to the above, the material should not be expensive, it should be nontoxic and it should be easy to prepare.

There is a large selection of materials that meet most of these requirements. However, BeO seems to be the material of choice and most TSEE research has been done with this material. Nevertheless, it is appropriate to list briefly other materials that may be usable.

1. *Alkali halides.* The most notable material from this group is LiF which has its main TSEE peak at about 120°C. Others in this class, which are not suitable, are LiCl, LiBr, NaF, NaBr, NaI, KCl and KF. All of the materials were ruled out because all had TSEE emission peaks below 100°C.
2. *Alkaline earth halides.* The primary representatives are CaF_2, BaF_2 and SrF_2. In all of these materials, a linear relationship between exposure and response has been observed between about 2 to 100 R.
3. *Alkaline-earth sulfates.* Materials considered were $CaSO_4$, $BaSO_4$ and $SrSO_4$.
4. *Metal oxides.* The primary material is BeO. However, this material has the disadvantage of being highly toxic in the low-fired state. By heating the material to more than 1,000°C, there is an increase in grain size and a change in the surface toxicity and an increased TSEE sensitivity. Other materials in the group were CaO, MgO, Al_2O_3 (with and without Cr), SiO_2, ZnO, NiO and CuO.
5. *Metal sulfides.* Most commonly studied materials in this class are ZnS, CaS and SrS, with and without Pb or Cu. The characteristics of these materials depend strongly on the impurities present and the method of preparation.
6. *Complex compounds.* These include limestone, $Li_2B_4O_7$, glasses such as metaphosphate and quartz, complex ferroelectric materials, and even hemoglobin and egg shells.

It is clear that there are a wide variety of TSEE detector materials available. However, the effects of temperature, preirradiation history, impurities and preparation techniques on the TSEE characteristics are poorly understood. The exact phenomenon is usually considered to be relatively complex.

Detector Characteristics

In addition to the general criteria, the sensitivity of the detector depends on the concentration and size of the emitting surface. If a constant percentage of electrons is counted the response is a linear function of the surface area. For LiF, the counts per unit exposure increased linearly with the surface area over two orders of magnitude.

The lower detectable dose limit using TSEE may be determined more by counting statistics than by counter background. It is important to remember that electrons are detected in the GM counter. Standard techniques such as shielding, use of small volume counters, and anticoincidence techniques can be used to reduce the background. With $CaSO_4$ Kramer measured doses of about 3 μR. This is equivalent to an exposure of a few minutes to the natural radiation background in many areas of the world. He estimated that he could measure 10^{-7} R by optimizing detectors and the reader. The response as a function of exposure for most TSEE materials is linear over a wide range of exposures; Figure 6.26 shows data

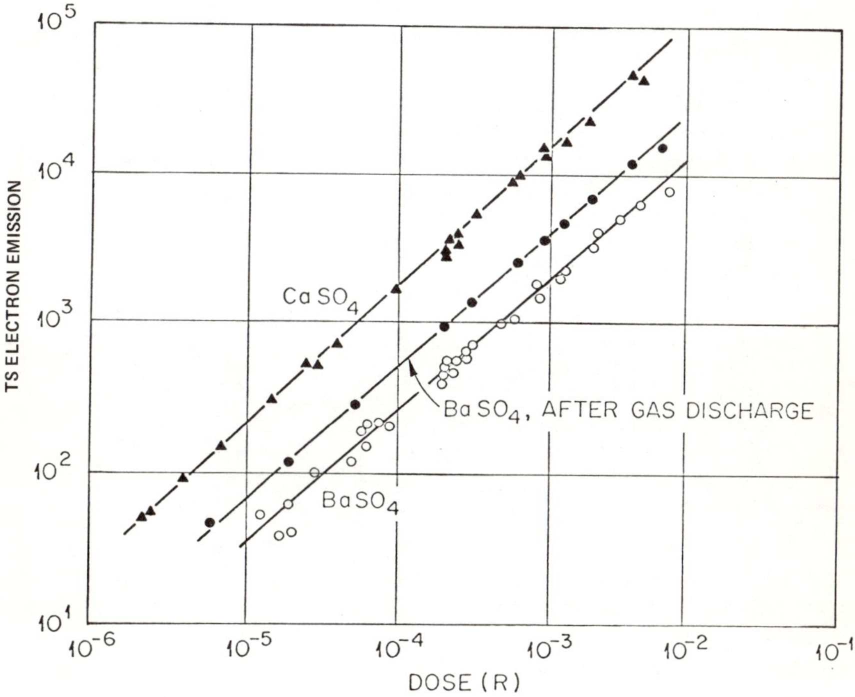

Figure 6.26 Relative TSEE response for $CaSO_4$ and BaSO4.

for $CaSO_4$ and $BaSO_4$. An important point shown here is the high sensitivity of the $CaSO_4$; exposures near 10^{-6} R or below appear to be a reality.

The effects of LET of the radiation and the storage temperature on fading can be seen in Figure 6.27. In BeO it appears that fading is LET-dependent and a strong LET dependence has been noted also in $CaSO_4$. Fading is highly dependent on the storage temperature and on exposure to light. No fading has been observed in BeO exposed to the standard red or yellow darkroom lights.

The photon energy dependence of a mixture of BeO and graphite is shown in Figure 6.28. Also shown in this figure are the ratios of the mass absorption coefficients for BeO and C to those for air.

No dose-rate-dependence data are available for TSEE materials. However, it is usually assumed that since TLD and RPL materials do not exhibit significant dependence, neither will the TSEE materials.

TSEE materials exhibit a small fast-neutron response. This response can be enhanced by surrounding the TSEE dosimeter with a polyethylene

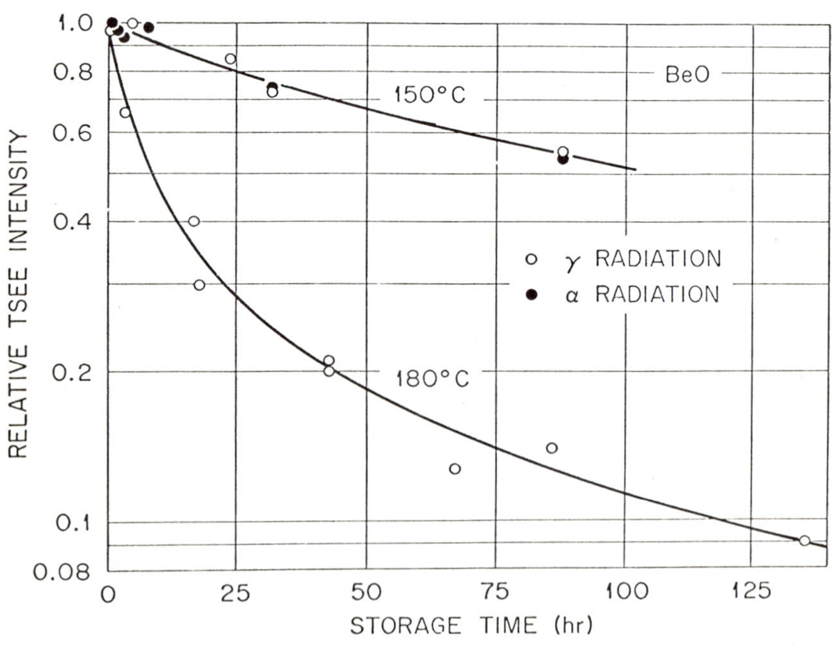

Figure 6.27 Effect of storage temperature and LET on BeO.

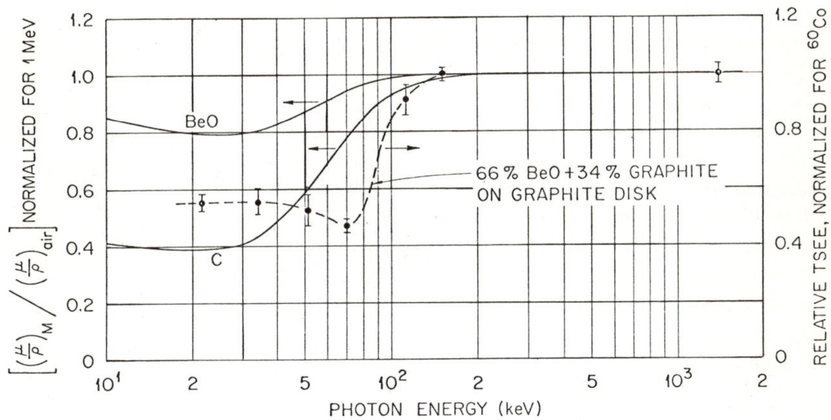

Figure 6.28 TSEE energy response for a composite BeO and graphite dosimeter.

radiator. For a $Li_2B_4O_7$ crystal which was exposed to fission neutrons, for samples exposed both with and without a polyethylene radiator, a more than twofold increase in count rate at 150°C was observed with the polyethylene.

Advantages of TSEE Dosimetry Techniques

There are several advantages of the TSEE technique. The main advantages include a high inherent sensitivity (10^{-6} to 10^{-7} R) and a linear response over a wide range of exposures (~9 orders of magnitude). Other obvious advantages, at least in using certain materials, include a wide choice of materials, and the use of some inexpensive "off-the-shelf" materials.

There are several areas which require further study. These include adjusting the energy response by mixing materials, the use of a high-temperature TSEE peak in applications where TLD is not suitable, the use in mixed radiation fields, and the applications of TSEE (a surface phenomenon) to microdosimetry.

Thermally stimulated exoelectron emission is a sensitive technique which may be applied to radiation detection. Its high sensitivity makes it the method of choice for studies of low-level radiation exposure, low-energy beta emitters, etc. However, there are many questions to be answered as to the exact characteristics of the dosimeters under a variety of exposure conditions. Thus, at this time, TSEE appears to be more a novelty than a well-defined dosimetry technique.

REFERENCES FOR FURTHER READING

Attix, F. H. Ed. *Luminescence Dosimetry*, CONF-650637, USAEC Symposium Series No. 8 (Washington, DC: U.S. Atomic Energy Commission, 1967).

Attix, F. H., and W. C. Roesch, Eds. *Radiation Dosimetry* (New York: Academic Press, Inc., 1966).

Auxier, J. A., K. Becker and E. M. Robinson, Eds. *Proceedings of the Second International Conference on Luminescence Dosimetry,* CONF-680920 (Washington, DC: U.S. Atomic Energy Agency, 1968).

Advances in Physical and Biological Radiation Detectors, IAEA (Vienna: International Atomic Energy Agency, 1971).

Becker, K. *Solid State Dosimetry* (Cleveland, OH: CRC Press, Inc., 1973).

Bertolini, G., and A. Coche. *Semiconductor Detectors* (Amsterdam: North-Holland Publishing Co., 1968).

Cameron, J. R., N. Suntharalingam and G. N. Kenney. *Thermoluminescence Dosimetry* (Madison: The University of Wisconsin Press, 1968).

Dearnaley, G., and D. C. Northrop. *Semiconductor Counters for Nuclear Radiations,* 2nd ed. (New York: Wiley, 1966).

Heath, R. L. "Ge(Li) and Si(Li) Spectrometry," (Rept. ANCR-1000), in *Gamma-Ray Spectrum Catalogue,* 3rd ed., 3 vols. Aerojet Nuclear Co. (1974).

PROBLEMS

6.1 The specific loss per ion pair in silicon is W = 3.6 eV. Compare the number of ion pairs produced by a 6-MeV α particle in silicon with that produced in a scintillator or a gas-filled ion chamber.

Assuming comparable collection times and a Gaussian distribution, compare the pulse-height resolution in the three cases.

6.2 Calculate the depletion depth in pure Si and Ge when an electric field of 1,000 V/cm is applied. (Dielectric constants of Si = 12.5, of Ge 15.8.)

6.3 Estimate the minimum time for complete charge collection in a 4-mm thick fully depleted Si detector.

6.4 Explain the charge collection in a coaxial Ge(Li) detector. If the drifted volume is increased 30% in diameter, what effect would this have on the resolution? Why?

6.5 Compute the theoretical resolution, in keV and percent, for the 0.66-MeV Cs-137 γ-ray in a planar Ge(Li) detector.

6.6 Calculate the energy loss of a 3-MeV proton passing through a 0.02-mm-thick dE/dx silicon detector. Neglect losses in the gold window layers.

6.7 Estimate the energy width per channel for the Co-60 spectrum in Figure 6-8. What would be the accuracy of the peak energy determination if the source strength had been only 1/100 of that shown or the pulse-height analyzer had 400 instead of 4,000 channels?

6,8 What is the maximum acceptable thickness for the window of a surface-barrier detector if the dead-layer loss should be less than 5% in determining 0.8-MeV protons, or 5-MeV α particles?

6.9 Identify the optimum photomultiplier responses (Figure 5-3) required to match TLD emissions for $CaSO_4$:Mn and LiF dosimeters.

6.10 Compare the neutron sensitivities of 25-mg LiF TLDs consisting of natural Li or 99% Li-6 assuming equal excitation probabilities for the α particles produced.

6.11 A certain TLD reader passes the response peak 10 min after starting the heater cycle with a 6-amp heater current. What would be the effect of operating with an 8-amp heat current? How would the integrated light emission from an exposed TLD compare in both cases?

6.12 Analyze the advantages and disadvantages of converting a personnel dosimetry program from film badges to TLDs.

6.13 Estimate the relative size (and cost) for comparable sensitivities in a scintillator γ-ray spectrometer and a Ge(Li) system exposed to 0.5-2 MeV γ-rays.

6.14 Calculate the trap energies above the valence bands associated with the main light emission peaks in CaF_2:Dy.

CHAPTER 7
TRACK DEVICES

INTRODUCTION

Few things are as convincing to the uninitiated, when first confronted with new concepts in nuclear physics, as a picture showing the actual path taken by a nuclear particle in a medium or the personal observation of a track being formed. The existence of many types of particles was first demonstrated by observing their *tracks* and the first pictures of the emission of alpha particles from a plane source, or the divergent tracks of positive and negative electrons in the magnetic field of a Wilson cloud chamber, have been among the landmarks in science. The analysis of tracks in bubble chambers, spark chambers and photographic emulsions is still one of the most powerful techniques employed in nuclear physics research. Track-etch dosimetry, where the radiation damage spike accompanying the track of an energetic particle is made visible, is a related method that is finding increasing applications.

In all these cases a method must be devised to make visible, either momentarily or permanently, the density and location of the streak of ions produced by the passing energetic particle by means of some phenomenon that effectively amplifies the ionization process without inducing any movement of the ions. The Wilson cloud chamber and the simple photographic plate have been gradually superseded by bubble chambers and a wide range of special nuclear and radiographic emulsions; these will be among the detectors discussed in this chapter.

For many purposes, measurement of track density, track deviation and range in photographic emulsions still remains the reference method for particle identification and energy determination. Independently, the simple spark discharge occurring between an electrified point and an opposing plate, when the intervening air is ionized, has been the starting point for the development of a host of large-volume spark chambers with sophisticated computerized data treatment analyzing the tracks and characteristics of a variety of high-energy events.

PHOTOGRAPHIC EMULSIONS

A photographic emulsion consists of a fine dispersion of silver halide grains, mainly AgBr, in a gelatin layer, mounted on a backing material such as cellulose acetate or glass. When the halide grains are excited by ionizing radiation they are left in a long-lived excited state, which makes them "developable," that is, those grains are easily reduced to elemental silver by a suitable developer solution whereas the nonexcited grains are not. The latter can be subsequently washed out by a "fixer" solution. Some diffusion of the Ag^+ ions also occurs within the grain facilitating this reduction. If the plates are warmed or left undeveloped for a long time, recombination of the excitons occurs and the *latent*, that is, undeveloped, image is said to be *fading*. The sensitivity of the emulsion in terms of the amount of energy that has to be absorbed to make a grain developable depends on grain size, grain density and the addition of certain impurities, such as AgI or S, that may modify the excitation-level structure. For radiographic and track emulsions, the resolution of an image depends also on grain size, grain density and uniformity in distribution; a larger number of finer grains may result in better image definition, but may also require a larger dose deposition to make a sufficient number of grains developable. Because the developed image is permanent, photographic emulsions are important for record-keeping purposes. Specialized photographic emulsions have been produced for three types of radiation detection: (1) radiography and autoradiography, (2) nuclear track recording, and (3) dosimetric applications.

Nuclear track emulsions permit observation of individual tracks under a microscope or on a projection screen. They usually contain a high density (up to 82% AgBr by weight) of fine halide grains (~0.3 μm) in an emulsion that may range from 25 to 600 μm in thickness. Such thicknesses require very careful control of development conditions, especially temperature control and presoaking of the emulsion, to ensure uniform development at all depths. The tracks are viewed individually and measurements can be made on their length, direction, angle of dip, if any, and changes in ionization capability (LET). Genetic relationships in complex decay chains can be followed and evaluated, provided most of the tracks lie within the emulsion. Figure 7.1 shows several types of tracks to illustrate the type of image obtained. It is evident that the alpha particles leave relatively short, easily recognized tracks, with the heavily ionizing end of the track corresponding to the peak in the Bragg

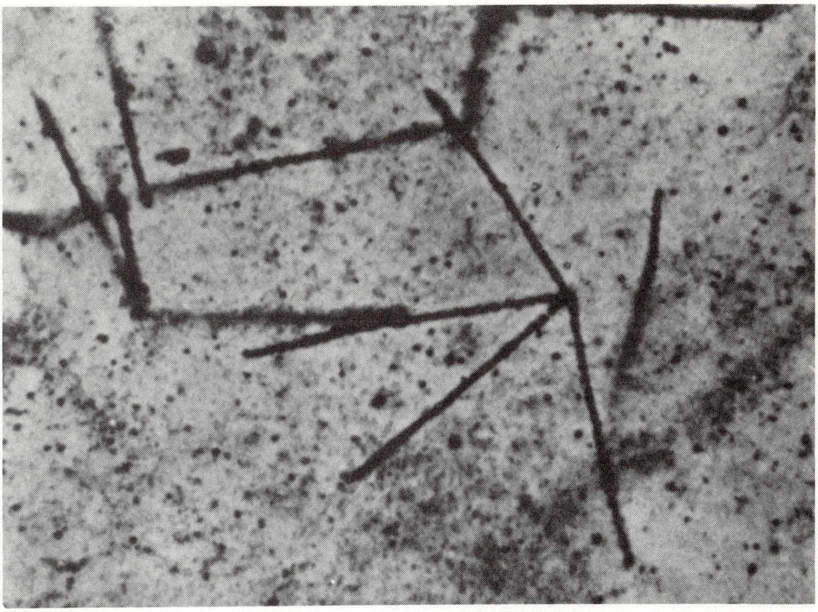

Figure 7.1a Particle tracks in nuclear emulsions (after Powell and Occhialini): radiothorium stars.

curve. The long, widely spaced tracks of the protons are characteristic of high-energy particles, and the grain density can be used to determine the particle energy.

The range is largely influenced by the composition of the emulsion, in which the gelatin constitutes 17-45% of the weight but a much higher proportion of the total volume. The composition of the gelatin varies and has to be known for specific applications; typical concentrations are C 36-42%, O 38-50%, H 6-8% and N 9-11%, with minor impurities. Radiographic emulsions have a much lower silver halide content ($\sim$40% by weight) and the grains tend to be coarser than in track emulsions.

In *dosimetric and radiographic emulsions*, the latent image consists of large areas that are "blackened," that is, their grains have been made developable, by the uniform absorption of energy over a greater area. This process, which involves no specific particle tracks, represents the

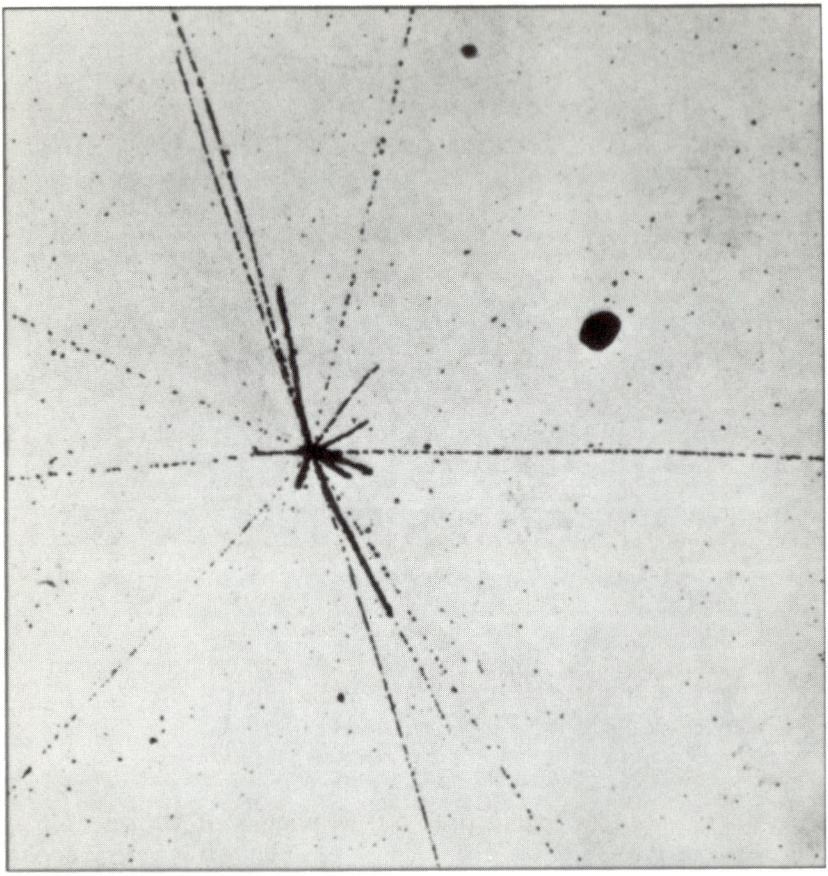

Figure 7.1b Particle tracks in nuclear emulsions (after Powell and Occhialini): tracks of particles of great energy.

condition when photons interact with the AgBr grains, mainly by Compton interactions, leaving behind a latent image. The quality of this image depends on the uniformity of response of the grains to the absorption of comparable amounts of energy; hence, uniformity of grain size and dispersion is important. The degree of blackening, or the number density per unit area of developed grains, is usually measured by photometry of

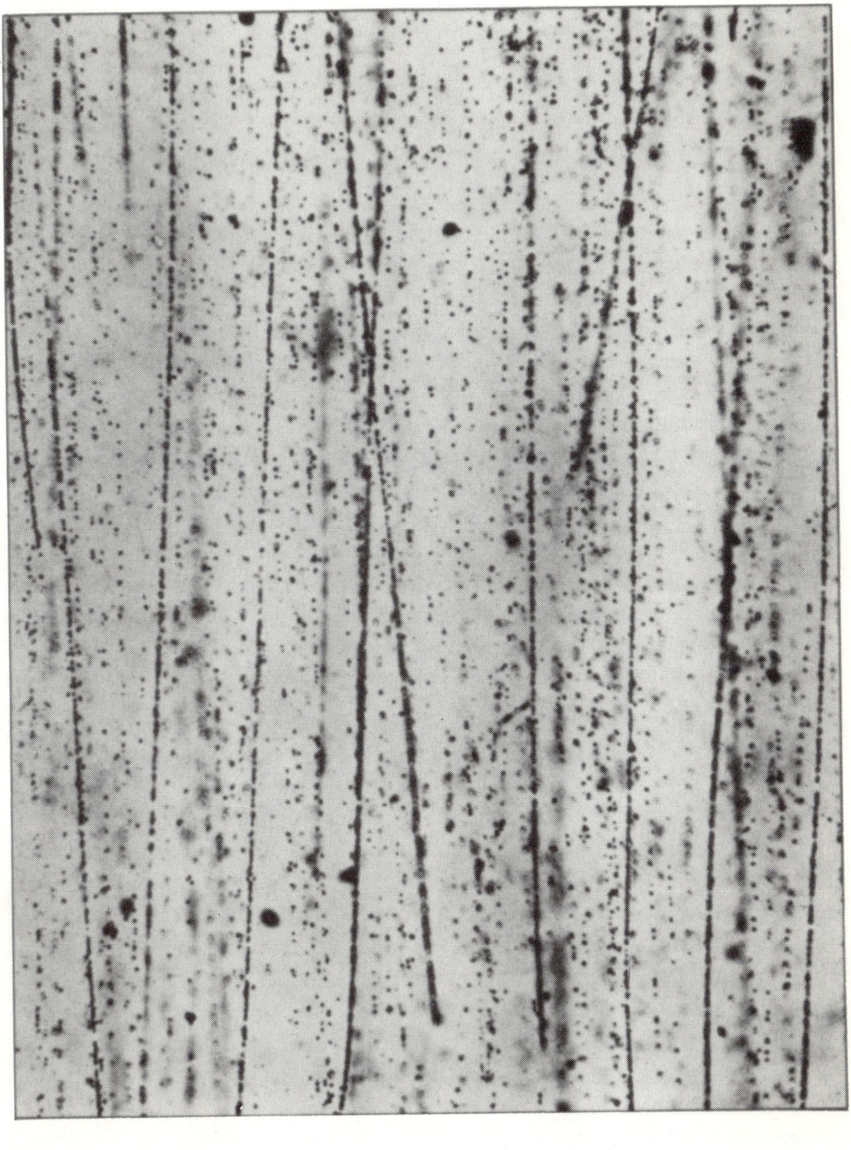

Figure 7.1c Particle tracks in nuclear emulsions (after Powell and Occhialini): explosive disintegration of a nucleus.

relative intensity of light transmitted through the developed film. This is a logarithimic relationship of the form

$$\text{optical density } S = \log \frac{I_o}{I_x} \qquad (7\text{-}1)$$

where I_o = light intensity through unexposed (developed) film

I_x = light intensity through exposed film

For a given type of radiation, the blackening is proportional to the radiation dose, but not linearly so. With some films it is possible to cover a dose range of 10^2-10^3 to 1, but this requires gray adaptability; high contrast and a wide range of sensitivity tend to be mutually exclusive properties. *Film contrast* expresses the ability of a film at a particular density and under particular conditions of irradiation to distinguish between two nearly equal incident fluences. Since contrast depends on density and hence on fluence, it is not a constant parameter and is often determined by the use of a "gray wedge," a stepped absorber with differences in thickness that are accurately known so that the difference in fluence reaching the film can be determined precisely.

A single ionizing particle may deposit sufficient energy in a grain to establish a latent image, so that the duration of exposure is unimportant. With photons, however, several photon interactions are required to form a latent image, with the precise number needed depending somewhat on the rate of arrival. In general, though, the "reciprocity law" applies which states that the photographic effect of a particular radiation fluence is independent of the rate of incidence:

$$S \propto I \times t \qquad (7\text{-}2)$$

Departure from reciprocity occasionally arises from insufficient control of development conditions or from overexposure. Overexposure results if many subsequent particles interact with grains that were already made developable by previous particles and depends on the total dose and the emulsion sensitivity. Figure 7.2 shows a set of exposure curves plotted as a semilog plot. The useful response region is the near-linear portion up to the horizontal line. Overexposure can also lead to an apparent shifting of contour boundaries, arising apparently from a diffusion of free silver ions from a region of high concentration into a low exposure zone (Eberhard effect).

An effect of importance in low-level dosimetry is fading of the latent image. Fading arises from ion recombination, oxidation of the latent image sites by atmospheric oxygen, and by temperature and humidity conditions. Fading is reduced with increasing grain size, increasing LET

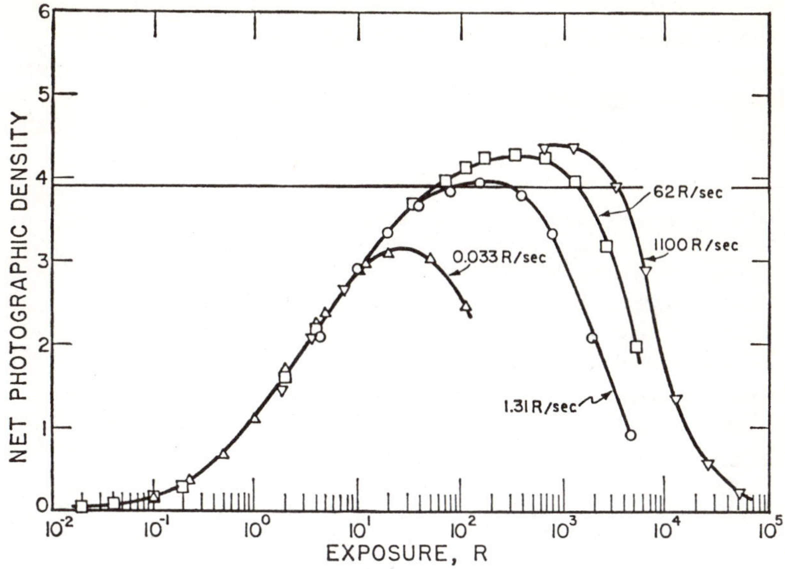

Figure 7.2 Net density vs exposure curves for various exposure rates. DuPont 502 Film irradiated by 50-kV X-rays (Ehrlich, 1963).

of incident particles, decreasing humidity, and higher emulsion pH. It can be minimized by storing exposed film in a dry, cool environment. For X-ray films at room temperature, fading may be a few percent per day initially, rather less after a few weeks. In the routine use of large numbers of film dosimeters, such fading effects have to be taken into account. For very fast films, fading effects may be severe unless films are stored carefully in a low-humidity atmosphere.

Fast films also show an increase in background fog from cosmic rays, thermal effects and random ionization effects. For nuclear emulsions it is important to guard against fallout contamination of raw materials and accidental inclusion of natural radioactive impurities (^{40}K,U,Th).

Nuclear Track Emulsions

Nuclear track emulsions are used for the study of nuclear reactions, particularly those that occur at a slow rate so that the continuous sensitivity of photographic films is helpful in providing a cumulative record. This has been particularly useful in studying cosmic ray phenomena and space events, where the small weight, compactness and absence of electronic components proved helpful. On the other hand, scanning of plates visually through a microscope is tedious and fatiguing and various systems for computerized scanning of projected images have been developed.

Items of interest are:

1. range in the emulsion, provided the track ends within the emulsion. A correction must be applied for angle of dip and for shrinkage of the emulsion during development;
2. scattering angle, a function of charge and energy;
3. grain density along the track, which depends on LET and may identify a particle;
4. star formation, indicating simultaneous or consecutive emission of particles from a given parent, recoil effects and decay chains (see Figure 7.1a); and
5. deflection of tracks in a magnetic field. This can indicate charge or mass of a particle, but usually requires careful registration of stacks of overlying emulsions, with air gaps, to identify track continuity.

The use of range-energy curves in emulsions to determine particle energy is one of the few direct energy determination methods available, particularly for fast neutrons, above about 0.5 MeV. Monoenergetic neutrons produce recoil protons in the emulsion with a uniform distribution of energy from zero to E_n. By selecting only proton tracks of maximum length within $\pm 5^\circ$ in the forward direction, one can obtain a good statistical sample for "bull's-eye" collisions. Since a "track," to be identified as such, must consist of at least 3-4 lined-up developed grains and a four-grain track is produced by stopping a 0.43-MeV proton, 0.5 MeV is the approximate lower limit for neutron energy determinations by the recoil method. Figure 7.3 gives the range-energy relations for charged particles in a nuclear emulsion; more detailed curves for various emulsions have been published by Brabner and Barkas.

The absorbed dose contributed by the (n,p) reaction in tissue and emulsion is proportional to the number of recoil events and their average energy. For fast-neutron dosimetry such track analysis is very laborious, but again can be facilitated by semiautomatic scanning and computer evaluation. The number of tracks per cm^2 produced by an absorbed dose of 1 rad (10 mGy) in fast-neutron dosimeter film depends on the angle of incidence. In the range of 0.6-14 MeV, track densities of the order of 1.4×10^4 tracks/cm^2/rem have been reported. This also will be affected by any major change in hydrogen concentration in the emulsion and by surface contamination from film-packing material.

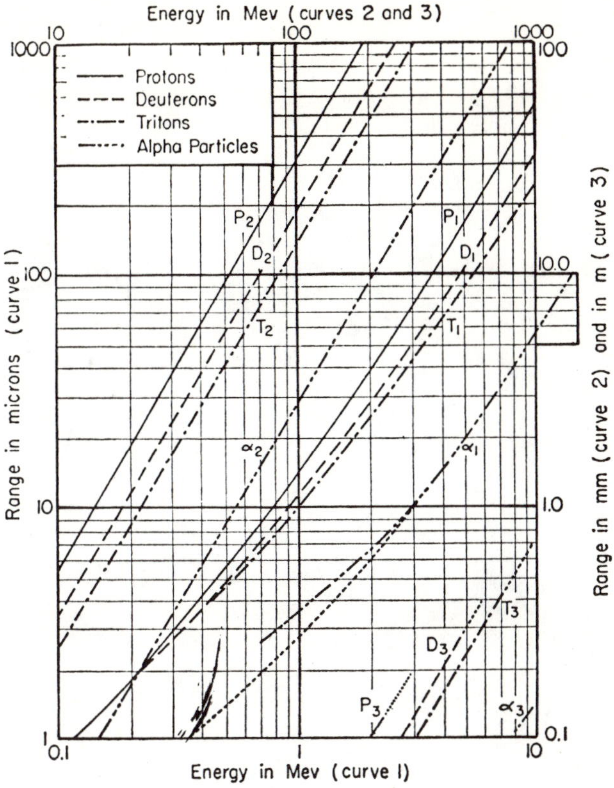

Figure 7.3 Range-energy relationships for protons, deuterons, tritons and alpha particles in dry Ilford C2 Emulsion (Vigneron, 1953).

An alternative method of neutron detection is by the use of spiked or "loaded" emulsions. In that case, the emulsion is soaked in a weak solution of the element of interest to incorporate those atoms in low concentration in the emulsion. Table 7.1 lists some of the reactions of interest and the characteristic tracks, including several stars, that will be observed. Such loaded emulsions must be calibrated in a known neutron fluence.

Radiography

The use of photographic film in medical and industrial radiography is well established. In such cases a point source of X-rays or γ-rays is

Table 7.1 Various Methods of Neutron Detection in Nuclear Emulsions

Reaction	Energy Range	Cross-Section (barns)	Recorded Particle: Type	Energy	Range
$^{14}N(n, p)^{14}C$	$n_{th} + n_{ep}$	1.82	p	0·6 MeV	7 μm
$^{6}Li(n, \alpha)^{3}H$	$n_{th} + n_{ep}$	950	α ^{3}H	2·06 MeV 2·74	6 μm 36 μm } 1 track
$^{10}B(n, \alpha)^{7}Li$	$n_{th} + n_{ev}$	4,010	α Li	1·47 MeV 0·9	5 μm 2 μm
$^{235}U(n, f)$	$n_{th} + n_{ep}$	590	F.P.		
$^{1}H(n, p)$	0·5-20 MeV	6-1	p	$0{\cdot}3\text{-}E_n$	4-1,000 μm
$^{12}C(n, n')3\alpha$	>12 MeV	0·3	3α		Three-track star
$^{6}Li(n, \alpha)^{3}H$	0·1-5 MeV	0·1	α ^{3}H	2·06 MeV 2·74 MeV	6 μm 36 μm
$^{10}B(n, 2\alpha)^{3}H$	0·2	0·2	2α ^{3}H	$\Sigma E_1 = E_n$	Three-track star
$^{10}B(n, \alpha)2\alpha$			2α α	1·3 MeV each	"Hammer" track
Inelastic Scattering	20-400 MeV	0·5	p α Nuclear fragments		Multitrack stars

used to produce a shadow picture on a photographic film after the radiation passes through the object whose structure will absorb the radiation to a varying extent (see Figure 7.4). To obtain sharp images, scattered radiation must be eliminated from the beam path as much as possible and the radiation should enter the film as near normally as possible. For this reason the source-to-object distance should be large, to ensure a near-parallel incident beam, and the film should be as close and as

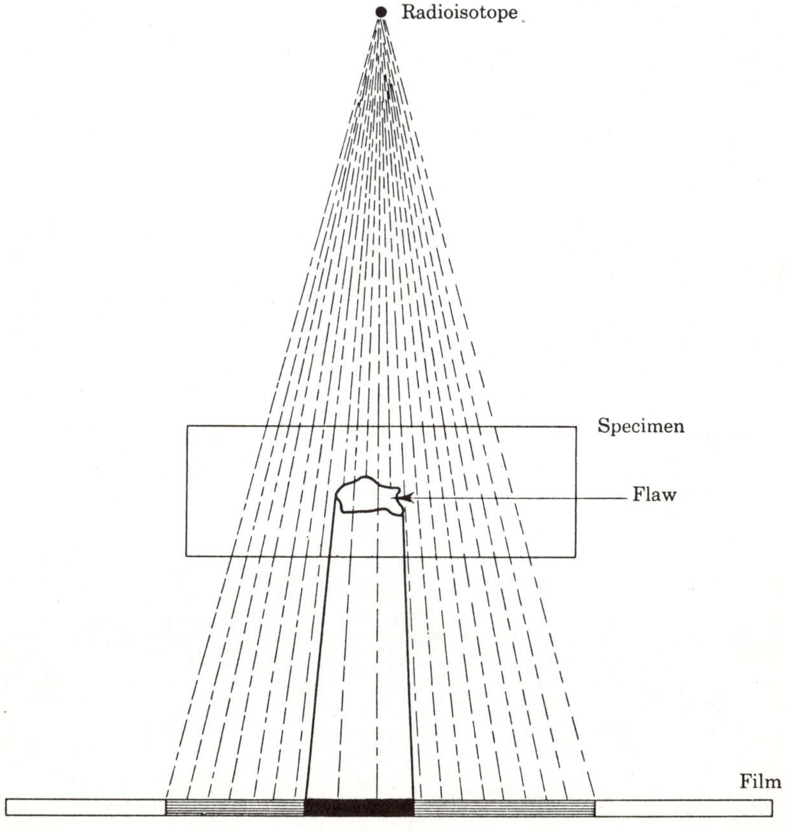

Figure 7.4 Fundamentals of a radiographic exposure. The dark region of the film represents the more penetrable part of the object, the light regions the absorbing portions.

parallel to the emergent surface as possible. In general, the reciprocity law applies so that for a point source of output A at distance d

$$\text{film exposure} \propto \frac{A \cdot t}{d^2} \tag{7-3}$$

Figure 7.5 shows exposure graphs for some representative industrial radiographic films. Similar charts are available for all commercial medical

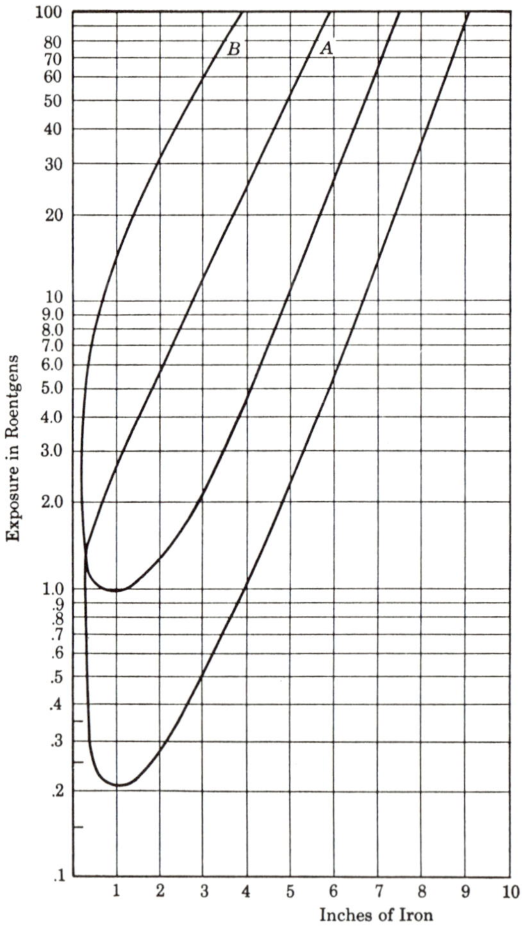

Figure 7.5 Two-percent sensitivity curves for cobalt-60: (A) Kodak Type K Film; (B) DuPont Type 506 Film (exposed between 0.25 mm front and 0.58 mm back Pb screens).

radiography films. The *sharpness* of the picture is affected by scattering from surrounding objects, distortion from the geometrical configuration of target and film, and by an irregularity in outline, volume and transmission path of intervening absorbers. This can be expressed by the coefficient of unsharpness U

$$U = \sqrt[3]{U_{sc}^3 + U_{geom}^3 + U_{film}^3} \tag{7-4}$$

To improve the sensitivity of detection it is often convenient to use *intensifying screens* and *masks*. Intensifying screens are thin layers of fluorescent material, deposited on a thin plastic substrate, that are excited by the incident radiation and will fluoresce strongly in the visible or near-visible range to which the emulsion is sensitive. Use of such screens modifies effective film sensitivity and special radiographic films have been developed for this procedure.

Masking involves the use of thin lead absorbers covering all regions of the film not in the direct beam. This reduces the scattering contribution to the image and, hence, the background fogging, thus improving overall film sensitivity.

The resolution of radiographic film, that is, the ability to respond to abrupt spatial variations in intensity, is limited by grain density, emulsion thickness and the energy of the incident radiation. Use of twin point sources with known spacing and application of a grey wedge absorber can define the attainable resolution. This is particularly important in autoradiography, where the emulsion is used to measure the distribution and location of radioactive material on a surface in contact with or closely spaced to the emulsion. (Contact must be avoided if the emulsion would chemically interact with the surface material.) This is a widely used technique in localization studies by means of radioactive tracers in biological systems and in metallurgy. In general, resolution is better for beta emitters than for gamma-rays because the shorter range permits closer correspondence between the source of emission and registration in the film.

Autoradiography is also used in the transfer technique of neutron radiography. In that process the neutrons transmitted through the radiographed object impinge on a thin screen of gadolinium oxide. The gadolinium is activated and, when the screen is placed on a photographic film, a transfer image of the neutron radiograph is formed by the beta radiation emitted by the activated gadolinium.

Film Dosimetry

In film dosimetry, the total blackening of the emulsion under X- or γ-ray irradiation is used to determine exposure; with suitable converters and selective absorbers exposure to beta rays or neutrons can also be recorded. For quantitative dosimetry, standardized and carefully controlled development conditions are essential; in addition, use of blank control samples helps to eliminate background effects during storage or shipping or accidental exposure of a given batch of films.

The films are evaluated by photometric densitometry. The lower level of significant dose observable is set by fogging and background conditions at about 40 mR for most films. The sensitivity of the film to photons depends on their energy as shown in Figure 7.6. The shape of the curve depends on grain size and grain spacing, with larger and sensitive grains

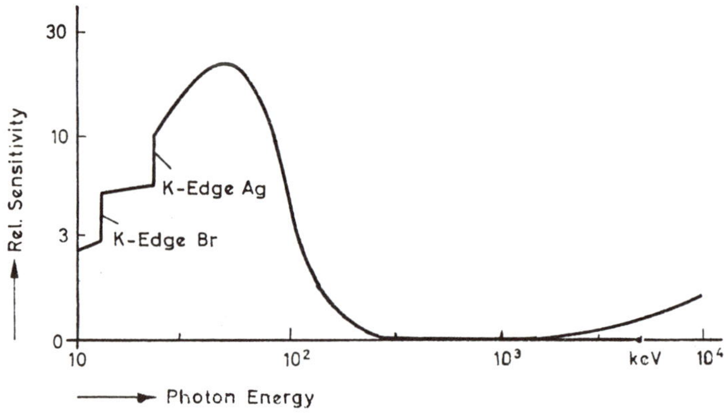

Figure 7.6 Relative sensitivity of a photographic film as a function of photon energy (after Becker).

favoring detection at higher energies. Most commercial films aim at a flatter response curve. One way of achieving this is by the use of filters which absorb the lower-energy photons preferentially. This is shown in Figure 7.7 which presents the response curve for a film behind a filter consisting of 0.71 mm tin plus 0.31 mm lead. Without filtration, sensitivity at 100 keV would be seven times higher, at 40 keV 27 times higher than at 1.25 MeV. Obviously this effect depends on the angle of

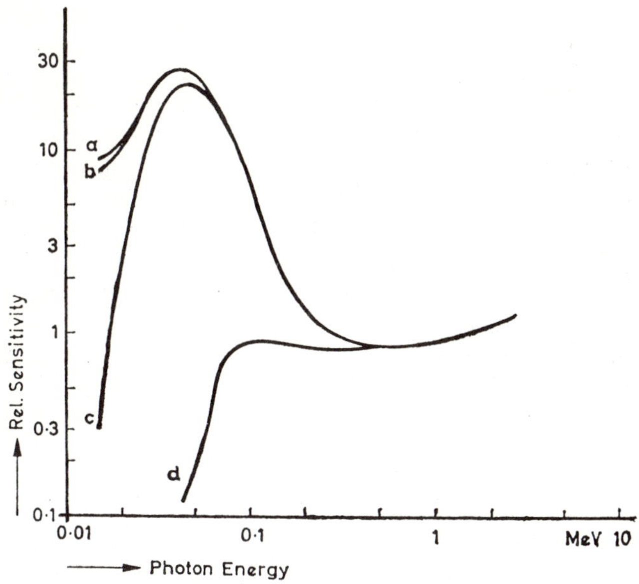

Figure 7.7 Effect of filters on the energy response of a photographic film (Kodak RM film) [after Heard].

incidence since both the path length in the filter and in the film vary with angle.

By suitable packaging, different portions of the film can be used to record exposure to either different energy ranges or different types of radiation. Table 7.2, after Kiefer and Maushart, summarizes some of these configurations that have been devised to obtain fairly flat dose-energy readings and their relative advantages. In general, film dosimeters have the advantage of relatively low cost, portability and the availability of a permanent record. Their disadvantages include dependence on development conditions, problems of standardization, insufficient sensitivity at low dose levels, and fading effects.

TRACK-ETCH DOSIMETERS

Introduction

As seen above, track registration has been used for a long time in photographic emulsions. However, nuclear emulsions have a number of

Table 7.2 Examples of Various Methods for Obtaining Energy-Independent Dose Recording with Film Dosimeters

	Filter Combination	Filter Combination: Ehrlich and Fitch (1962)	Film Combination van Stekelenburg (1958)	Fluorescence Compensation: Hoerlin (1953)	Fluorescence Compensation: Becker (1957)	Filter Analysis Dresel (1956)
Film dosimeter						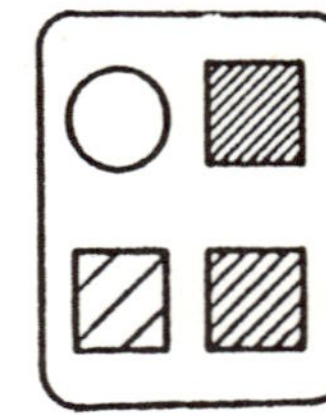
Combination	Kodak PM1 1.2 mm Sn	Du Pont 510 8.25 mm Bakelite 1.07 mm Sn 0.3 mm Pb	Ilford PM 3 2 mm Sn Ilford PM 1 2 mm Sn	Special film *p*-terphenyl pressed emulsions	Agfa-AGEPE *p*-terphenyl emulsion 1 mm Cu	Adox film, filter-free region 0.05 mm Cu 0.5 mm Cu 1.2 mm Cu
Energy Range	0.100-1.2 MeV ±20%	0.115-10 MeV ±21%	0.030-1.2 MeV ±20%	0.040-1.2 MeV ±15%	0.110-1.2 MeV	0.030-1.2 MeV energy-dependent
Advantages	Simple arrangement	Electron equilibrium up to 10 MeV	Energy-independent	Nondirectional, dose sensitivity up to mR-range		Energy determination (in homogeneous radiation, etc.)
Disadvantages	Direction dependence	Direction dependence	Direction dependence Evaluation of two films	Intensity-dependent fading		Direction dependence, complicated handling because of graphical evaluation and 3 dose fractions

drawbacks. These include fogging by unwanted types of radiation, fading, limited shelf-life, need for a darkroom processing facility, shrinking of the sensitive layer during processing which modifies the original track length, tedious microscopic evaluation, and limited flexibility in composition. For these reasons, there has been a significant amount of research on other methods of registering tracks in condensed media. One of these is the "track-etch" method which depends on the formation of a damage region in plastic foils along the path of an ionizing particle.

Track-Etch Mechanism

The basic mechanism of track formation is as follows. Radiation impinging on a solid foil causes damage along the radiation tracks. These damaged regions are attacked preferentially by chemicals (that is, they are "etched") and become visible either microscopically or to the human eye. The number of tracks produced per unit area can be related directly to the amount of radiation incident on the material, and the absorbed dose.

Track detectors fall into two categories: inorganic crystals and glasses, and organic polymers. Only particles that lose energy at a rate greater than 15 $MeV/mg/cm^2$ can be detected in inorganic materials. Organic materials are more sensitive; all organic polymers which have been tested record particles of less than about 4 $MeV/mg/cm^2$ and some cellulose derivatives will record particles which lose energy at a rate as low as 1 $MeV/mg/cm^2$.

Because these materials have very different physical and chemical properties, no unified theory of track formation has been agreed upon. Currently, in minerals and inorganic glasses, the "ion-explosion spike" model is predominant. This theory states that a region of positive charge is produced along the track of the particle as it dislodges (ionizes) electrons. There is a repulsion due to all the like charges which results in lattice dislocations. Thus, a region of imperfections is created in the crystal which is more easily attacked by the etching reagent than is the surrounding material.

Two other theories used to explain the process are the "thermal spike" theory which says the damage is due to local heating in the lattice and the "displacement spike" theory which says that the damage region is due to elastic collisions in the crystal lattice. The actual mechanism may be a combination of all three processes, each with a relative importance depending on the material.

The accepted theory of track formation in polymers is based on a radiochemical damage mechanism. According to this theory, etchable tracks are formed by radiolytic breaking of long polymer chains into shorter fragments. The theory also takes into account the production of reactive, low molecular-weight radiolytic products that are more easily etched than the surrounding, undamaged bulk plastic. Figure 7.8 illustrates the formation of an etched track.

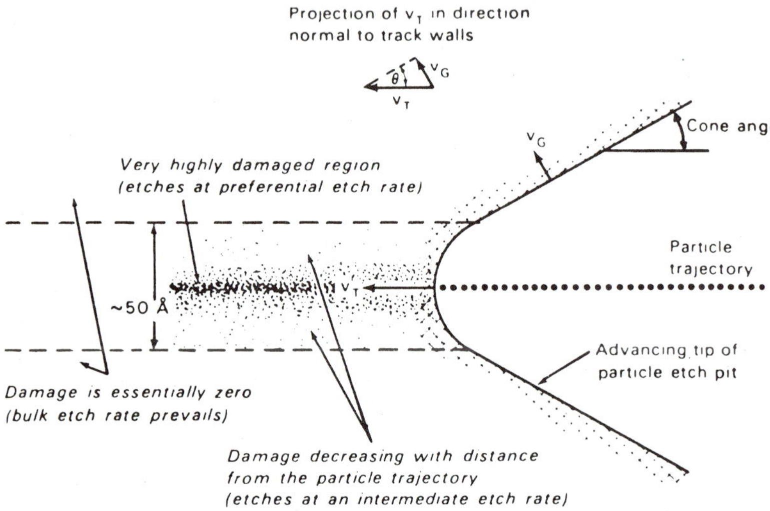

Figure 7.8 Formation of particle tracks in a plastic foil.

The track-etch technique is very simple in its general characteristics. A solid material is exposed to a radiation source and interactions either in the material itself or in a "radiator" placed close to the material produce charged particles that leave damage tracks in the solid. The number of tracks is proportional to the radiation flux and thus to the absorbed dose. The tracks are enlarged by placing the solid in a solution which selectively attacks the damaged regions around the tracks; the tracks can then be seen and counted under a microscope.

The "Passive" Track-Etch System

To understand more clearly the general characteristics of a track-etch system, consider an organic polymer system. Most current research is concerned with these polymers; these systems have wide application limited only by the ingenuity and resourcefulness of the researcher.

One of the first considerations in selecting a track detector is the stability of the latent tracks in the solid, one of the recognized limitations of nuclear track emulsions. After etching, the visibility of a track is limited only by the melting, softening or chemical decomposition of the detector. However, before etching, the damage zone may be partially or completely annealed at lower temperatures. In the discussion of TLDs the word "anneal" was used to indicate the "dumping" of traps at higher temperatures so that the detector could be reused. In track-etching, the word "anneal" usually means the reduction in size or the complete loss of a damage track. This is most easily understood by considering what happens to a material when it is taken to a temperature near its melting point. If tracks are present, there may be sufficient plastic flow so that the damage is repaired or obscured such that no etching takes place. This is, in general, what is meant when annealing is mentioned in discussions of track-etching techniques.

If the melting point of the detector material is high, there is a good possibility that the detector will remain stable with little or no fading for long periods of time. For example, no fission-track fading has been observed in SiO_2 after one hour at a temperature of 1,000°C. Mica and quartz can be used for measurements at temperatures greater than several hundred degrees Celsius. However, not all glasses have these qualities. For example, microscope slides have exhibited considerable fading within several days at 65°C.

In polymers, no fading has been observed at temperatures below about 50°C. However, if the softening temperature of the plastic is approached rapidly, annealing of the tracks occurs. As might be expected the fading rate is a function of the storage temperature. However, it is probably not a linear function. In Figure 7.9 fading begins at about 60°C and is rapidly accelerated above 100°C. Note the effect of "extended etching," which means that the material is left longer in the etching solution. In this case, some tracks seem to reappear, probably due to a surface closing of the tracks since there must be some distribution of temperature through the material which causes the tracks to close from the surface inward.

There are other effects which affect the fading, and many of the observed effects are not easy to explain. These effects include intense ultraviolet light which has been shown to induce 50% fading of α-tracks in cellulose nitrate within about 30 min. High-LET tracks from fission fragments tend to fade less rapidly than those produced by α-tracks in the same material. Fading is accelerated in some materials by high relative humidities. In addition, more fading has been noticed in those foils in which the damage track penetrates completely through the foil, than in thick foils where the track opens a channel only on one side of the material. Figure 7.9 shows the fading of fission fragment tracks in a foil of polycarbonate, 10 μm thick which was stored at 60°C and a relative humidity of 77%.

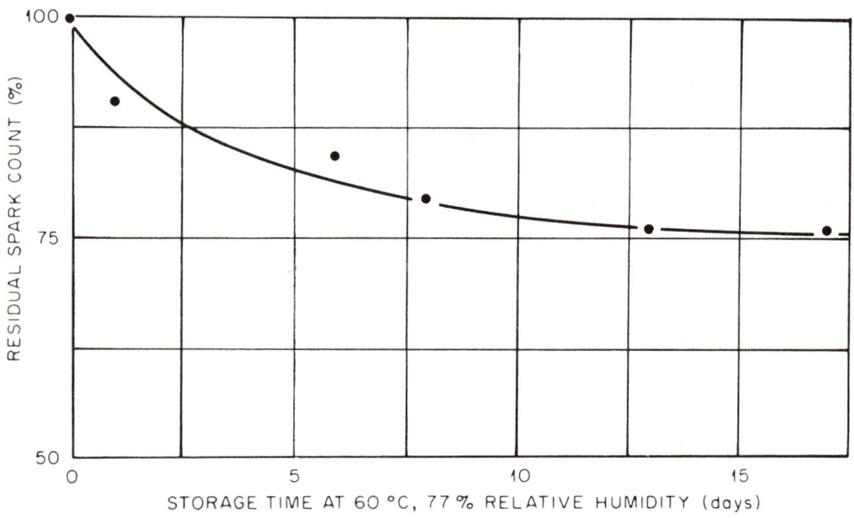

Figure 7.9 Fading of latent fission fragment tracks in a 10-μm Kimfol foil (Sohrabi and Becker, 1972).

Table 7.3 represents a summary of the characteristics of some inorganic materials. There are more than 10 inorganic materials and over 20 organic polymers used in track-etching techniques. Many of the materials are manufactured by more than one company and the characteristics of the material are not necessarily the same even though the material has the same chemical formula. Therefore, it is necessary that optimized etching conditions be established empirically for each detector material. Any chemical that attacks it at a sufficient speed can, in principle, be used as an etchant. Table 7.3 gives a partial list of etching conditions for glasses and organic polymers that have been found to

Table 7.3 Etching Conditions for Fission Fragment Track Registration (after Becker)

Material	Etchant Composition and Temperature
Glasses	
Mica	15% HF, 20 sec, 50°C
Quartz	KOH, 3 hr, 150°C 48% HF, 24 hr, 23°C
Borate Glass	H_2O, 1 min, 23°C
Phosphate Glass	48% HF, 5-20 min, 23°C
Soda-Lime Glass (Microscope Slide)	48% HF, 5 sec or 5% HF, 2 min
Organic Polymers	
Cellulose Acetate (Triafol A)	28% KOH, 30 min, 60°C $NaCO_3$, 30 min, 100°C
Cellulose Acetate Butyrate (Triafol B)	28% KOH, 60 min, 60°C 6.25N NaOH, 12 min, 70°C
Cellulose Nitrate	28% KOH, 30 min, 23°C 6.25N NaOH, 2 min, 70°C
Cellulose Triacetate (Triafol TN)	28% KOH, 60 min, 60°C

work best for those materials. The parameters which influence the etching speed are:

- Type and Velocity of Charged Particle
- Chemical Composition of the Etchant
- Temperature and Concentration of the Etchant
- Chemical and Physical Pre- and Postirradiation Treatment of the Detector Material
- Chemical and Physical Condition of the Material During Irradiation

The optimum etching time is a function of the track density in the foil. This is illustrated in Figure 7.10 which presents results for fission fragment tracks in polycarbonate foil. Note that at low track densities the etching time is not a sensitive parameter; however, as the track density increased, the etching time becomes extremely important. In some cases, the etching time reaches a plateau. Figure 7.11 shows the data for α-particle tracks in four kinds of polycarbonate foil. The size of the developed track increases with etching time and reagent strength and is also affected by the presence of air.

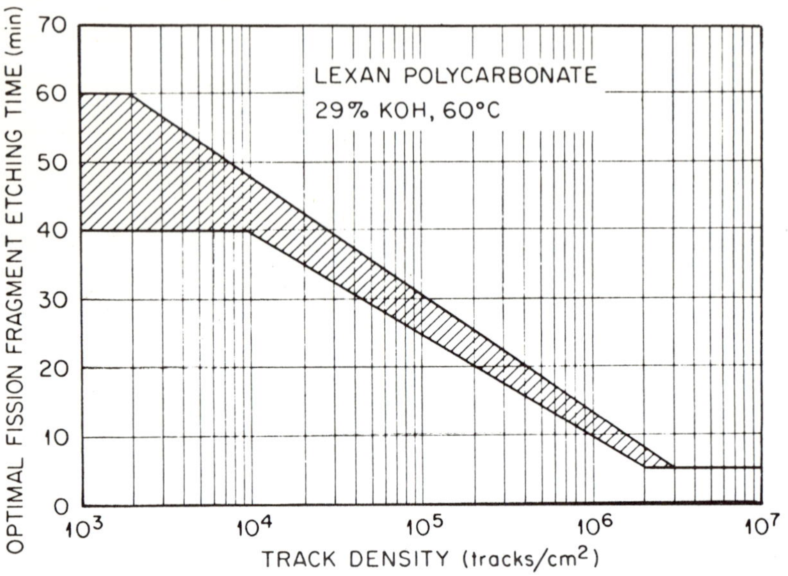

Figure 7.10 Optimum etching time for fission fragment tracks as a function of track density (Prêtre *et al.*, 1968).

If a detector is exposed to fission fragments, most of the tracks become visible at about the same time. Further etching simply increases the track diameter but does not affect the number of visible tracks. If the detector is exposed to a wide energy spectrum of particles with a relatively lower LET, then the number of tracks becomes a complex function of the etching time. First, the tracks due to low-energy, high-LET particles become visible. The other tracks and their visibility during the etching procedure are dependent on the detection threshold of the material. Secondly, there is a variation in the number of visible tracks because some low-LET particles must slow down in the material before reaching the detection threshold. Thus, they produce tracks at a depth in the material which will only become accessible to the etchant after the surface material has been etched away.

A basic disadvantage of track etching is that only tracks at or near the surface that are accessible to the etchant can be made visible. Another disadvantage is the need to use microscopic techniques to evaluate the number of tracks. In general, the thickness of the foil is not an important factor when visual track counting is used. If the

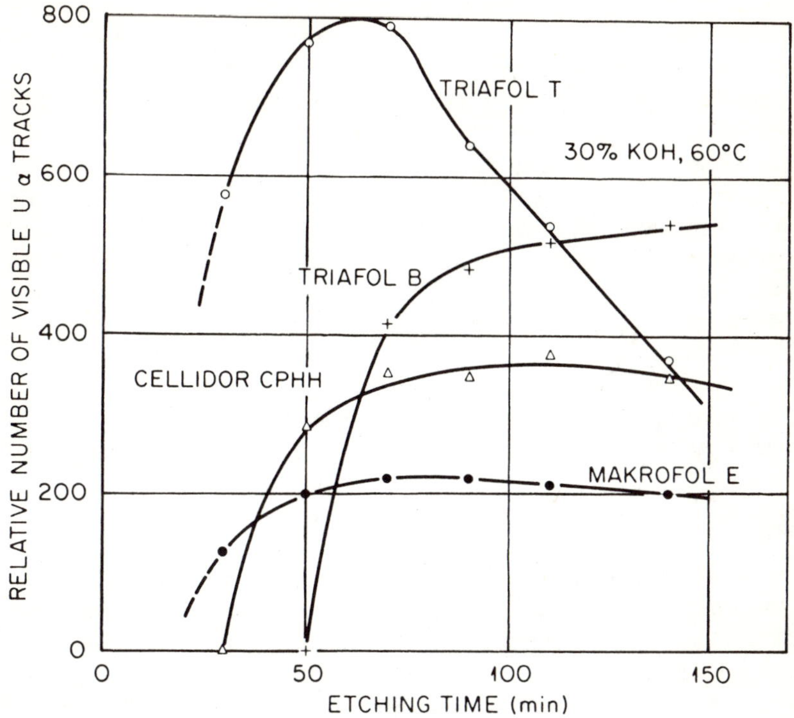

Figure 7.11 Relative number of alpha tracks in various materials as a function of etching time in 30 % KOH at 60°C.

evaluation process is based on perforating the material, then the foils should be thin, say 6-20 μm thick, and uniform thickness of the foils becomes important in order to obtain reproducible results.

Most dosimetric applications of these solid-state particle detectors require that the user determine the number of etch pits per unit area of detector surface. Visual track counting in a microscope is the simplest way of doing this. However, using a microscope to evaluate a foil is a slow and tedious process; it is also an inaccurate process and is subject to systematic errors, especially when the track density is low so that large areas of the detector must be scanned. Other factors which affect the accuracy are high track densities where the tracks overlap and situations where the tracks are small, making them hard to see.

Visual track counting can be made a little easier by such techniques as projection, adjustment of etch time to match expected track density, or special techniques of optical magnification. One technique is to mount the track-etch material on another material that is not as sensitive to the etchant. A red cellulose nitrate foil is available which is mounted on a clear base. When etched, the tracks appear as clear holes on a red background. The reverse process is also used where the track holes or pits are filled with a red dye and stand out clearly against the clear background. Use of a dark-field microscope has been tried, also. In this case, the tracks stand out as white dots or "stars" against a black background. Nevertheless, any visual method using a microscope is awkward, especially when speed, high accuracy and the evaluation of a large number of foils is required.

Many other methods have been tried that do not use a microscope. The most popular, and one which has all the characteristics necessary to provide speed and accuracy, is the spark-counting method of Cross and Tommasino. Spark counting represents the least complicated, quickest and most accurate method for determining low and medium track densities over a relatively large area. Spark counters are easy to build and, therefore, are fairly inexpensive.* Their use is as follows (see Figure 7.12):

The foil is exposed and etched according to the procedure found to be the best for the particular detector material. It is removed from the etching solution, washed quickly in tapwater, then in distilled water, and dried. Usually the foil is attached either to a small ring by gluing or one uses a double-ring arrangement (similar to a miniature embroidery hoop) which clamps the foil rigidly in place. In some cases, these rings are used routinely to hold the foils, making it easy to locate and handle them. The foil is then placed over an electrode of the spark counter. Next, it is covered with a piece of aluminized Mylar with the aluminized side facing the foil. The piece of Mylar must be sufficiently large to be in contact with the other electrode of the counter. Usually a weight made of a nonconducting material is placed on top of the Mylar to ensure good contact through the entire system.

When a voltage is applied to this system, sparks "jump" through the etched holes in the foil. Each spark causes the evaporation of the aluminum in a small spot on the Mylar film. The area of this spot is much larger than the area of the etched hole so that only one spark will jump through each hole. Thus, the number of holes in the foil is obtained by observing the number of sparks which jump. The sparks are counted by using a standard laboratory scaler coupled to the spark counter through a simple GM-type quenching circuit. In this way, each spark is recorded as one count on the scaler.

*Note: The term spark counter as used here for track evaluation should not be confused with the spark counters and chambers described later in this chapter.

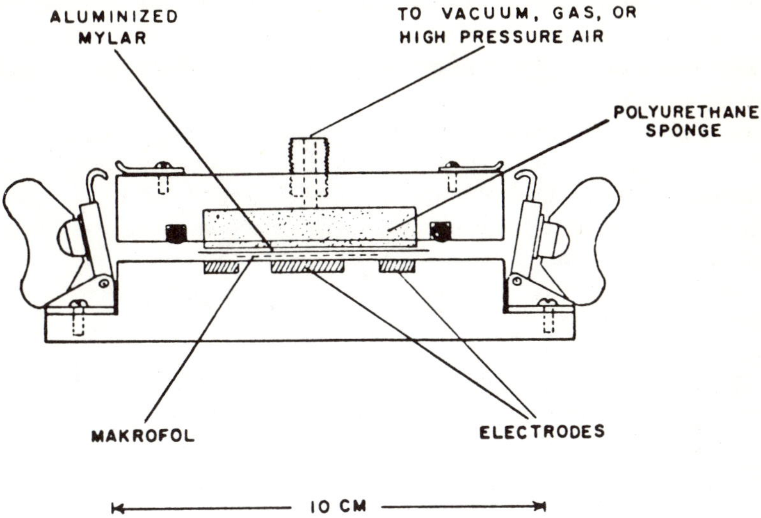

Figure 7.12 Simplified diagram of track-etch spark counter.

The exact operating voltage of such a system must be determined empirically because it depends on the counting circuit, the foil thickness, the etching conditions and the aluminum thickness. Some of the tracks may not be etched all the way through the foil and after sparking through the existing holes there may be a "spark-punching" effect which completes other holes through the foil. So, usually a first count is done at a slightly higher voltage, and all subsequent counts are done at a lower voltage. In a typical system there is a sparking voltage plateau which may extend from 500 to 800 volts.

One minor problem with this system is that evaporated aluminum begins to coat the electrode and also may be partially coated out on the foil itself. This can be overcome by wiping the electrode with a cotton swab dipped in alcohol and by rinsing the foil in a NaOH or KOH solution.

The advantages of this system are obvious. First, it eliminates the use of a microscope in evaluating the foils. In addition, the aluminized mylar, after sparking, provides a replica of the foil which may be stored as a permanent record of the count. Thirdly, the foil may be sparked many times by simply changing the mylar between counts. Thus, it is possible to obtain a set of data for each foil which may be evaluated by standard statistical methods. This is a great advantage since quantitative errors associated with any sort of visual evaluation method are usually not obtained directly. The linearity of the spark-counting system is good up to about 3,000 tracks/cm^2 and the upper limit on the useful dynamic range is about 5,000 tracks/cm^2. The reproducibility of repeated counting of the same foil in routine work is about 3%. The detection efficiency for fission fragments incident normal to the surface of the foil is nearly 100%. However, for isotropic incidence, the detection efficiency drops to about 70%.

Fission Foil Technique

The most widespread dosimetric application of track-etching has been its use in neutron dosimetry by fission-fragment registration. The technique is very simple. An ordinary microscope slide contains enough trace uranium to allow the measurement of the neutron fluence in a nuclear reactor. The tracks can be etched by dipping the microscope slide in an HF solution for a few seconds. The track density can be related to the neutron fluence by a constant, weighted by the fission cross section of fissile material.

The threshold detector technique can be applied to fast-neutron dosimetry by placing unencapsulated fission foils close to the track-etch material. The fissile foil and the track-registration material may be separated by a thin spacer and it is possible to place several such sandwiches inside a cadmium container and surround the container with a ^{10}B sphere. This arrangement is essentially identical to the arrangement used for fission-foil activation detectors. Instead of the complicated counting procedure used to evaluate the activity of fission foils, the very simple spark-counting technique is used. In addition, the track registration does not require that the foils be evaluated before the radioactivity decays. The tracks are in the material and will remain stable for long periods at room temperature.

For many applications the fission fragment registration technique of fast-neutron dosimetry has replaced the old technique of proton-recoil registration in photographic nuclear-track emulsions. The advantage of the track registration technique are:

1. the solid-state detectors are normally insensitive to high doses of x-rays, γ-rays and β-radiation. No changes have been noted in the etching characteristics of the foils for doses up to 10^6 to 10^7 rads.
2. There are no deterioration, storage and fading problems involved unless the temperature far exceeds room temperature. Most of

the time the detectors do not respond to disturbing influences such as light and humidity.

3. The development and evaluation are much easier than for photographic emulsions. The process can be automated and there is a potential for higher sensitivity since it is very easy to read large areas of the detector; in fact, it is simple to read the entire area of the foil.
4. It is possible to cover a very wide dose range with this technique. Two methods commonly employed are to dilute the fissile material to a different concentration or to use layers of the fissile material of different thicknesses.
5. The detectors can be switched on and off simply by separating the fissile radiator and the track-registration material. This allows the use of a neutron-fluence integrator of sorts which may be useful in certain applications.

There are some disadvantages also. The first and most important involves the neutron energy dependence of the detector. The cross sections for Np^{237}, U^{238}, etc., are not constant over a wide energy range even above the energy threshold.

The second disadvantage has to do with the radioactivity of the fissile material itself. Pu-239 is considered so radiotoxic that using it in a badge for personnel monitoring purposes is not desirable. Some of these materials have a high spontaneous-fission rate and, therefore, produce a high background reading. For example, it has been calculated that a foil of U-238 would produce a background reading in the track-etch material equivalent to 44 mrad/month. Another disadvantage is that most of the fissile materials emit α-particles. These α-particles can change the etching characteristics of the foil and give spurious readings.

Another major disadvantage stems from the fact that these materials emit β-particles and γ-rays which irradiate the wearer. Thus, the detector is not only useful to measure chronic exposure it may be instrumental in producing a chronic exposure! Finally, the fissile radiators are not easy to prepare; there is a hazard that the material may rub off producing a severe α-contamination problem if the samples are not handled properly. These disadvantages for routine personnel dosimetry are largely overcome by the electrochemical etching technique described below.

There are several other applications of the track-etch technique, especially the registration of fission fragment and α-particle tracks in thin foils. These include the detection of natural uranium in tapwater, providing information on the spatial distribution of fissile material in bones or other samples, and the monitoring of radon progeny in mine atmospheres.

Electrochemical Etching Technique

One technique that has been developed recently is the method of electrochemical etching. It involves the use of an electrochemical procedure which amplifies fast-neutron-induced recoil particle tracks in polymer foils. The method was introduced by Tommasino in 1973. It is based on the conductive energy loss along the tracks when applying a high-frequency ac voltage, having square or sinusoidal waveforms, in an etchant at room temperature. This results in a preferential local etching along the track at a rate much greater than the bulk etching rate, leading to very large tracks usually round in cross section. This technique has resulted in better tracks than those obtained by conventional etching methods, especially when polycarbonate foils were used for recoil particle track registration.

The setup for electrochemical etching is different from that used in the passive etching process, which is to use an etchant at an elevated temperature to enlarge the tracks. This requires that the etching solution be maintained at a constant temperature; this is usually done by placing the entire system in a large low-temperature oven. In electrochemical etching the etchant is usually maintained at room temperature. The etching chamber in this case consists of a central region to hold one or more foils to be etched and two electrodes, one at either end of the chamber. After the foils to be etched are placed in position, the two sections of the chamber are assembled and the etching solution is introduced. At this point a high voltage at a frequency of the order of 2 kHz is applied across the etching chamber.

If the potential across the chamber was due to direct current (dc), then there would be a steady ion movement across the chamber. However, since the potential is due to alternating current (ac), the direction of motion of the ions is changed twice each cycle. It is thought that this local oscillation of the ions at the damage sites is a controlling factor on the sensitivity of the technique. The major advantage is that this technique provides, in most cases, clearly defined damage tracks of a reasonably large diameter which makes visualization much easier.

To illustrate the kind of data which may be obtained with the electrochemical etching system, Figure 7.13 shows the depth-dose distribution in an elliptical phantom irradiated by fission neutrons. The detectors were bare 250-μm polycarbonate foils which registered recoil tracks from neutron interactions in the phantom. The distribution of neutron dose at a bone-tissue interface is shown in Figure 7.14. These data again were obtained with 250-μm polycarbonate foils placed at 6 cm depth in a tissue-equivalent phantom. In neither case was there any need to

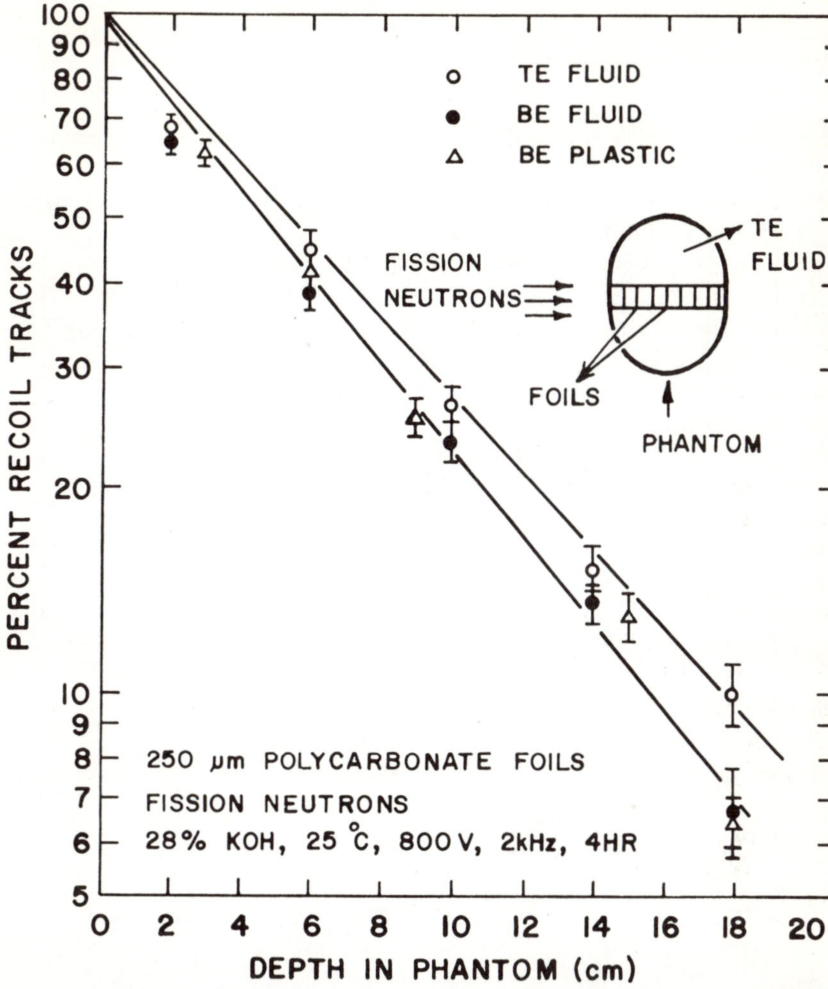

Figure 7.13 Depth dose measurement for fast neutrons irradiating an elliptical phantom with electrochemically etched polycarbonate foils (after Sohrabi).

use a fission-foil converter. As a result, electrochemically etched polycarbonate foils appear to be fully acceptable as personnel dosimeters.

Finally, Figure 7.15 compares the neutron sensitivity of the polycarbonate foil as a function of energy with data of the ICRP and that obtained by passive etching techniques.

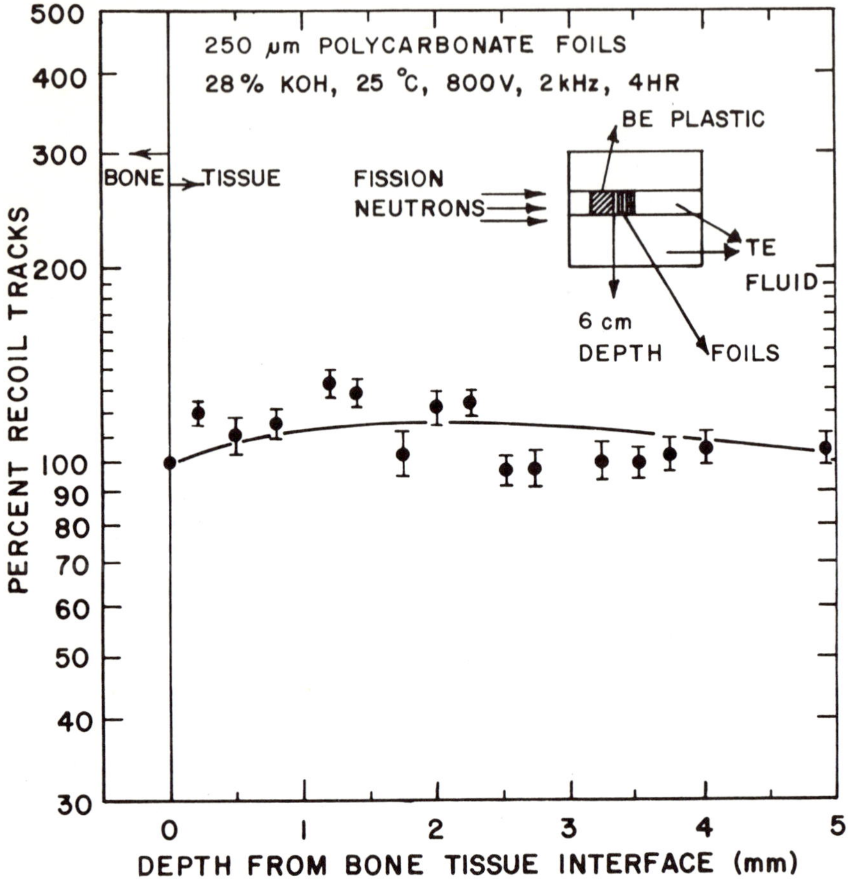

Figure 7.14 Neutron dose measurements at the bone-tissue interface (after Sohrabi).

In summary, track-etch dosimetry, whether utilizing passive etching or electrochemical etching techniques, is extremely simple and of moderate cost. There are a wide variety of materials which exhibit the necessary characteristics. The range of radiations detected by the materials is large and can be selected based on the radiation to be detected. Whatever the radiation to be detected or the material to be used, it is necessary that the system be calibrated carefully before extensive use.

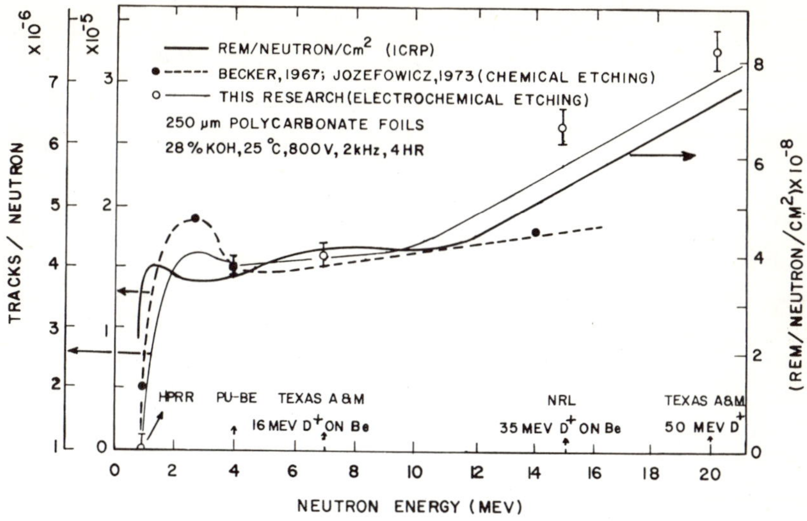

Figure 7.15 Neutron response curve for electrochemically etched polycarbonate foils compared with the ICRP-rem curve (Sohrabi, 1976).

SPARK COUNTERS AND SPARK CHAMBERS

Introduction

From studies on lightning and spark discharges it has long been known that actual passage of charge in air between two strongly charged surfaces is facilitated by the presence of ionized molecules. Conversely if a region of ionized molecules has been created, spark or arc discharges can be made to follow a path through such a region. The movement of an intense current of ions through the air can result in secondary ionization and excitation events, and the light emitted when the excited ions return to the ground state can itself give rise to further photoelectric events, thus causing a spread of the affected air volume and giving rise to multiple discharge strokes.

Even before a discharge stroke occurs, ionization of the gas exposed to intense electric field around a charged point or sharply curved conductor can give rise to a glow phenomenon known as a *corona*. The corona region can be expanded towards the oppositely charged electrode, forming a virtual equipotential layer, by increasing the voltage or ionizing additional molecules: a violent spark discharge occurs when the effective voltage gradient exceeds the dielectric breakdown strength of the intervening neutral gas. This sequence is illustrated in Figure 7.16.

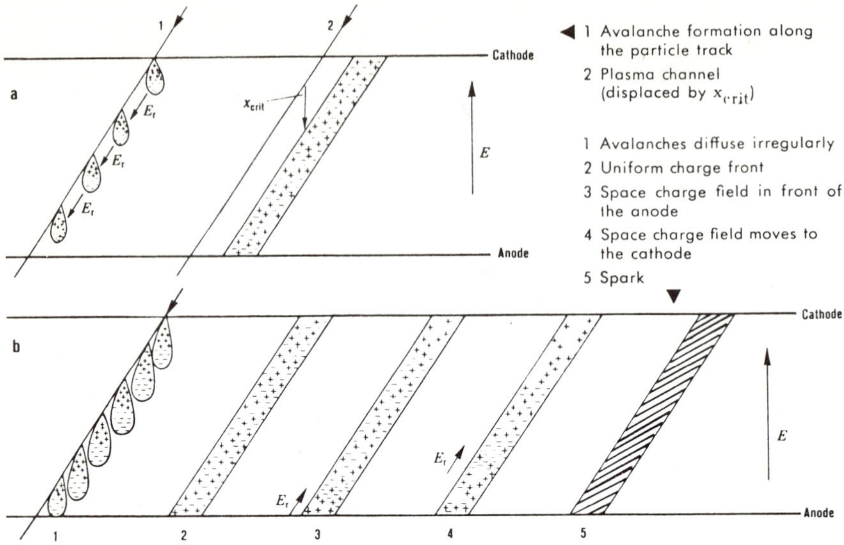

Figure 7.16 Two models of formation of a spark track in a track spark chamber (after Allkofer).

Spark discharges can be triggered at lower voltages if free ions are introduced into the gap region by a (lower-voltage) trigger electrode, such as are used in certain arc rectifiers, or by exposure to ionizing radiation. It is this latter process which has led to the development of a number of devices for the detection and tracking of ionizing particles by observation of sparks. Such systems have the advantage of simplicity, good spatial resolution and large, easily detectable phenomena: the passage of a spark is visible to the naked eye, it causes a large, readily counted electric pulse, and the noise accompanying its motion even makes acoustic localization possible.

Early spark detectors were very simple devices. This simplicity has led to the stacking of such detectors in a three-dimensional volume where correlation of track lengths and directions permits discrimination between different types of nuclear particles. Such three-dimensional spark systems are referred to as *spark chambers.* They represent a major tool in modern high-energy physics research since the magnitude of the signals and their correlation in space lend themselves to fast online computer evaluation of the data. A brief description will be given of *spark counters*, that is, detection devices based on this process, and spark chambers, the more sophisticated three-dimensional track recording systems derived from it.

Spark Counters

Greinacher showed in 1931 that a spark discharge would occur when a spark gap consisting of a point facing a flat plate electrode was exposed to alpha particles. Chang and Rosenblum showed that a larger sensitive volume could be obtained if a charged wire was used instead of a point, and Eichholz, in 1952, described the first multiwire spark counter, shown in Figure 7.17. In this counter a number of parallel wires faced a flat or corrugated plate. On applying a sufficient voltage between wires and the plate to establish a corona regime near the wires, the gap region as a whole becomes sensitive to charged particles and discharges occur whenever a single alpha particle enters the gap region. Controlling the voltage to minimize secondary events and photoinduced events, the spark can be spatially correlated with the path of the incident particle. The magnitude of the spark current must be controlled to ensure speedy recovery of the supply voltage, which is momentarily lowered by the flow of charge. In some systems, a quenching pulse is applied in reverse to remove residual gas ions and to speed up return to the normal operating voltage.

Although spark counters will operate in air, it is easier to obtain consistent performance in a noble gas atmosphere where the pressure, humidity and gas purity can be readily controlled. Helium or He-Ne mixtures provide a rapidly recovering filling gas, permitting operation at moderately low voltages. Argon and argon mixtures are more easily available; heavier gases can be used to obtain shorter ranges in the counter volume for very energetic particles. The spark pulses produced are of constant amplitude. Energy discrimination can be provided by interposing shielding foils or by varying the gas pressure, though that also changes the operating characteristics.

With a suitable converter material, spark counters can also detect neutrons. This is shown in Figure 7.18 which illustrates one type of neutron detector employing boron nitride converter disks. The alpha particles from the $^{10}B(n, \alpha)^{7}Li$ reaction trigger the spark discharges. Such a detector can operate over a wide temperature range and is insensitive to gamma-rays up to a level where gamma-ray ionization of the filler gas as a whole causes continuous discharges.

Spark Chambers

The most important extension of the spark counter occurred when multiwire layers were stacked and each wire used to provide spatial information about the location of successive path segments of a passing

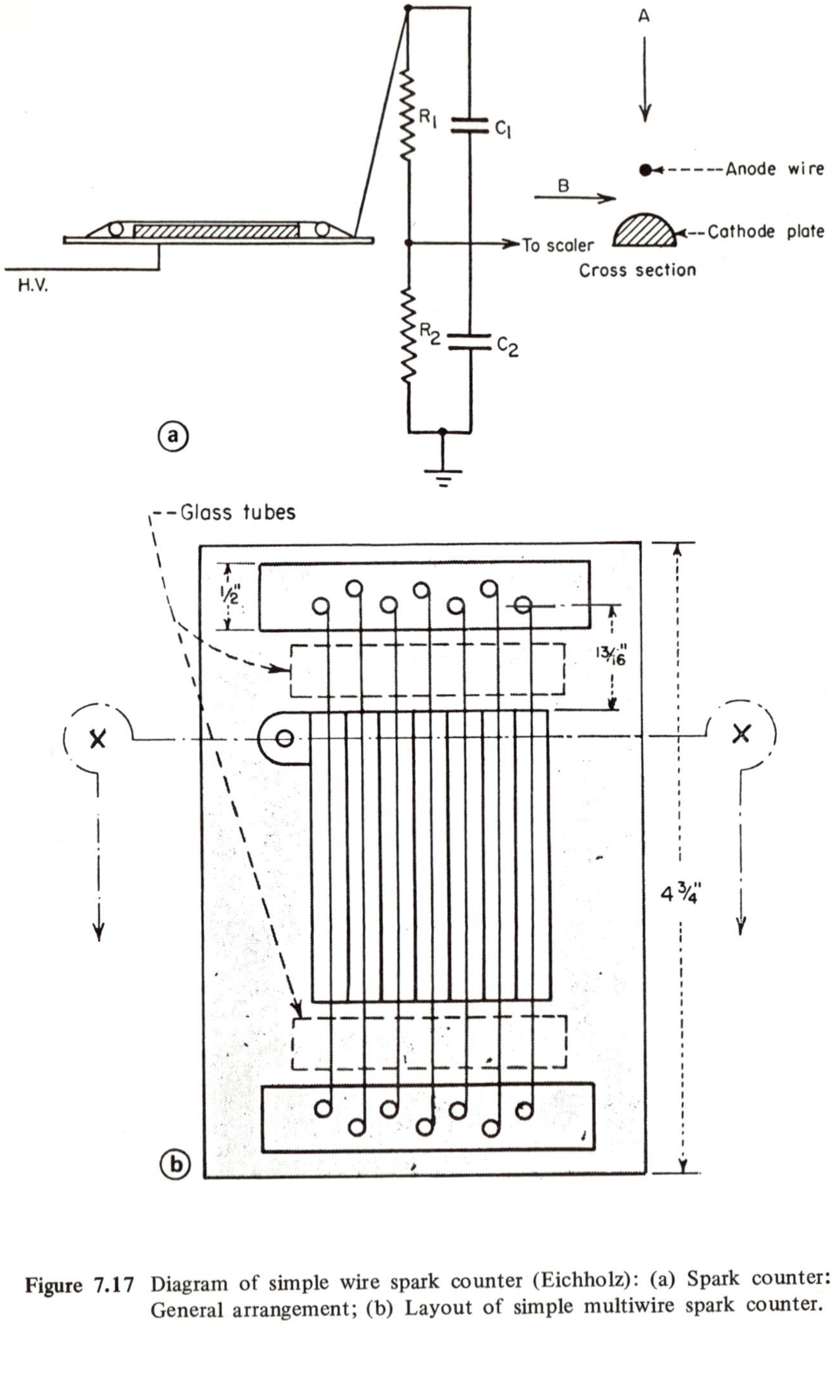

Figure 7.17 Diagram of simple wire spark counter (Eichholz): (a) Spark counter: General arrangement; (b) Layout of simple multiwire spark counter.

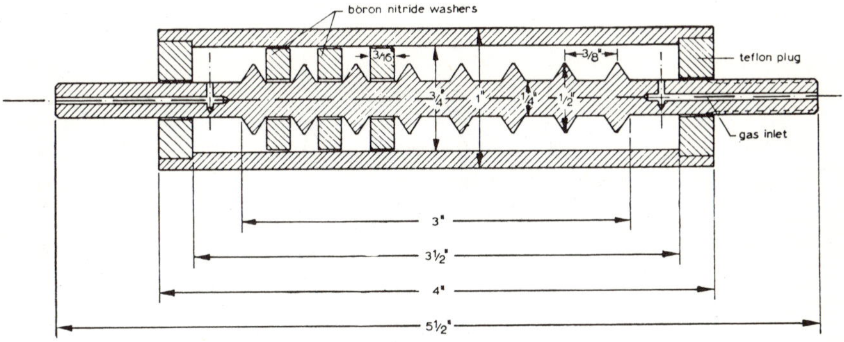

Figure 7.18 Cross section through cylindrical spark counter for neutron detection.

particle. Early spark chambers consisted of wire frames with the wires oriented at right angles in successive frames (Neumann, Reiffel). Each crossing of overlying wires then provides a coordinate point, so that a computer could determine the spatial angle and direction of a spark sequence affecting many counter layers. Such systems turned out to be difficult to control and suffered from a slow duty cycle, owing to microphonic effects and photoinduced processes.

Subsequent chambers use wire layers separated by oppositely charged thin solid plates; each wire is connected to a discriminator sorting circuit and events are collected during specific time intervals while a pulsed high voltage is applied to the electrodes. The sparks can be observed, photographed, counted electronically or detected acoustically; the latter method is adopted sometimes to reduce electronic interference from other sources. Because the sparks represent relatively large current pulses, a spark chamber can be a significant source of electrical interference and must be well shielded itself. In many cases cosmic-ray events or high-energy phenomena are selected on the basis of their angle of incidence. This is accomplished by connecting a scintillation counter or a set of GM counters above or below the spark chamber in coincidence with the chamber output, so that only incident particles arriving within a defined solid angle are recorded. Figure 7.19 shows an example of such a system. By operating in a dark room, on a slow pulse cycle, with an open camera shutter facing the transparent side of the chamber, excellent visual records of a passing energetic particle can be obtained. (Note that multiwire spark counters should be distinguished from multiwire proportional counters, which are widely used as energy-discriminating directional detectors).

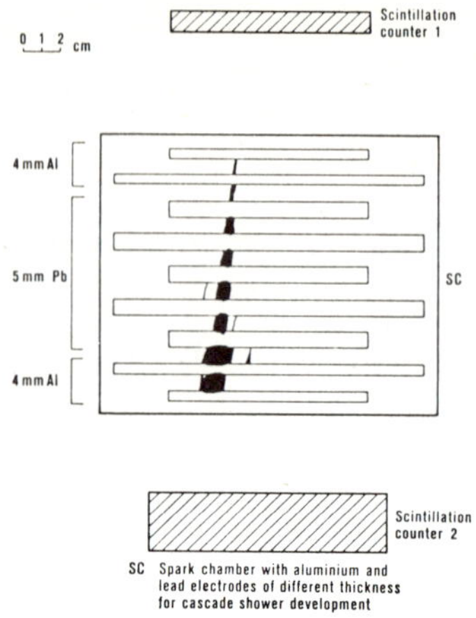

Figure 7.19 Diagram of a spark chamber "telescope" spectrometer.

As more powerful accelerators have been built, bigger and bigger chambers have also been designed, such as the Brookhaven chamber (1964) which used 93 Al plates 6 ft x 6 ft (about 2 m x 2 m) in size. The large number and size of the electrode plates tends to increase the chamber capacitance and hence slow down operation of the chamber. A number of ways to overcome this have been proposed: parallel-series connection of plate-wire sets, voltage reversal between pulse events to clear the field region, and fast-quenching gas mixtures. The recovery time of the chamber depends on the clearing field and on the amount of energy dissipated in the spark discharge. In a wire spark chamber for which only 1 millijoule per spark was used, recovery times on the order of a few μsecs were observed.

Spark chambers are classified as track chambers (wide gap, single plate-pair), streamer chambers, multiplate chambers, and wire chambers.

According to the method of recording and data analysis they may be classified as photographic, projection, image amplifier, sonic and digitized chambers, as well as some other methods that are less commonly used.

Wide-gap track chambers are used where it is desired to measure individual tracks without separation by electrodes. The gap width may be 6 to 60 cm, and the chamber records particles passing more or less normal to the plate surfaces. Under these conditions the spark is formed along the track of the particle, whereas in narrow-gap multiplate chambers the track follows strictly the direction of the electric field.

In a streamer chamber, particles moving parallel to the electrodes are made visible. By applying a very short (5-50 ns) high-voltage pulse, avalanches are produced locally along the track, streamer formation is initiated, but no spark is formed. During streamer formation, light is emitted and hence the track is visible as a series of luminous dots, a few millimeters in length. Photography of the tracks will usually require use of an image intensifier.

In multiplate chambers, a particle track is formed by combining successive sparks as the particle of interest crosses adjoining gaps. Gap widths of 3-10 mm are commonly used; electrode thicknesses can be varied to allow for the range of the particles and their decay products. Figure 7.20 shows examples of tracks observed.

Track resolution is an important characteristic of such chambers, representing the deviation in the spatial location of the sparks and the actual path of the particle. With 6- to 10-mm gaps, standard deviations of the order of 0.2 mm have been reported. Neon appears to give better spatial resolution than helium. Chambers have been designed to operate in magnetic fields up to 200 kG (20 T). The magnetic field affects the electron diffusion rate and hence the clearing time. It also introduces additional positional uncertainty, but permits separation of secondary particles by charge condition.

Most multiplate chambers require photographic recording, usually operating in the dark with the camera shutter left open. The brightness of the sparks depends on the chamber gas and the discharge current. Chambers that record mainly simple trajectories of particles, much the way a cloud or bubble chamber operates, are often called hodoscopes. Measurement of energy and momentum is done by reconstruction of the trajectory in three dimensions from stereoscopic pictures.

More precise track reconstruction, particularly where multiple particle production occurs, is possible in digitized chambers using online computers. This can be accomplished most easily with wire-spark chambers where planes of thin parallel wires are employed in place of solid plates as the ground electrode. The wires are threaded through tiny ferrite cores to provide spatial information. Less energy is required to flip the cores than to produce a visible spark so that the recovery time is much shorter than for plate chambers. After passage of the pulse, the cores can be interrogated for transfer of location data to the computer. The spatial accuracy of the wire chambers at low pulse energy depends on the wire spacing. To minimize vibration problems and photoionization effects, wire chambers have been constructed using printed circuit techniques.

An alternative method for hodography without photographic recording involves the use of microphones for acoustic location of sparks. Typical

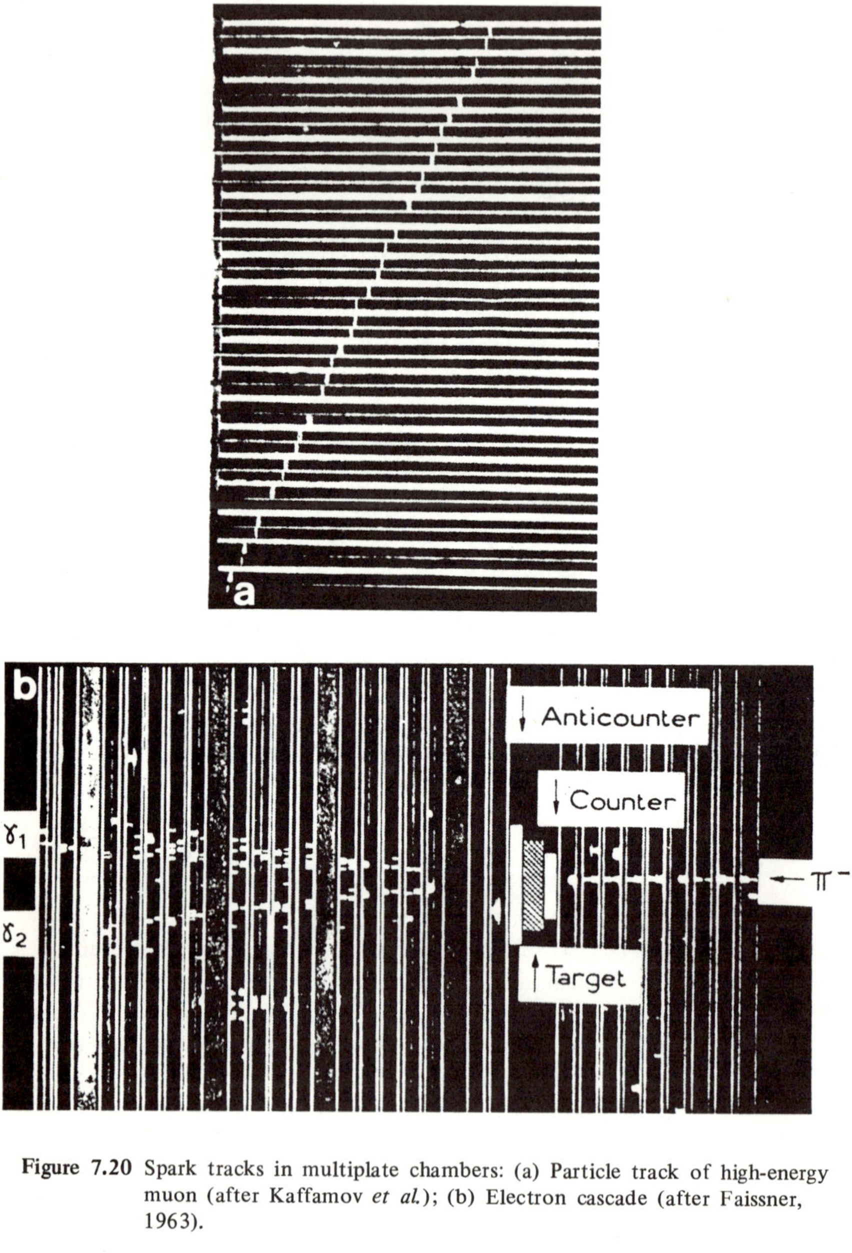

Figure 7.20 Spark tracks in multiplate chambers: (a) Particle track of high-energy muon (after Kaffamov *et al.*); (b) Electron cascade (after Faissner, 1963).

sound velocities in noble gases are 0.5 mm/sec and tracks can be localized with a rms deviation of 0.3 mm. The spark produces a shock wave which initially travels faster than ordinary sound waves. (The velocity of sound has a temperature coefficient of about 0.17% per °C.) Compact piezoelectric crystals, such as lead zirconate, can be mounted at the corners of the chamber and the track coordinates can be reconstructed from the arrival times and amplitudes of the acoustic pulses.

For large spark chambers online computers are essential to evaluate data continuously, to avoid accumulation of useless results, and to sort out events of different character.

CLOUD CHAMBERS AND BUBBLE CHAMBERS

C. T. R. Wilson in 1911 invented a system for the visualization of the tracks of ionizing particles in an expansion chamber filled with supersaturated moist air. If the volume of the chamber was expanded suddenly by dropping the floor of the chamber, the supersaturated water vapor condensed preferentially on any floating dust particle or free ion to form visible droplets. If an ionizing particle had crossed the chamber a brief moment before expansion took place, so that the ions formed along its track had not moved appreciably under gravity or by thermal action, then its ionized track would become visible as a necklace of tiny droplets that condensed on its ion trail. By suitable illumination, this string of droplets could be made visible and by stereoscopic photography the track could be reconstructed in space. Many historic discoveries have been made by cloud chamber photography: the positron, the neutron and nuclear disintegration. By mounting the cloud chamber in a magnetic field, positive and negative ions could be distinguished and the charge-to-mass ratio could be deduced from the curvature of their tracks.

Despite its long history of accomplishment, the Wilson cloud chamber suffered from a number of disadvantages. It was difficult to operate, the sudden movement of the piston tended to set up turbulence in the chamber gas, the pressure and expansion timing was difficult to control, and the flash photography introduced intense heat pulses that made the system very slow to recover and to recycle.

A simple modification of the cloud chamber that was introduced in the 1950s was the *diffusion cloud chamber.* In this case no expansion cycle was needed. One side of the chamber was permanently cooled by means of dry ice. Supersaturation was produced by spraying isopropyl alcohol into the air volume. If a radiation source was introduced close to the cooled floor of the chamber, alcohol would condense on any ionized air molecules when it diffused into the cooler region of the chamber. In this way tracks could be made visible in the light of an external light source in the cooler portion of the chamber volume whenever any particles passed through the sensitive chamber region. Ionized droplets would disappear by gravitational settling. Such chambers are widely used for lecture demonstrations of radioactivity; they do not lend themselves for large-volume research applications.

That function has been taken over by *bubble chambers* which were invented by Glaser in 1952. In the bubble chamber, ionized particles along the track of the incident radiation again serve as nuclei. However, instead of condensation of a vapor in the bubble chamber, ionization initiates boiling in a superheated liquid, giving rise to the formation of vapor bubbles along the path. In Glaser's experiments, diethyl ether was maintained at 20 atmospheres pressure (~ 2 MPa) and a temperature of 130°C. Periodically the pressure was lowered and, occasionally, spontaneous boiling was observed. Under the action of ionizing particles, tiny vapor bubbles form along the track and will grow rapidly in size until they can be photographed just prior to the onset of general boiling.

Figure 7.21 shows the pressure-temperature diagram for such a liquid. The liquid is in a stable state at any condition above the normal saturated vapor pressure P_∞. If the liquid is in the metastable state, *i.e.*, if it is superheated by lowering P at constant T or raising T at constant P from the equilibrium condition, bubble formation can be triggered. In a bubble chamber, sensitive conditions are established by lowering the pressure, keeping the temperature constant. Because liquids of low compressibility are employed, their temperature changes little when the pressure is dropped. The amount of superheat, S, determines whether bubbles are formed and how densely. Beyond a maximum superheat parasitic boiling occurs.

The liquid can stay in the metastable state only for a short time before general boiling occurs; during that time formation of bubbles occurs due to local evolution of heat due to the degradation of the energy lost by δ-rays (excited electrons) left behind by the passing particle. Several hundred electrovolts are required to form a bubble of critical radius and to do work against surface tension, σ, and the pressure of the surrounding liquid.

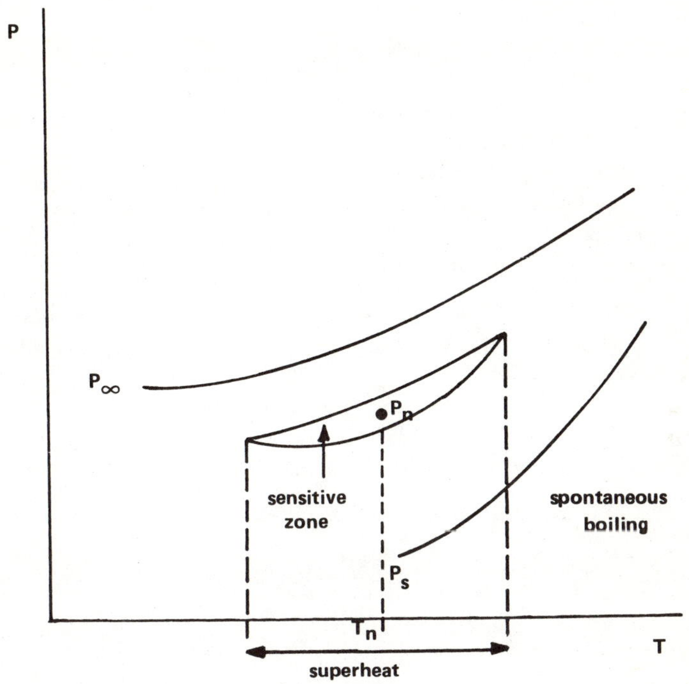

Figure 7.21 Thermodynamics of the bubble chamber.

When a bubble forms there is a change in thermodynamic potential

$$\Delta G = 4\pi r^2 \sigma \left(1 - \frac{2}{3} \frac{r}{r_k} \right) \tag{7-3}$$

where r = bubble radius
r_k = critical radius
$\approx 10^{-6}$ cm for most liquids

The energy required to form a bubble

$$W_o = (U' - U) \frac{4}{3}\pi r_k^3 \rho' + 4\pi r_k^2 \left(\sigma - T \frac{d\sigma}{dT}\right) + P \frac{4}{3}\pi r_k^3 \left(1 - \frac{\rho'}{\rho}\right) \tag{7-4}$$

where U and U' = specific internal energy of liquid and vapor
ρ and ρ' = densities of liquid and vapor

A bubble can form in a time of the order of 10^{-10} sec and grow almost explosively.

If a beam of charged particles passes through the liquid at relativistic speeds ($\beta = v/c$), the rate of energy loss

$$\frac{dE}{dx} = \frac{0.3 z_o^2 \; m_e c^2 \; Z\rho}{\mu\beta^2} \left[\log_e \frac{4 m_e c^2 4\beta^4}{I^2(1-\beta^2)} - 2\beta^2 \right] \tag{7-5}$$

where z_o = particle charge
Z = number of electrons in liquid molecule
I = mean ionization potential
= $13.5 \; Z_i$ eV
μ = molecular weight of liquid

A particle with $\beta^2 \sim 10^{-3}$ would lose about 100 eV over a distance of 10^6 cm in a liquid of density 0.1-1.0 g/cc. The number of secondary electrons formed

$$dn = \frac{\kappa(E,I)}{\beta^2} \left[\log \frac{2 m_e c^2 \beta^2}{I(1-\beta^2)} + \alpha_{(E,I)} - \beta^2 \right] dE \tag{7-6}$$

where α and κ depend on atomic states and a is small for low energies,

$$dn = 10 \frac{\kappa(E,I)}{\beta^2} dE$$

For propane

$$N = \Sigma \, dn = \frac{1.53 \times 10^5 z_o \rho Z}{\mu\beta^2 W}$$

where W = minimum energy to initiate bubbles.

For 18 MeV electrons, $\beta^2 \sim 1$

$$N = \frac{39.8 \times 10^3}{W};$$

if W = 400 eV, N $\gtrsim$ 100 cm^{-1}

Table 7.4 compares chamber conditions for different liquids.

The main characteristics that have made the bubble chamber such a versatile and almost universal detector for use with high-energy accelerators are as follows.

1. The instrument has high spatial resolution.
2. Its time resolution is good.
3. It can record tracks in a 4π solid angle.
4. The liquid used can be chosen according to the requirements of a particular experiment; for example, hydrogen and deuterium

Table 7.4 Bubble Chamber Operating Conditions for Different Liquids and Estimation of the Bubble Formation Energy (after Peyrou)

Liquids	H_2	He	C_3H_8	$CBrF_3$	SF_6	C_3F_8
Operating Temperature T	26.5°K	4.2°K	60°C	37°C	21°C	42°C
Saturated Vapor Pressure at Temperature T, p_∞(atm)	4.5	1.0	22	21.4	22.2	13.6
Expansion Pressure Drop $(p_\infty - p_e)$(atm)(p_e is the pressure of the expanded fluid)	2.5	0.28	10	8.4	6.6	4.8
Surface Tension at Temperature T, σ(dyn/cm)	0.9	0.10	3.6	2.6	2.05	2.16
Radius of Critical Bubble $r_c = 2\sigma/(p_\infty - p_e)$(Å)	72	83	73	63	62	90
Surface Free Energy $W_{s_1} = 4\pi r_c^2 \sigma$(eV)	3.7	0.54	15	8	6	13.5
$W_{s_2} = -4\pi r_c^2 \frac{T\partial\sigma}{\partial T}$(eV)	19.5	2.2	140	96	96	178
Vaporization Energy $W_v = \frac{4}{3}\pi r_c^3 (H/M)\rho'$(eV)	9	4.5	90	73	88	180
Expansion Energy $W_e = \frac{4}{3}\pi r_c^3 p_e$(eV)	2	0.94	11	8	10	17
Bubble Formation Energy $W_s' = W_{s_1} + W_{s_2} + W_v +$ W_c (eV)	34.2	8.2	256	185	100	390

can be used to provide free proton and (almost) free neutron targets, since the target and the detector are the same instrument.

5. A high density and, therefore, a low interaction and gamma-ray conversion length, can be obtained from heavy-liquid chambers using propane or Freon.
6. Large volumes can be scanned quickly for events of a particular type.
7. The cycle rate of bubble chambers is well matched to that of accelerators (about one per second).
8. The momenta of the charged particles can be measured by putting the chamber in a magnetic field.
9. Information on the velocity (β) of the charged particles can be obtained from the bubble density.

Since the invention of this technique, bubble chambers have evolved from small instruments containing a few cubic centimeters of liquids to huge machines such as the 12-m^3 French "*Gargamelle*" or the 15-foot bubble chamber at the Fermi Accelerator Laboratory shown in Figure 7.22, which has a diameter of 12.5 ft (3.8 m), with the nose protruding 2.5 ft (0.76 m) further. The chamber is mounted inside a large superconducting magnet with an outer diameter of almost 17 ft (5.2 m) producing a central magnetic field of 3 T (30 kG). The Big European Bubble Chamber (BEBC) at CERN, Geneva, consists of a cylinder 3.70 m in diameter, filled with 35 m^3 of liquid hydrogen. Such chambers can be used to observe events with 300-GeV protons or neutrino-induced events in a suitable chamber gas, such as neon. They are very complex and expensive to operate and a number of modifications are being tried. Among these is the vertex chamber where a compact visible target is introduced into the center of the chamber to produce more frequent events which would make it possible to operate at a higher repetition rate, possibly more than 50 times a second. Another possibility is the ultrasonic bubble chamber, where the expansion phase of a sound wave in the bubble chamber liquid is used to expand the liquid below the sensitive pressure, P_S, thus causing microbubbles to grow to a sufficient size before the next compression phase arrives. Such a system avoids the troublesome piston arrangement of conventional chambers. By using an ultrasonic transducer to produce standing waves in the liquids, bubbles have been observed to grow to visible size in 0.5 ms. Experimental chambers have been constructed working with both liquid helium and liquid hydrogen at frequencies of the order of 300 kHz, though no very large sonic bubble chambers seem to have been operated so far.

In summary, Table 7.5 compares the performance of various chambers and detectors that are used in high-energy physics research (after Kalmus).

REFERENCES FOR FURTHER READING

General

Holm, N. W., and R. L. Berry. *Manual on Radiation Dosimetry* (New York: Marcel Dekker, Inc., 1970).

Kiefer, H., and R. Maushart. *Radiation Protection Measurements* (Elmsford, NY: Pergamon Press, Inc., 1972).

Photographic Emulsions

Barkas, W. H. *Nuclear Research Emulsions*, 2 vols. (New York: Academic Press, Inc., 1963, 1973).

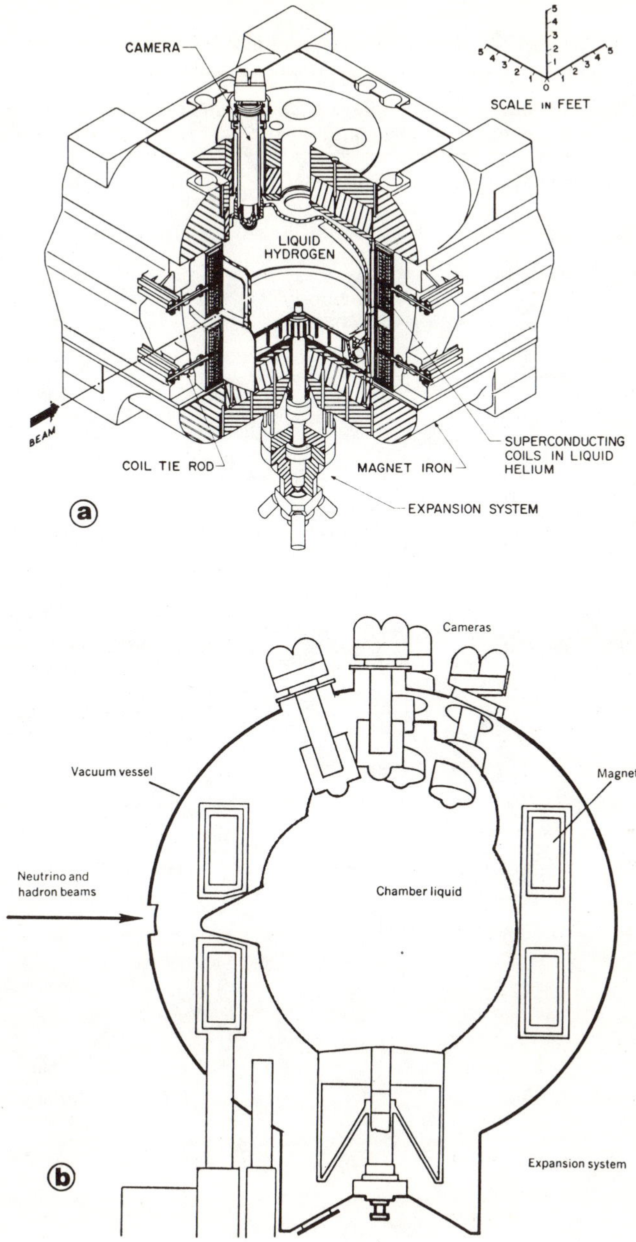

Figure 7.22 High-energy bubble chamber: (a) Cut-away view of large bubble chamber and its superconducting magnet; (b) Cross section through 15-ft bubble chamber at the National Accelerator Laboratory.

Table 7.5 Comparison of Various Experimental Techniques

Technique	Spatial Resolution	Time Resolution	Solid Angle (sr)	Target Material
Cloud chamber (expansion)	200-500 μm	100 ms	4π	Mixture of gases
Cloud chamber (diffusion)	200-500 μm	Bad (continuously sensitive)	$<4\pi$	Mixture of gases
Photographic emulsion	1 μm	None	4π	Photographic emulsion (Ag, C, H, O, etc.)
Counters (scintillation)	$\geqslant$1 cm	Good $\sim 10^{-8}$ s	Small ($\ll 4\pi$)	Separate target; can be any material including some containing polarized protons
Counter-spark chambers	300 μm	Good $\sim 10^{-6}$ s	Usually small ($\ll 4\pi$) but in general higher than counters	Separate target; can be any material including some containing polarized protons
Bubble chamber (H_2, D_2)	30 μm	$\sim$1 ms	4π	Liquid H_2 or D_2
Bubble chamber (heavy liquid)	30 μm	$\sim$1 ms	4π	Propane, Freon, etc.

Used in High Energy Physics (after Kalmus)

Density (g/cm^3)	Interaction Rate	Triggerable?	Comments
10^{-2} - 10^{-3}	Low: one interaction per 10^2-10^3 pictures in a typical chamber	Yes	(a) Cycle rate low; one expansion per 300 s (b) Distortion of tracks can be severe due to track motion
10^{-3}	Low: one interaction per 10^2-10^3 pictures	No	(a) Sensitive in a rather narrow vertical band. Therefore steep tracks quickly lost (b) High background
~4	High: one interaction per 30 cm of track	No	(a) Very slow to scan and measure (b) High multiple scattering—even sign of particle cannot be determined by bending in normal strength magnetic fields (c) Usually not possible to correlate V^0 to production vertices
0.1 upwards	High: 10^6 triggers/s with intense beams	Yes	(a) Details of interaction and/or decay vertices not available (b) Small solid angle and lack of spatial resolution means data are not bias free
0.1 upwards	Quite high: 20 triggers/s Limited by recovery time of chambers (see comment (c))	Yes	(a) Details of interaction and/or decay vertices not available (b) Lack of complete solid angle means that data are not bias free, but in general this problem is much less severe than for pure counters (c) The new multiwire proportional chambers allow much higher event rates than the normal wire spark chambers
0.06 (H_2) 0.10 (D_2)	~1 interaction per picture and 1 picture/s	No	(a) The expansion is not triggerable, but the light flash for taking the photograph is (b) In its untriggered (usual) mode, ideal for exploratory work since all interactions are observed (c) Lower event rates than either counters or spark chamber-counters for particular, well determined reaction types
~1	~1 interaction per picture and 1 picture/3 s	No (see H_2 bubble chamber comments)	All the comments under H_2, D_2 bubble chambers apply; also (a) Electrons are easily recognizable (b) γ rays have a high probability of converting to $e^{\pm}$ pairs (c) Momentum measurement by curvature worse than in H_2, D_2 chamber due to multiple scattering

Powell, C. F., and G. P. S. Occhialini. *Nuclear Physics in Photographs* (Oxford: Clarendon Press, 1947).

Track Etch Devices

Becker, K. *Solid State Dosimetry* (Cleveland, OH: CRC Press, Inc., 1973).
Fleischer, R. L., P. B. Price and R. M. Walker. *Nuclear Tracks in Solids* (Berkeley: University of California Press, 1975).

Spark Chambers

Allkofer, O. C. *Spark Chambers* (Munich: Thiemig Verlag, 1969).
Rice-Evans, P. *Spark, Streamer, Proportional and Drift Chambers* (London: The Richelieu Press, 1974).
Shutt, R. P., Ed. *Bubble and Spark Chambers*, 2 vols. (New York: Academic Press, Inc., 1967).

PROBLEMS

7.1 Explain the length, grain density and reasons for the general appearance of the "stars" shown in Figures 7.1 a and b.

7.2 A 5.3-MeV α particle has a range of about 23 μm in gelatin-AgBr and dissipates 0.23 MeV/μm of average path. If there were 2.7 x 10^9 AgBr grains per cm^2 of emulsion and the track contained 50 developed grains, estimate the energy required to make a grain developable.

Beta particles of 2.3-MeV maximum energy have an effective range of 2,800 μm in the emulsion. Would they leave developable tracks?

7.3 Explain the use of lead screens to enhance radiographic contrast. How thick should the screens be if used as front screens or as back screens?

7.4 In radiography a film has been calibrated with a source-to-film distance of 50 cm. If this distance has to be changed to 65 cm, how would this affect the exposure time?

Under what condition may reciprocity law failure occur?

7.5 An isotopic source, 5 mm in diameter, is collimated by a tubular channel 5 mm in diameter, 50 mm long. Calculate the focal spot dimensions and the penumbra on a film 1 m away.

7.6 Aluminum, copper and lead foils are used to flatten the response of some film dosimeters over a wide range of X-ray energies. Estimate the thicknesses required to balance response at 0.2, 0.5 and 0.8 MeV.

CHAPTER 8

MISCELLANEOUS DETECTORS

CHEMICAL DOSIMETRY

Introduction

In chemical dosimetry, radiation dose is determined from the chemical changes produced in an appropriate medium.

Chemical dosimetry systems are used primarily for measuring high doses beyond the range where ionization chambers are normally used. In some cases chemical dosimeters are used for calibrating other dosimetry systems and, in addition, in situations requiring maximum accuracy. These systems are also attractive because they consist of an aqueous solution, which, enclosed in a suitable container, approximates rather closely the atomic composition and density of many biological materials. Unfortunately, routine health physics monitoring requires a response below 1 rad and this sensitivity is beyond the limit of reliable chemical systems.

Listed below are the desirable characteristics of any chemical dosimeter; there are:

- sensitivity,
- adequate reproducibility,
- stability before and after irradiation,
- dose-rate independence,
- temperature independence,
- independence from the quality of the radiation (LET),
- linear relation between reaction product and dose, and
- simple preparation and analysis.

There are a wide variety of systems in use today. The most popular is the Fricke dosimeter which was developed in the period 1927 to 1929. The system is based on the oxidation of an aerated ferrous sulfate solution and is widely accepted as a standard in both radiation chemistry and radiation dosimetry. The Fricke dosimeter meets most of the above requirements. The main disadvantages of the dosimetry system are its limited normal dose range and the dependence of yield on the LET of the radiation. An additional disadvantage is a demonstrated dose-rate dependence at very high dose rates.

Basic Principles

In dilute aqueous solutions, which include chemical dosimetry systems, chemical reactions arise from the action of the radiation on the water, resulting in free radicals and molecular products. The free radical group includes hydrogen atoms, hydroxyl radicals and solvated electrons. Hydrogen and hydrogen peroxide are the main molecular products. The radiolysis of water has been studied for a number of years and the equations describing the formation of free radicals and molecular products have been published widely.

As with most other radiation detectors the chemical changes produced in the dosimeters are due to excitation and ionization caused by charged particles produced in the solution. It is assumed that the number and kind of energetic states formed (per unit absorbed energy) are independent of the type of radiation. These energetic states degenerate rapidly, forming chemically active species. In water this occurs in about 10^{-10} to 10^{-11} sec.

For a given chemical system, the production of radicals in a solution can be established through several measuring techniques. The primary techniques are chemical titration and absorption spectroscopy. In this manner, the chemical system can be calibrated as a radiation dosimeter. The yield of the reactions is expressed in terms of the number of molecules or radicals formed/100 eV of absorbed energy. The yield is usually designated by the symbol G and it is common to refer to the "G-value" for a certain chemical system.

Calculation of Absorbed Dose

If the radiation chemical yield is assumed to be a linear function of the absorbed dose then:

$$G(x) = \frac{\Delta n}{\Delta E} \cdot 100 \qquad (8\text{-}1)$$

where G(x) is the radiation chemical yield expressed as the number of molecules or ions produced per 100 eV of absorbed energy, Δn is the number of molecules of x formed per unit volume, and ΔE is the amount of energy (in eV) absorbed per unit volume. Usually, ΔE is established by a standard measurement technique, *e.g.* calorimetry, and Δn can be determined by chemical analysis.

If G(x) is known, the absorbed energy (expressed in electron volts per unit volume) is given by:

$$\Delta E = \frac{\Delta n \cdot 100}{G(x)} \tag{8-2}$$

If, in Equation 8-2, the concentration of x is expressed in moles/liter (M), then,

$$\Delta E = \frac{(\Delta M)\,(6.02 \times 10^{25})}{G(x)} \tag{8-3}$$

where the absorbed energy is in electron volts per liter (eV/l). Using the conversion that 1 rad equals 100 ergs/g and 1 eV equals 1.602×10^{-12} ergs, the absorbed dose can be obtained:

$$D = \frac{(\Delta M)\,(6.02 \times 10^{25})\,(10^{-3})\,(1.602 \times 10^{-12})\,(10^{-2})}{G(x)\rho}$$

$$D = \frac{\Delta M\,(9.64 \times 10^{8})}{G(x)\rho}\ \text{rad} = \frac{\Delta M}{G(x)\rho}\,(9.64 \times 10^{6})\ \text{Gy} \tag{8-4}$$

where ρ is the density of the irradiated system. The above equation, because it is general, may be applied to any type radiation. However, G(x) must be measured for each radiation to which the particular system is to be applied.

General Experimental Techniques

All chemical systems have one common requirement; the reagents must be scrupulously pure. In many cases, commercially available reagents are repurified. For example, triply distilled water is required for the more sensitive dosimetry systems. Even slight traces of contaminants can produce loss of reproducibility and stability.

Glassware and irradiation cells must also receive careful consideration, including not only the geometry and dimensions of the cells, but also proper cleaning and handling procedures. These latter requirements are due to the extreme sensitivity of radiolytic reactions to trace impurities. The cells made of silica or Pyrex usually give reliable results; except in very special cases, plastic cells should be avoided. For mixed radiation fields, especially for high-level irradiations, high-purity quartz cells are recommended in order to reduce the effects of induced activity.

Ferrous Sulfate System (Fricke Dosimeter)

Because its atomic composition and density are close to biological systems of interest, water is a popular substance for dosimetry. In addition, there are high-purity reagents and analytical reactions available and the yields and mechanisms of many aqueous radiation-chemical reactions are known. Perhaps the most popular chemical dosimetry system, and certainly the best-known, is the ferrous sulfate or Fricke dosimetry system. The range of this dosimeter is from 4,000 to about 40,000 rads. In some cases, with special techniques, the range can be extended beyond both that low- and high-dose limit. This dosimeter consists of an aerated dilute solution of ferrous sulfate in 0.8 *N* sulfuric acid. This concentration was initially selected to match the mass absorption coefficient for X-rays in air.

The standard Fricke dosimeter consists of 0.001 *M* $FeSO_4$ or $Fe(NH_4)_2(SO_4)_2$+0.8 *N* H_2SO_4 (air-saturated), and is prepared from triply distilled water. To maintain greater stability, stock solutions should be stored in a dark place and, if possible, refrigerated.

A second or modified Fricke dosimeter is also in use. This consists of 0.001 *M* $FeSO_4$ or $Fe(NH_4)_2(SO_4)_2$, 0.8 *N* H_2SO_4. (air-saturated), and 0.001 *M* NaCl. The NaCl is used to desensitize the system to the effects of organic impurities so that water of a lower quality may be used.

When the Fricke dosimeter is irradiated, ferrous ions are oxidized to ferric ions. The number of ions oxidized is directly related to the absorbed energy due to X- or gamma-radiation. The simplified reactions are:

$$Fe^{++} + OH \rightarrow Fe^{3+} + OH^-$$

$$H + O_2 \rightarrow HO_2$$

$$Fe^{++} + HO_2 \rightarrow Fe^{3+} + HO_2^-$$

$$HO_2^- + H^+ \rightarrow H_2O_2$$

$$Fe^{++} + H_2O_2 \rightarrow Fe^{3+} + OH + OH^-$$

All the above reactions occur rapidly with the exception of the reaction of the ferrous ion with hydrogen peroxide. However, even this reaction occurs with a half-life of about 14 sec. The total G-value, *i.e.*, $G(Fe^{3+})$, is 15.6.

The dosimeter may be evaluated by chemical titrations to determine the ferric ion concentration. Potentiometric titrations have been used in the dose range 4,000 to 50,000 rads. These methods have the advantage that they are absolute.

Usually, it is more convenient to use absorption spectroscopy to determine the ferric ion concentration. This method is fast and accurate and is expedient for the analysis of low ferric ion concentrations and small quantities of solution. A spectrophotometer can be used to determine the amount of ferrous ions oxidized by measuring the absorption change at the 5,100-Å peak or the ferric ion absorption change at the 3,040-Å peak. At this latter wavelength, the optical absorption coefficient of the ferrous ion is less than 0.05% of that of the ferric ion.

In an extensive dosimetry program, calibration curves of optical density as a function of ferric ion concentration should be obtained using a spectrophotometer designated for the purpose. For this calibration it will be necessary to prepare a series of ferric ion solutions by oxidation of ferrous sulfate in 0.8 *N* acid with a standard hydrogen peroxide solution.

The absorbed dose can be calculated by:

$$D = \frac{N\,\Delta(OD)\,100}{\Delta\epsilon\,10^3\,G(Fe^{3+})\,f\rho l} \tag{8-5}$$

where N = 6.02×10^{23} molecule/mole
$\Delta(OD)$ = difference in optical density between irradiated and control solutions
$\Delta\epsilon$ = difference in molar extinction coefficient between ferric and ferrous ions, at the wavelength used for the optical density measurements
ρ = density of solution (for 0.8 *N* H_2SO_4 = 1.024)
l = optical path length (cm)
f = 6.24×10^{13} eV/rad

The Fricke system is relatively stable; however, even in the absence of radiation, the aerated ferrous sulfate solution will oxidize slowly. It is standard procedure to correct for this oxidation by using one unirradiated sample of the same solution as a control.

It has been demonstrated that the ferric ion formation in the dosimeter is directly proportional to the absorbed dose as long as some oxygen remains in the solution. The sulfuric acid concentration will affect this linearity of response of the chemical system. Concentrations below 0.1 *N* H_2SO_4 result in a loss of reproducibility and linearity. However, at 0.1 *N* there is only about a 2% drop in the G-value.

The G-value is not completely independent of photon energy. It rises from about 13.8 at 10 keV to 15.6 at 600 keV and is relatively constant up to ^{60}Co energies (1.25 MeV). This variability illustrates the necessity of knowing the effective energy of the radiation to which the dosimeter is exposed.

The Fricke dosimeter has been shown to be dose-rate independent from 0.1 to 4,000 rads/sec. In experiments utilizing a linear accelerator, the response was demonstrated to be independent of dose rate up to 2×10^8 rads/sec for 2-μsec pulses.

The G-value changes with temperature, increasing gradually with increasing temperature of irradiation. The exact value of this temperature dependence is not completely known. Three separate authors reported the following; no temperature dependence in the range 20 to 45°C, 0.09%/°C in the range 2 to 65°C, and 0.04 ± 0.03%/°C in the range 0 to 70°C.

The usable range of the Fricke dosimeter can be extended outside its normal limits by suitable modifications both to the composition of the system and to the method of analysis. If the concentration of the ferrous ion is raised to 0.05 *M* the range may be extended to 10^6 rads. However, air or oxygen bubbling is required to maintain an adequate oxygen concentration during irradiation.

The range of the dosimeter may be lowered to about 400 rads simply by increasing the length of the absorption cell in the spectrophotometer to 10 cm. Doses in the range below 100 rads may be surveyed by using a ^{59}Fe tracer and extracting the ferric ion from the irradiated solution. The lower limit of this method is about 10-20 rads.

Ceric Sulfate Dosimetry

Ceric sulfate is used to measure absorbed doses above the general range in which the Fricke dosimeter is used, *i.e.*, 4 x 10^4 rad to about 10^8 rad (400-10^6 Gy). The system consists of an aerated solution of ceric sulfate or ceric ammonium sulfate in 0.8 *N* sulfuric acid. The principal reaction is the reduction of ceric to cerous ion, accompanied by the formation of oxygen and hydrogen gases. This dosimetry system requires careful calibration because of its high sensitivity to impurities at low ceric ion concentrations.

In this particular chemical system, the ceric ions are reduced by hydrogen atoms, hydroperoxy radicals and hydrogen peroxide. The cerous ion is oxidized by the hydroxyl radicals. The equations describing the mechanism are:

$$H + O_2 \rightarrow HO_2$$

$$Ce^{4+} + H \rightarrow Ce^{3+} + H^+$$

$$Ce^{4+} + HO_2 \rightarrow Ce^{3+} + H^+ + O_2$$

$$Ce^{3+} + OH \rightarrow Ce^{4+} + OH^-$$

$$Ce^{4+} + H_2O_2 \rightarrow Ce^{3+} + H^+ + HO_2$$

The total G value for these reactions, $G(Ce^{3+}) = 2.34$.

For dosimetry above 40,000 rads, the concentration of ceric sulfate in 0.8 *N* sulfuric acid is varied according to the dose range of interest. For example, in the range 0.1 to 0.3 Mrad the recommended concentration is

0.001 *N*, whereas in the range 6 to 15 Mrad, the concentration is 0.05 *N* (see Table 8.1).

Table 8.1 Recommended Ceric Sulfate Concentrations for Several Dose Regions

	Dose	
N Ce^{4+}	(megarads)	(kilograys)
0.001	0.10-0.30	1-3
0.004	0.30-1.20	3-12
0.02	1.20-6.0	12-60
0.05	6.0-15.0	60-150

Ceric sulfate is very sensitive to light, both ultraviolet and visible, especially in dilute solution. It is recommended that stock solutions be stored in darkened bottles. In addition, care must be exercised to avoid excessive exposure to light before, during and after irradiation. The use of control solutions, which are treated as nearly like the irradiated solutions as possible, is recommended as a simple solution to this problem.

The irradiated solutions may be analyzed by titration or by a spectrophotometer. In the megarad range, titration with ferrous sulfate using orthophenanthroline as an indicator is the usual procedure. The ceric-ion concentration can be measured much more conveniently by using a spectrophotometer and measuring the absorption at 3200 Å. If great care is taken in the preparation, irradiation and analysis of solutions, it is possible to obtain a precision of ± 1%.

The rate of reduction in the ceric sulfate solution varies linearly with dose at concentrations below about 50 mM Ce^{4+}. This relationship holds essentially until the ceric ion is completely reduced. Over the ceric ion concentration range, 0.1 to 50 mM, the yield $G(Ce^{3+})$ decreases at high ceric ion concentrations.

Over the concentration range from deaerated to aerated solutions, oxygen has no effect on the yield. In addition, for sulfuric acid concentrations between 0.8 to 2.0 *N* the cerous ion yield is constant.

The ceric sulfate system has not shown a dose-rate dependence up to about 500 rads/sec. An increase in the cerous ion yield has been observed at doses in excess of 10^8 rads/sec delivered in accelerator pulses of 1- to 2-μsec duration. Between 15 and 30°C the yield is practically independent of irradiation temperature.

Ferrous Sulfate - Cupric Sulfate System

Another chemical system designed to extend the range of the Fricke dosimeter is based on the addition of cupric sulfate to the ferrous sulfate. This system can be used in the range 40,000 to 10^7 rad (400-10^5 Gy) with only slight modifications in the analysis. This extension into the megarad region is possible because the G-value drops from 15.6 to 0.66 when cupric sulfate is added to the ferrous sulfate solution at pH 2. In addition, higher concentrations of ferrous and cupric sulfate can be used since no oxygen is consumed in the reactions.

The standard dosimeter solution consists of 0.001 *M* $FeSO_4$, 0.010 *M* $CuSO_4$, and 0.010 *N* H_2SO_4 (air-saturated). The solution can be made with analytical-grade reagents because the dosimeter is relatively insensitive to organic impurities. However, cupric ions lessen the thermal stability of ferrous sulfate. In routine situations fresh solutions should be made up daily.

The ferrous-cupric dosimeter is evaluated spectrophotometrically at a wavelength of 3,040 Å. There is a strong absorption of the cupric ion at 2,240 Å and, therefore, measurements are impractical in this spectral region.

The G-value changes little with minor variations in acid concentration around 0.01 *N*. This is due to the fact that a 50% change in acid concentration changes the yield less than 5%. If the sulfuric acid concentration is increased to 0.8 *N*, the G-value for this system changes from 0.66 to 4.0.

The ferrous-cupric dosimetry system may be used in a dose range which is about a factor of 24 higher than the conventional Fricke dosimeter. No increase in the ferrous sulfate concentration is necessary for measurements in the dose range 10^5 to 10^6 rads. The practical upper limit is about 8×10^5 rad (8 kGy) with a reproducibility of about ± 0.3%. In order to extend the dose range to 10^8 rad (10^6 Gy) an increase in the ferrous and cupric ion concentration is necessary.

The G-value for the ferrous-cupric dosimeter is LET-dependent. For gamma-rays the G-value is 0.66 ± 0.02; it increases to 3.4 for 0.02-MeV alpha particles. This increase in G with increasing LET is the opposite of that demonstrated in the Fricke dosimeter. Using these two dosimeters in tandem offers the possibility of obtaining measurements in a mixed radiation field.

Primary uses of the ferrous-cupric dosimeter have been in underwater monitoring of fuel elements and ^{60}Co gamma-ray food-irradiation sources.

Oxalic Acid System

This system is intended primarily for use in the high-dose range and is considered to be a supplement to the ceric sulfate system.

It has shown a high thermal stability before and after irradiation. In addition, it is insensitive to organic impurities. The primary use of the system is for in-pile measurements because neutron activation is neglibible compared to chemical activation in the ferrous sulfate and ceric sulfate dosimeters. The major disadvantage of the system is its poor reproducibility, which has been reported to be about ± 10%.

The oxalic acid chemical system is based on the radiation decomposition of oxalic acid, in the presence of oxygen, into [CO_2], [H_2] and [H_2O]. The initial concentration of oxalic acid is varied between 0.025 and 0.60 *M* according to the dose range to be measured; its sensitivity range is 10^6 to 2 x 10^8 rads. Some typical concentrations and the applicable dose range are shown in Table 8.2.

Table 8.2 Recommended Oxalic Acid Concentration for Several Dose Regions

$H_2C_2O_4$	Dose	
(*M*)	(megarads)	(kilograys)
0.025	0.3-1.6	3-16
0.050	1.6-8	16-80
0.100	3.2-16	32-160
0.600	16-20	160-1600

The oxalic acid solution is irradiated in a glass or quartz cell. The cell is stoppered or flame-sealed and a minimum of air is left above the solution. The container must be able to resist the pressure of hydrogen and carbon dioxide which are produced during radiation. Quartz containers are used to avoid neutron activation during reactor irradiations.

The amount of oxalic acid consumed during radiation can be related to the absorbed dose. This is accomplished by measuring the optical absorption of a cupric benzidine complex. The complex is formed in oxalic acid by adding an acetic acid solution containing recrystallized benzidine and cupric acetate. The absorption is measured with a spectrophotometer at 2,460 Å. The molar extinction coefficient at this wavelength is 2,490 ± 250 M^{-1} cm^{-1} at room temperature. The G-value for the reaction is about 4.9 and the density of the dosimeter is 1 g/cm^3. The solution has exhibited a temperature coefficient of -0.7%/°C.

Two other analytical methods have been employed with the oxalic acid dosimeter. The analysis of the carbon dioxide liberated is a very sensitive indication for low doses. For high doses, an acid-base titration with standard NaOH gives reproducible results.

The amount of oxalic acid which is decomposed is a linear function of dose up to about 30% decomposition. The G-value is independent of dose rate up to 0.5 x 10^6 rads/hr. Temperature during irradiation has been found to have essentially no effect in the range 20 to 80°C.

Chlorinated Hydrocarbon Systems

Chlorinated hydrocarbons, such as chloroform, trichloroethane and tetrachloroethane, have been utilized as radiation dosimeters for certain applications. These solutions, irradiated in the presence of oxygen, produce high yields of hydrochloric acid. The system has been developed for low-level dosimetry (10-20 rads) because of several attractive characteristics. Most notable of these characteristics are the high yield of the reaction and the high sensitivity of indicators for measuring small changes in pH near neutral.

The G-value for the production of hydrochloric acid may be as high as 10,000 in pure compounds. This fact coupled with the ability to measure accurately acid concentrations of 10^{-8} N should allow a detection limit of a fraction of a rad. However, during irradiation the chlorinated hydrocarbon systems show poor reproducibility, temperature-dependence and dose-rate sensitive yields. These characteristics can be overcome by adding a free-radical chain retarder to the solution. However, a loss of sensitivity must be accepted for this gain in stability.

The acid produced in solution can be measured by standard methods such as titration or electrometric determination of pH or conductivity. In some situations, especially in low-dose usage, it is advantageous to use an acidic dye. The color density of the indicator can be read without opening the capsule and, in some cases, an absorbed dose of 10 rads (0.1 Gy) can be determined visually.

There are two types of chlorinated hydrocarbon dosimeters used frequently. The single-phase, sometimes called water-equivalent single-phase, dosimeter consists of an aqueous dye solution saturated with a chlorinated hydrocarbon. The solution is made with triply distilled water and a retarder is usually added. The two-phase dosimeter consists of the chlorinated hydrocarbon overlaid with a layer of aqueous dye solution and retarder.

The single-phase is not as sensitive to radiation as the two-phase dosimeter. However, it is much simpler to use, it is more stable and it has a better spectral response when used for the dosimetry of low-Z materials. In both dosimeters there is a linear relation between dose and the amount of acid liberated over a wide dose range. The single-phase system, with modifications, can be used over the dose range 20 to 2 x 10^6 rad (0.2-2 x 10^4 Gy) while the two-phase dosimeter ranges from 2 to 2 x 10^6 rad. Both dosimeters are temperature-independent from 0 to 55°C and no dose-rate effects have been noted for dose rates up to 4,000 rads/min.

After exposure, the dosimeters may be evaluated in a number of ways. For example, simple visual observation of color change in the single-phase dosimeter is sufficient to measure doses of about 100 rads (1 Gy) with an accuracy of 10-20%. However, in more precise dosimetry, a spectrophotometer is used. The indicators used have two widely separated absorption peaks in the visible spectrum. These belong to the acid and basic forms of the solution. Normally, the dosimeters are evaluated by determining the ratio, R, of the transmittance of the two bands rather than just a single band. The dose is obtained from a plot of the log of the abosrbed dose versus the log of the difference, ΔR, before and after irradiation.

Summary

Chemical dosimeters have a wide range of possible applications. The dosimeters are economical when prepared in large numbers and they can be made in a range of sizes. The primary application of chemical dosimetry systems is in the high-dose range. However, some systems (*e.g.*, chlorinated hydrocarbons) have the potential for measuring doses of less than 1 rad.

The major system in use is the Fricke dosimeter which utilizes the oxidation of an aerated ferrous sulfate solution. The characteristics of this system have been studied extensively.

CALORIMETRY

Introduction

This method of detection represents one of the few absolute methods of determining the energy deposited by ionizing radiation when it interacts in an absorbing medium. The calorimeter is a device which measures the amount of heat produced in a medium due to the absorption of ionizing radiation. Since this absorbed energy is ultimately degraded to heat, the calorimeter is a fundamental instrument of measuring absorbed dose.

In general, calorimeters have been used in many areas of physics and medicine. However, the primary use of the calorimeter as a dosimeter has centered around the measurement of high-intensity beams of radiation, *e.g.*, a radiation therapy beam. Calorimeters may be designed (1) to measure energy locally absorbed, or (2) to measure an energy flux density. Both types of calorimeters will be discussed briefly in this section.

Before discussing the design of specific types of calorimeters, it is instructive to consider the temperature rise to be expected in a material from a given absorbed dose. In order to keep the discussion simple, consider an absorbed dose of 1,000 rads (10 Gy) in water. The temperature rise in water can be calculated easily:

$$1{,}000 \text{ rad} = 10^5 \frac{\text{ergs}}{\text{g}} = 10^{-2} \frac{\text{joule}}{\text{g}} = 10 \frac{\text{joule}}{\text{kg}} \tag{8-6}$$

since 4.18 joules is equal to one calorie, we find that

$$1{,}000 \text{ rad} = 0.239 \text{ cal/g} \tag{8-7}$$

The specific heat of water is one calorie per gram per degree. Thus, we see that an extremely large dose of radiation will cause a temperature rise slightly greater than 0.2°C. Obviously, this temperature rise is so small that very sensitive methods are required to measure this absorbed energy.

The development of calorimeters progressed rapidly after the introduction of the thermistor. A thermistor is a metallic oxide semiconductor with an electrical resistance which changes rapidly with temperature; that is, it exhibits a high negative temperature coefficient (3-5%/°C at 20°C is typical). The resistance change in the thermistor can be measured with a Wheatstone bridge and a galvanometer. However, a dc amplifier and a galvanometer are also used in certain situations.

Measurement of Energy Flux Density

In therapy it is usually necessary to determine the radiation output of a source or device in terms of the energy flux density (*i.e.*, $J/m^2/sec$). For measurements of this type, a total-absorption calorimeter is used. A schematic drawing of such a detection system is shown in Figure 8.1a. Basically, the radiation beam impinges on an absorber (usually lead) which is sufficiently large to absorb the radiation completely. Usually, the calorimeter has a thin-walled aperature to allow the radiation beam to enter essentially unperturbed. Thermal insulation and a vacuum are used to isolate the absorber from its surroundings. In some cases a temperature-controlled oil bath is used to surround the calorimeter, except for the entrance port, in order to maintain the device in a constant-temperature environment. The absorbed energy in the material causes a rise in temperature which is measured by a thermistor embedded in the absorber. Since the thermistor is attached to one arm of a Wheatstone bridge the temperature rise in the absorber is indicated by an unbalance of the bridge. The temperature rise is easily determined by measuring this unbalance.

Measurement of Absorbed Energy

The measurement of absorbed energy is of more interest from the radiological point of view. In this case the absorber is much smaller and is of a low-atomic-number material similar to tissue. In many cases graphite is the material

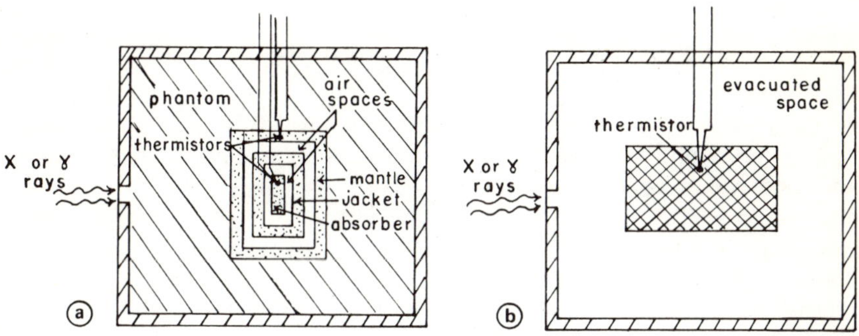

Figure 8.1 Schematic diagrams of calorimeters designed to (a) measure energy flux density and (b) measure energy locally absorbed.

of choice. This device is shown schematically in Figure 8.1b. The absorber is placed inside a shell of the same material. This arrangement allows the build-up of thermal and electron equilibrium. The entire arrangement is insulated from its surroundings, usually by a vacuum. In addition, the absorber and the equilibrium material are insulated from each other. When the radiation beam impinges on the calorimeter, only a small fraction of the beam is absorbed in the thin absorber. The temperature rise in the absorber is determined by use of a thermistor as previously described. Calorimeters of this type are sometimes called "low perturbation calorimeters." In addition to measuring absorbed energy, they have been used to determine electron stopping-powers of thin foils.

Both the calorimeters described above are classed as adiabatic calorimeters; that is, the calorimeter jacket and the calorimeter are maintained at the same temperature. The radiation energy input can be measured easily in terms of temperature rise of the calorimeter.

Two other types of calorimeters have been used; these are the twin-differential and the isothermal calorimeters.

In the twin-differential calorimeter there are actually two essentially identical calorimeters. In one calorimeter, heat is added at a known rate while the other calorimeter is used to determine the unknown heat input. The rate of known heat input is adjusted in the first calorimeter until the temperature difference between the two is zero.

In the isothermal calorimeter it is allowed to come to temperature equilibrium with the heat-removal rate equaling the heat input due to the absorption of radiation. One common method of accomplishing this is to flow water through the calorimeter and measure the temperature rise in the liquid.

A calorimeter of the twin-isothermal type has been designed which is a tissue-equivalent calorimeter (Bewley *et al.*, 1974). In principle, such a device can be used for the measurement of absorbed dose due to neutrons. However, since a calorimeter can measure only the total amount of heat produced, a separate measurement of one of the components of the mixed radiation field is necessary in order to evaluate the absorbed doses separately.

Summary

Calorimeters allow the measurement of absorbed energy in absolute units. This is their greatest advantage. However, their use requires an extensive facility and is usually confined to use in well-equipped laboratories. Their low sensitivity, size and complexity are the primary factors that restrict widespread usage. In addition, calorimetric measurements often require more time and analysis than do other dosimetry systems.

REFERENCES FOR FURTHER READING

Attix, F. H., and W. C. Roesch, Eds. *Radiation Dosimetry*, Vol. II (New York: Academic Press, Inc., 1966).

Auxier, J. A. "Special Methods in Radiation Dosimetry," in *Principles of Radiation Protection*, K. Z. Morgan and J. E. Turner, Eds. (New York: John Wiley and Sons, Inc., 1967).

Bewley, D. K., E. C. McCullough, B. C. Page and S. Sakata. "Neutron Dosimetry with a Calorimeter," *Phys. Med. Biol.* 19:831 (1974).

Holm, N. W., and R. J. Berry, Eds. *Manual on Radiation Dosimetry*, (New York: Marcel Dekker, 1970).

Johns, H. E. *The Physics of Radiology*, (Springfield, IL: Charles C. Thomas, 1964).

Laughlin, J. S., and S. Genna. "Calorimetry" in *Radiation Dosimetry*, Vol. II, F. H. Attix, W. C. Roesch and E. Tochilin, Eds., (New York: Academic Press, 1966).

Morgan, K. Z., and J. E. Turner, Eds. *Principles of Radiation Protection*, (New York: John Wiley and Sons, Inc., 1967).

PROBLEMS

8.1 Calculate the dose received by a 1-cm cell containing a ferrous sulfate solution if the spectrometric transmittance values at 2240 Å were 0.052 and 0.47 before and after irradiation. [($\epsilon(Fe^{3+})_{25°C}$ = 4,565 M^{-1} cm^{-1}, $\epsilon(Fe^{++})_{25°C}$ = 20 M^{-1} cm^{-1} at 2,240 Å)]

8.2 Outline a procedure for calibrating a TLD dosimeter by means of a Fricke dosimeter.

Explain in what sense the Fricke dosimeter provides an absolute calibration.

8.3 Compare the detection efficiencies for 1-MeV gamma-rays of a chemical dosimeter and an ionization chamber, making reasonable assumptions regarding sizes and dimensions, concentrations and other operational parameters.

8.4 Calculate the rate of gas evolution if pure water is exposed to a dose rate of 10^5 gray per hour [($G_{(H_2+O_2)}$ = 0.575 for ^{60}Co γ-rays)].

8.5 Copper-constantan thermocouples have a response constant of 40 μV/°C. If such a thermocouple is used in conjunction with a well-insulated calorimeter filled with 100 ml of deaerated water, estimate the minimum X-ray dose that can be determined if the thermocouple indicator reads in microvolts.

8.6 In the previous example, if the calorimeter was of the twin differential type, how many watts of electrical energy will be needed for balance in a 1-MR/hr field?

CHAPTER 9
NEUTRON DETECTION

INTRODUCTION

Neutrons form a major component in the cosmic radiation background, but the main reason for the interest in their detection lies in their association with nuclear reactors, fusion devices and certain high-voltage accelerators, as well as with certain industrial and research applications employing spontaneous fission neutron sources (californium-252) or sources utilizing the Be(α,n) reaction such as Am:Be, Pu:Be or Ra:Be neutron sources. In most of the cases the source strength must be determined (usually in terms of neutron flux), or the neutron dose (preferably in terms of the dose equivalent in units of rems or sieverts).

The problem with precise neutron measurement is twofold: because neutrons interact only with the nuclei of target atoms, rather than their electronic structure, cross sections for interactions and, hence, detection efficiencies, tend to be much lower than for charged particles; also, because neutron energies may be rapidly modified by scattering events and most of the cross sections vary substantially with energy, both detection and dosimetry depend critically on a detailed knowledge of the energy spectrum of the neutron field. This may vary from point to point unless it has been deliberately thermalized, *i.e.*, has reached thermal equilibrium at the ambient temperature. Additional problems arise from the fact that the actual detection process may arise from a secondary effect, such as decay of an activated nuclide, with its own half-life and decay scheme, or from ionization effects of a recoiling nucleus in a hydrogenous or gaseous medium. In all such cases the overall detection efficiency will depend both on the energy-dependent cross section of the primary neutron interaction as well as on the efficiency of detection for the secondary products, be they hydrogen ions, beta particles or any number of other excitation products. This multiplicity of possible reactions provides an opportunity for selecting the most appropriate detection method, but may also introduce some unexpected neutron-induced background events, *e.g.*, through the $^{14}N(n,p)^{14}C$ or $^{16}O(n,p)^{16}N$

reactions, or by activation of the detector medium itself or its shielding, giving rise to a slowly decaying background activity.

Because capture cross sections are high in the thermal neutron region and it is relatively easy to conduct measurements at thermal energies (kT $\sim$ 0.025 eV at 20°C), many methods have been developed for the detection of thermal neutrons or for measurements of fast neutrons that have been slowed down to thermal energies to make them more detectable. However, in many cases, particularly in complex spectra, the slowing-down process introduces additional and unacceptable uncertainties, and it is necessary to employ methods that are capable of observing energetic neutrons directly. For purposes of description neutron energies are frequently classified as:

- slow or thermal neutrons, up to 1,000 eV in energy
- epithermal or intermediate energy neutrons, 1-100 keV
- fast neutrons, 0.1-20 MeV
- high-energy (relativistic) neutrons, $>$ 20 MeV

The slow-neutron region is often also defined by reference to the cadmium absorption edge, that is the rather steep drop in capture cross section at 0.18 eV (see Figure. 2-17) which makes elemental cadmium virtually opaque to neutrons below that energy and increasingly transparent above it. For this region "thermal" neutrons are often considered to be those that are stopped by a cadmium foil, whereas those that are not are assumed to be epithermal or fast neutrons. The intermediate-energy regions are the range of energies where capture resonances occur in many nuclides (Figure 2-16), while fast energies, especially above 0.6 MeV lie above the threshold energies for fission (n,f) and charged particle emission [(n,α), (n,d), (n,p) and (n,t) reactions] in many nuclides. Fast neutrons also are capable of ejecting atoms from crystal lattices and liquid molecules and a number of detection methods have been based on such phenomena.

Neutron detection methods can be classified according to:

1. energy (slow, intermediate, fast)
2. type of information desired:
    - number flux
    - energy spectrum
    - dose equivalent
3. principal interaction mechanism:
    - activation (neutron capture), leading to beta decay, fission, or secondary nuclide emission
    - scattering (recoil production)
    - damage tract production (spike formation)

4. detection process employed:
   - ion collection
   - track indicator
   - paired detector (time-of-flight methods)
   - rise-time discrimination
   - sandwich (double-layer) systems
   - activation foils
   - fission converters and related devices

Some of the above classes overlap and the distinction between methods cannot always be maintained sharply. Monoenergetic neutrons are rarely encountered and, hence, most neutron detectors must be calibrated in a neutron spectrum of reproducible character. The detected neutrons must be similar in spectral distribution or corrections must be made, usually by computer programs, to relate the experimental measurements to reference spectra. The principal source of nearly monoenergetic neutrons is provided by high-voltage machines, such as a Van de Graaff generator, bombarding a thin beryllium target with deuterons or photons. In this way the cross-section values in the fast range and the resonance regions have been determined accurately for many target nuclides.

The reference energy spectrum for thermal neutrons is not monoenergetic but represents a Maxwellian distribution around the most probable energy value, kT, for a well-moderated neutron population and to that extent may be considered reproducible. In many practical cases it is necessary to check its "purity" by means of cadmium absorbers. Other neutron sources emit "known" spectra that may be rapidly degraded by scattering from intervening air molecules, floor, walls and experimental equipment; great care must be taken to ensure reproducibility. This is particularly important when calibrating foils or dosimeters in neutron fluxes of "known" energy from portable sources, *e.g.*, Pu:Be, nuclear reactors with a fission spectrum clustered in the 2-3 MeV region, and neutron generators employing the $^3H(d,n)$ reaction with a peak energy of 14.2-14.6 MeV.

A further criterion that often arises in choosing a neutron detector is its ability to discriminate against other radiations, such as the gamma-ray field usually present around most neutron sources.

THERMAL NEUTRON MEASUREMENTS

Thermal neutrons are most easily detected by means of capture reactions that have high cross sections in this energy range. Three types of reactions can be distinguished:

1. prompt capture reactions, leading to charged particle emission. These include the $^{10}B(n,\alpha)^{7}Li$ reaction (σ_{th} = 3836 barns), the $^{6}Li(n,\alpha)^{3}H$ reaction (σ_{th} = 940 barns), and the $^{235}U(n,f)$ reaction (σ_{th} = 586 barns).
2. delayed activation reactions, employing thin foils or converter materials that will emit detectable radiations with a convenient half life and energy. Tables 9.1 and 9.2 list a few nuclides that are useful for this purpose, usually as thin metal foils. Activation materials are used primarily for flux determinations, for fluence (burnup) monitoring and for self-powered detectors.
3. calorimetric methods, measuring the temperature rise in a capture medium.

All of the neutron detection media can be incorporated as neutron-to-ionizing particle converters in many of the detection systems mentioned in previous chapters.

Gas-Filled Detectors

Because of their inherent simplicity gas-filled detectors, in the form of ionization chambers or proportional counters, are widely employed for neutron monitoring or simple dosimetry. The neutron absorber can be introduced into the sensitive volume either as a constituent of the filling gas or as a solid converter foil which can emit ionizing particles to interact with the counter gas.

The most widely employed thermal neutron detector is a proportional counter using boron trifluoride (BF_3) as the filling gas, preferably in the enriched state with the boron containing 96% ^{10}B. The $^{10}B(n,\alpha)$ reaction can also be employed by applying a layer of B_2O_3 or BC_4 on the walls of the detector. Such a boron-lined counter has an inferior plateau compared with the BF_3 counter since the α particles lose part of their energy in the lining, whereas in the BF_3 gas practically all the alpha particles are absorbed in the counting gas. Since these detectors are proportional counters, it is possible to discriminate against gamma radiation. However, in high gamma flux densities enough ions may still be produced to lead to a "pile-up" of pulses, resulting in a loss of discrimination and even complete loss of sensitivity. For this reason gas-filled detectors are best employed in low-level neutron fields. BF_3 detectors are normally cylindrical in shape, filled with 96% enriched BF_3 at a pressure of about 12 torr (1.6 kPa). Since the counting gas detects alpha particles from the $^{10}B(n,\alpha)$ reaction, it is fairly easy to dimension the volume so that the entire energy is dissipated within it. This results in a fairly flat operating-counter plateau, extending typically over 200 volts for an operating voltage of the order of 1,400 volts for a 6 in. (15 cm) long counter tube.

Table 9.1 Materials for Thermal Neutron Activation Detectors

Elements	$N/\rho \times 10^{-22}$ (g^{-1})	Element		Isotope		Radioactive Isotope (Half-Life)
		σ_a (barn)	σ_s (barn)	(Abundance in %)	σ_{act} (barn)	
Mn	1.097	13.2 ± 0.1	2.3 ± 0.3	Mn^{55} (100)	13.2 ± 0.1	Mn^{56} (2.58 h)
Co	1.022	37.1 ± 1.0	7 ± 1	Co^{59} (100)	16.9 ± 1.5	Co^{60m} (10.4 min→5.28y) (99%)
					20.2 ± 1.9	Co^{60} (5.28y)
Cu	0.948	3.81 ± 0.03	7.2 ± 0.6	Cu^{63} (69.1)	4.41 ± 0.20	Cu^{64} (12.87h)
				Cu^{65} (30.9)	1.8 ± 0.4	Cu^{66} (5.14 min)
Ag	0.558	64.5 ± 0.6	6 ± 1	Ag^{107} (51.35)	45 ± 4	Ag^{108} (2.3 min)
				Ag^{109} (48.65)	3.2 ± 0.4	Ag^{110m} (253 d)
					105 ± 15	Ag^{110} (24 sec)
In	0.525	194 ± 2	2.2 ± 0.5	In^{113} (4.23)	56 ± 12	In^{114m_1} (49 d)
					2.0 ± 0.6	In^{114} (72 sec)
				In^{115} (95.77)	160 ± 2	In^{116m_1} (5.12 min)
					42 ± 1	In^{116} (14.1 sec)
Dy	0.371	940 ± 20	100 ± 20	Dy^{164} (28.18)	2,000 ± 200	Dy^{165m} (1.3 min)
					800 ± 100	Dy^{165} (140 min)
Au	0.3055	98.5 ± 0.4	9.3 ± 1.0	Au^{197} (100)	98.5 ± 0.4	Au^{198} (2.695d)

Table 9.2 Materials for Threshold Activation Detectors (after Beckurts and Wirtz)

a. Substances for (n,p) Threshold Detectors

Element	Target Nucleus (Abundance)	Reaction Product (Half-Life)	E_s[MeV]	σ_s [mbarn]	E_{eff} [MeV]	$\overline{\sigma}$ [mbarn]	Practical Detector Substances	Emitted Radiation	Minor Activities (Threshold Energy [MeV] and Half-Life)
P	P^{31} (100%)	Si^{31} (157 min)	0.716	140	3.0	30	Phosphor Containing Glasses	β^-	P^{30} (12.7; 2.6 min) Al^{28} (2.0; 2.3 min) Al^{29} (13.6; 6.6 min) P^{32} (thermal; 14.2 d)
S	S^{32} (95%)	P^{32} (14.2 d)	0.954	350	3.2	66	Tablets of elemental sulfur	β^-	P^{33} (8.7; 25 d) Si^{31} (1.32; 157 min) S^{35} (thermal; 87 d)
Ni	Ni^{58}(68%)	Co^{58} (72 d)	– (Q = 0.399)	600	3.45	100	Metallic Ni metal foil	γ, (EC), β^+	Ni^{57} (12.0; 36 h) Co^{61} (0.5; 1.65 h) Ni^{56} (thermal; 2.56 h)
Fe	Fe^{54}(5.82%)	Mn^{54} (290 d)	– (Q = 0.094)	400	3.75	53	Fe metal foil	γ, (EC)	Cr^{51} (–; 27.8 d) Mn^{56} (2.95; 154.5 min)
Al	Al^{27}(100%)	Mg^{27} (9.393 min)	1.89	80	5.30	3.5	Al metal foil	γ, β^-	Al^{28} (thermal; 2.3 min) Na^{24} (3.26; 14.97 h)
Fe	Fe^{56}(91.6%)	Mn^{56} (154.5 min)	2.95	130	7.70	0.97	Fe metal foil	γ, β^-	Cr^{51} (–; 27.8 d) Mn^{54} (–; 290 d)
Mg	Mg^{24}(78.8%)	Na^{24} (14.97 h)	4.9	200	8.00	1.2	Mg metal foil	γ, β^-	–
b. Substances for (n, n′), (n, α) and (n, 2n) Threshold Detectors									
Rh	Rh^{103} (100%) (n, n′)	Rh^{103m}(57 min)	0.040	1500	0.9	1093	Rh metal foil	γ, e^-	Rh^{104m} (thermal; 4.4 min) Rh^{102} (9.5; 210 d)

In	In^{115} (n, n′)	(95.8%)	In^{115m}	(4.5 h)	0.335	350	1.65	171	In metal foil	γ, e⁻	In^{116m_1}	(thermal; 54.1 min)
											In^{113m}	(0.393; 1.7 h)
											Ag^{113}	(9.5; 5.3 h)
											Ag^{112}	(–; 3.4h)
Al	Al^{27} (n, α)	(100%)	Na^{24}	(14.97 h)	3.26	120	8.15	0.61	Al metal foil	γ, β⁻	Mg^{27}	(1.89; 9.39 min)
Cu	Cu^{65} (n, 2n)	(30.91%)	Cu^{64}	(12.8 h)	10.1	1000	11.7	0.31	Cu metal foil	γ, β⁺, β⁻	Co^{61}	(10.5; 1.65 h)
											Ni^{65}	(–; 2.56 h)
											Cu^{64}	(thermal; 12.8 h)
Cu	Cu^{63} (n, 2n)	(69.09%)	Cu^{62}	(9.8 min)	10.9	800	13.2	0.073	Cu metal foil	β⁺	Co^{60m}	(–; 10.5 min)
											Cu^{66}	(thermal; 5.1 min)
											Co^{62}	(–; 13.9 min)
											Ni^{65}	(–; 2.56 h)
											Co^{61}	(–; 1.65 h)
Ni	Ni^{58} (n, 2n)	(68.0%)	Ni^{57}	(37 h)	12.0	80	14	4×10^{-3}	Ni metal foil	γ, β⁺	Ni^{56}	(thermal; 2.56 h)
											Co^{61}	(0.5; 1.65 h)

c. Fission Detectors as Threshold Detectors

Isotope	E_s[MeV]	σ_s[mbarn]	E_{eff}[MeV]	$\bar{\sigma}$ [mbarn]
U^{234}	≈ 0.3	1500	0.62	1200
Np^{237}	≈ 0.4	1500	0.87	1100
U^{236}	≈ 0.7	850	1.25	520
Th^{232}	≈ 1.3	140	1.40	28
U^{238}	≈ 1.3	606	1.55	310

The efficiency for the detector is defined as the fraction of neutrons entering the counter volume that result in a measurable counting event. This, or course, depends on the shape of the counter volume and the direction of the incident neutron. If a neutron passes a distance x through the counter, the probability of absorption is

$$(1 - e^{-\Sigma_a x}) = (1 - e^{-N\sigma_a x}) \quad (9\text{-}1)$$

where Σ_a = macroscopic absorption cross section
N = number of boron-10 atoms per cm^3
σ_a = absorption cross section (3,836 barns at 0.024 eV)

For thermal neutrons travelling parallel to the axis in a 6-in. (15 cm) cylindrical detector, efficiencies of about 22-25% can be attained.

Because of the (1/v) energy dependence of the capture cross section, the BF_3 detector is relatively insensitive to fast neutrons. Such detectors can be employed in "slow-fast" detector units by pairing a "bare" BF_3 tube, detecting slow neutrons, with another one that is surrounded by a paraffin moderating layer, which is sheathed on the outside with cadmium, so that the second tube will detect only fast neutrons, after thermalization. Boron trifluoride detectors are convenient and widely employed in portable detectors.

Another type of gas-filled detector is an ionization chamber with a uranium-235 foil inserted in the center of the chamber or used as a wall lining. Such an ionization chamber is often referred to as a *fission chamber* and depends on the ejection of highly ionizing fission products from the enriched uranium foil or layer, which has to be thin enough to minimize self-absorption of the fission products before they reach the counter gas. The upper limit for coating thicknesses is about 1 mg/cm^2; for thicknesses less than 0.1 mg/cm^2 fission pulses are about twice as large as other pulses, facilitating pulse-height discrimination. Argon is used as a filling gas in many commercial chambers, especially for in-core reactor use, while helium is a preferred gas if neutron-to-gamma discrimination is important. Figure 9.1 illustrates a commercial fission chamber. Dimensions for the electrode spacings are of the order of 0.25-4mm.

The sensitivity of fission chambers depends on their size (converter surface area), gas pressure and applied voltage. They are relatively insensitive to temperature variations. Boland has quoted thermal neutron sensitivities of the order of 2×10^{-17} amp/n/$cm^2 \cdot$ sec for a gamma sensitivity of the order of 10^{-4} amp-hr/rad for in-core fission chambers, subject to a maximum thermal neutron flux of 10^{14} n/$cm^2 \cdot$sec. In higher fluxes overloading occurs.

Both types of gas-filled detectors have a limited usable life due to deterioration of the gas pressure as a result of leaks, chemical interaction with

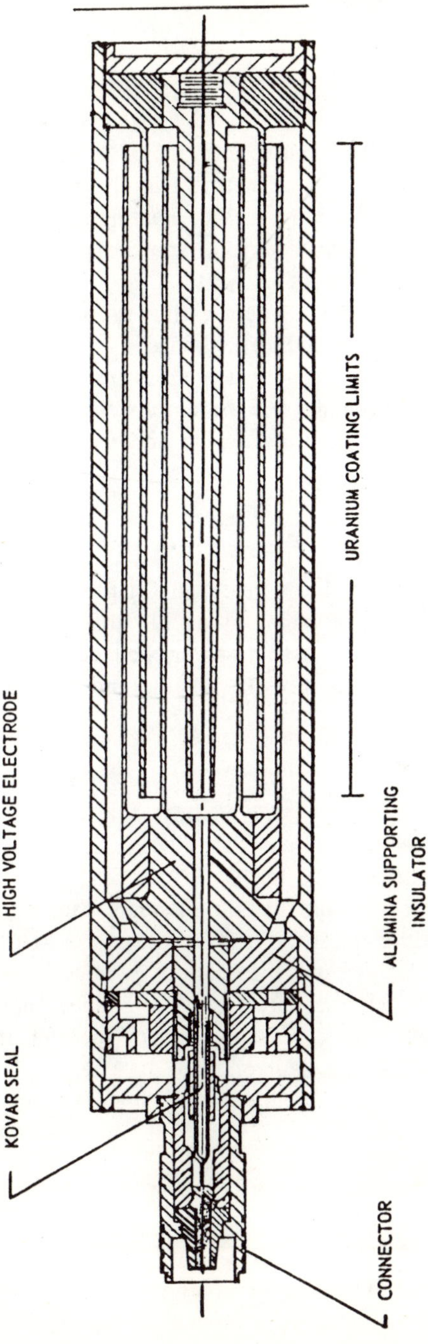

Figure 9.1 Fission ionization chamber for neutron flux density measurements (after Baer and Bayard, 1953).

wall materials and buildup of impurities. In addition, the sensitivity will diminish with use as the converter nuclide is consumed ("burnup").

Autoradiography

In neutron radiography and photographic dosimetry, photographic emulsions are employed to measure neutron intensity by means of indirect conversion. This can be done by surface coating the film with boron or lithium compounds, by incorporating appropriate compounds in the emulsion (see Table 7.1) or the use of converter foils. The latter method is widely used in neutron radiography where a thin gadolinium screen is exposed and used to store the neutron-induced "image". It is then placed on a photographic emulsion where the activated areas produce a developable image. In a similar way film badges can be employed with and without a cadmium screen to register neutron-induced beta activity from a copper foil (see Table 7.2).

Thermoluminescent Dosimetry

Reference has already been made, in Chapter 6, to the use of LiF TLD's for neutron dosimetry. By employing LiF made of natural or Li-7-enriched material and comparing it with Li-6-enriched detectors, separation of neutron- and gamma-produced exposures is possible and practical.

ACTIVATION FOILS

Activation foils can be employed both for thermal and fast-neutron detection, according to the material selected, and it is expedient to treat these applications together. The basic principle is simple: a thin, preferably metallic, foil is selected that has an adequate capture cross section in the neutron energy region of interest. It is exposed to the neutron field for a known period of time and the induced radioactivity is counted in an appropriate beta or gamma detector. The induced activity depends on the number of exposed atoms, *i.e.*, the mass of the foil, its composition and thickness, the bombarding neutron flux, its energy spectrum, and the capture cross section σ of all the nuclide species involved, which is a function of the energy spectrum.

Hence,

$$\text{Activity } A = \phi_{(E)} N_o \sigma_{(E)} (1 - e^{-\lambda t_e}) e^{-\lambda t_c} \qquad (9\text{-}2)$$

Where N_o = the number of target atoms

λ = the time constant of the induced activity

$\phi_{(E)}$ = the effective neutron flux
$\sigma_{(E)}$ = the effective capture cross section
t_e and t_c = the exposure and cooling times, respectively

In practice, attempts are made to use only highly pure foil materials and to employ elements where only one predominant isotope contributes to the induced activity. This cannot always be taken for granted and often both irradiation time and cooling times must be selected to discriminate against unwanted activities or to permit them to decay away before counting. It is also usually assumed that the neutron flux is uniform across the area of the foil; this too may have to be verified and encourages use of small-area foils or wires.

Corrections must be applied for decay of activity between exposure and counting, and during the counting interval, for self-absorption and for detector sensitivity and geometry. In practice this means that the detector must be calibrated in a reproducible geometry, with a foil that has been exposed in a known flux of neutrons of known energy, not always an easy condition to meet. Tables 9.1 and 9.2 list some of the elements used for detection foils. These include cobalt, manganese, silver, indium, dysprosium and gold for thermal neutrons; indium, gold, tungsten, lanthanum, and manganese for resonance neutrons; and plutonium, neptunium, nickel and sulfur as threshold detectors for fast neutrons. For thermal neutron measurements, $\phi_{(E)}$ is assumed to represent a fully thermalized neutron flux and the applicable cross section $\sigma_{(E)}$ is the listed one (Table 9.1). To verify this, it is often advisable to compare bare and cadmium-covered foil measurements.

If the cross section is fairly large, a correction for self-shielding may have to be applied unless the foil is thin compared with the effective mean free path $1/\Sigma_a = 1/N\sigma_a$. This becomes significant when dealing with such materials as indium, cadmium, gold, gadolinium or fission foils. When the induced beta activity is counted, corrections may also be necessary for beta self-absorption in the foil. For indium the optimum thickness lies at 120 mg/cm^2; a decrease in detected activity occurs for both thinner and thicker foils because of the above-mentioned effects.

If the foil introduces a significant neutron absorber into the neutron field, its very presence may result in flux depression and the measured flux may not be representative of the unperturbed neutron field. Correction curves for flux depression by indium foils can be found in Beckurts and Wirtz.

For epithermal and fast neutrons the simplifying assumptions regarding an effective average neutron energy are no longer valid; foil materials must be selected that either have accurately known cross section curves over the energy

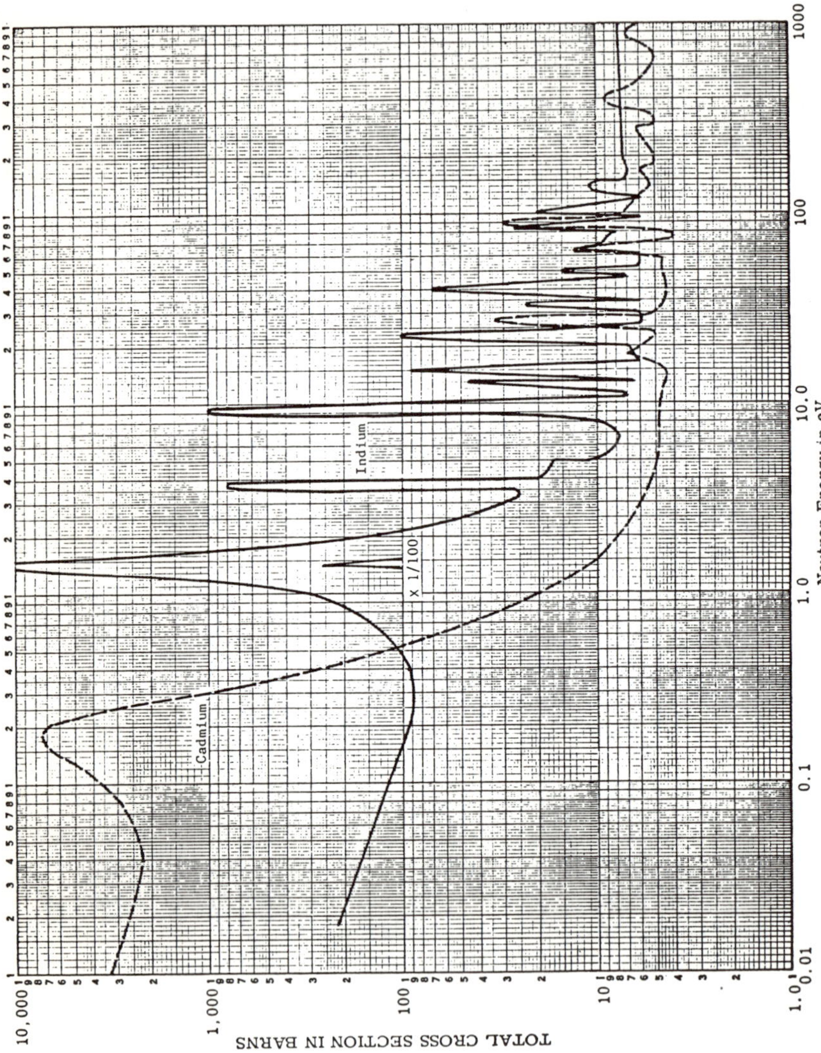

Figure 9.2 Total neutron cross sections for indium and cadmium.

range of interest, or respond by threshold reactions that make them insensitive to energies below the effective threshold. By piecing together responses for specific energy regions the total neutron spectrum can then be reconstructed. The practical difficulties in this procedure arise from problems of calibrating foils in a "known" neutron field of reproducible energy, especially in the epithermal region.

If one wishes to determine solely the total neutron flux associated with each region one can employ suitable foils, subtract the thermal contribution by means of a cadmium shield and the neutrons below about 10 eV with an indium shield. Figure 9.2 shows the two cross section curves. Comparing the count rate $C = C_{th} + C_{epi}$ for the bare foil, with that of the cadmium-covered foil C_{cd}, one finds that they are related by a cadmium correction factor F_{cd} which depends on the thickness of the cadmium cover, the foil substance, the moderator substance and temperature. Figure 9.3 shows values of F_{cd} for gold and indium foils as a function of thickness. The purely thermal activation count of the foil then is obtained as

$$C_{th} = C - F_{cd} C_{cd} \tag{9-3}$$

Alternatively, to avoid flux depression by the presence of the cadmium, the thermal and epithermal contributions can be separated by irradiating simultaneously two foils, one made of a material with a large resonance integral, such as gold, and the other of a material whose thermal absorption is predominant, such as copper or manganese.

Threshold Foils

Certain nuclear reactions, such as (n,2n) reactions and most (n,p), (n,d), (n,α), (n,f) reactions occur only when the incident neutrons have enough energy, of the order of several MeV, to overcome binding energies and the Coulomb barrier to permit ejection of a massive nuclide from the compound nucleus. In the form of foils, such threshold reactions can be used to determine the flux, energy distribution and relative dose due to fast neutrons. The cross sections for such reactions are typically of the order of tens of millibarns, occasionally higher. The threshold energy, when the reactions can first occur, usually corresponds to a situation where there is barely enough excess energy for the product particle to escape and the cross section is hardly measurable. Usually the cross section rises rapidly with energy above the threshold, reaches a maximum and then falls off gently at yet higher energies. The energy at which the cross section attains a value close to the average over a broad energy spectrum is defined as the "effective threshold energy" and the corresponding cross section is the *effective* cross section. Table 9.2 lists some of the useful threshold detectors; it is evident that the effective threshold energy, E_{eff}, in some cases lies several MeV above the nominal threshold energy, E_s.

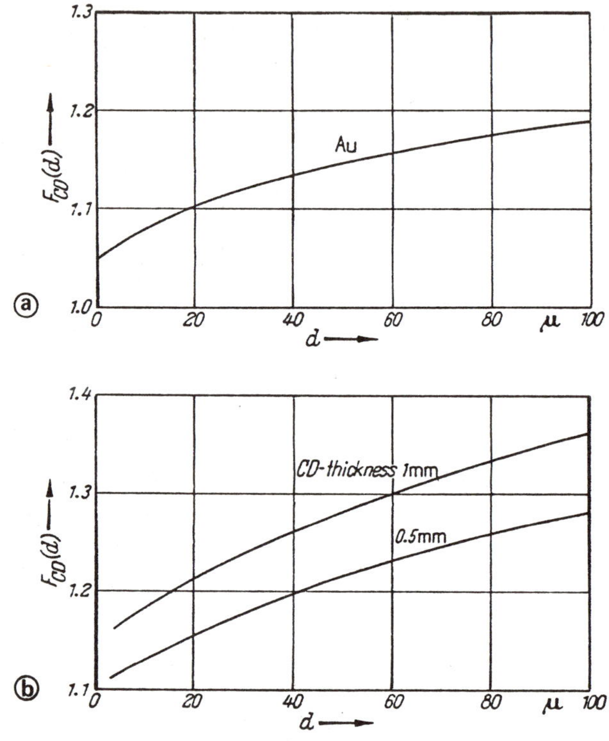

Figure 9.3. Cadmium correction factors in an isotropic neutron field as a function of foil thickness: (a) gold; 1-mm Cd filter; (b) indium, 0.5- and 1-mm Cd filters.

Sulfur-32, used in the form of pressed pellets, is most important as the threshold detector for the 3-14 MeV region. Between 0.6 and 2 MeV fission detectors are particularly useful, but their fabrication and handling is more complicated and expensive than that of other threshold detectors. Nevertheless, if coupled with a suitable computer program, they provide one means of reconstructing a fast-neutron spectrum. For precise neutron spectrum determinations other methods, such as proton recoil evaluation and time-of-flight methods, must be employed.

TIME-OF-FLIGHT MEASUREMENTS

One way of determining the energy spectrum of neutrons is by utilizing their finite velocity. A thermal neutron at room temperature travels at an average speed of about 2,200 m/sec, well within the range of measurable time intervals. If the starting time of a bunch of neutrons can be determined accurately, their time of arrival at the end of a carefully measured flight path yields a direct measure of their velocity, and, hence, of their energy.

If the neutron detector is at a distance d from the source, the arrival time for neutrons of energy E is

$$t = \frac{d}{v} = \frac{d}{\sqrt{2E/m}} \tag{9-4}$$

In practice the detector is connected to a multichannel analyzer, in which each channel registers events occurring in incremental time intervals. If the flight path d is expressed in meters, the flight time in μsec, and the neutron energy in eV, then

$$t = \frac{72.3d}{\sqrt{E}} \tag{9-5}$$

The accuracy of the energy measurement depends on uncertainties in the flight time determination, Δt, and in the flight path, Δd, though the latter uncertainty is usually negligible.

Then the uncertainty in energy ΔE can be expressed as

$$\begin{aligned} \Delta E &= \left(\frac{\partial E}{\partial t}\right) \Delta t = 2E \frac{\Delta t}{t} \\ &= 0.028 \frac{\Delta t}{d} E^{3/2} \end{aligned} \tag{9-6}$$

for the conditions in Equation 9.4. For high energies, flight paths of several hundred meters are required, ideally in evacuated pipes, with energy uncertainties ΔE/E as low as 5×10^{-4} for 100-eV neutrons.

A great deal depends on the means of generating precisely a burst of neutrons. Usually this is done by means of a rotating shutter, called a "chopper," or by means of a sharply pulsed accelerator. A chopper typically consists of a cylinder pierced by a series of cadmium-lined channels, preferably curved for good energy resolution, which is rotated rapidly while producing a synchronous timing pulse for the detector system. Figure 9.4 illustrates such a system. Alternatively, a rotating single crystal may be used as a monochromator.

For fast neutrons the time-of-flight method is limited mainly by the availability of unobstructed flight paths and by difficulties in precise timing of emerging neutron pulses.

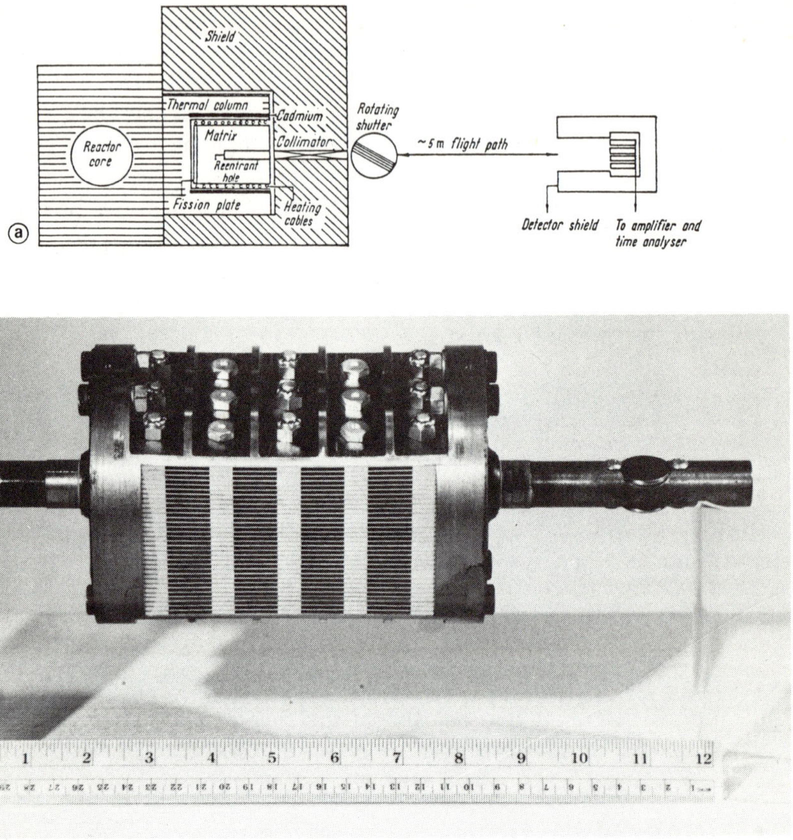

Figure 9.4. Neutron chopper systems: (a) Stone and Slovacek's set-up; (b) Details of chopper rotor assembly (Lake, 1973).

MEASUREMENT OF FAST NEUTRONS

As mentioned before, one easy means of measurement of fast-neutron fluxes is by slowing them down consistently and then measuring the moderated neutrons. If that approach is not acceptable, fast-neutron reactions can be employed in otherwise conventional systems, though at a lower efficiency imposed by the generally lower cross sections.

These methods include:

a. fission ionization chambers,
b. helium-3-filled proportional counters,
c. plastic scintillators and Hornyak buttons,
d. track-etch detectors, and
e. semiconductor diodes.

Figure 9.1 has illustrated a fission ionization chamber for reactor flux measurements. Its energy response depends on the fissile material chosen. Np-237 with a threshold at 0.32 MeV, U-238 at 1.3 MeV and Th-232 at 1.5 MeV are widely used. Chambers are generally filled with argon and operated in the ionization chamber regime. Because their sensitivity is limited they are often employed in high-intensity reactor fields where other detectors would overload and they offer fair discrimination against gamma radiations. Their response is fairly flat over the spectral range of fission neutron energies.

For spectrum determinations in the 200 keV to 2 MeV range, proportional counters filled with helium-3 yield pulses whose height is proportional to E + Q, where Q = 764 keV is the reaction energy of the $^3He(n,p)^3H$ reaction.

Helium-3 has some ideal characteristics for neutron detection.

1. The absorption cross section, 5,400 barns at 0.025 eV, varies smoothly over the energy range 0.001 eV to 0.04 eV, showing no resonance. (See Figure 9.5.) The (n,p) reaction has a cross section of 900 mb at 1-2 MeV.
2. Very high neutron sensitivity can be achieved by increasing gas pressure, which does not increase required operating voltage.
3. Of significant consideration regarding helium-3-filled counters is the effect of neutron reactions which occur near the counter walls (wall effects). Since reaction products travel on paths 180° apart and since helium-3 has a relatively low stopping power, many of the reactions near the counter wall do not contribute their full energy to the output pulse. These pulses do not come under the peak in the pulse spectrum from the counter. There are three means of overcoming this effect: (a) use of a large diameter counter in which the proportion of events near the counter wall is small; (b) increase in helium-3 pressure, thereby improving stopping power; (c) addition of a small percentage of a gas such as krypton, which has a far higher stopping power and is far less costly than helium-3.
4. In improving the stopping power of the gas, pulse rise time is decreased, while gamma sensitivity and jitter time are increased. Low jitter time (the time required for all the charge produced to be collected at the anode) is frequently a major requirement for time-of-flight applications.
5. Figure 9.5 shows the increased efficiency of a 1-in. (2.5-cm) diameter detector with increased helium-3 pressure. It should be noted that, for thermal neutron detection, filling pressure beyond 4-5 atmospheres does not provide appreciable increase in efficiency because of the self absorption of neutrons. This is not true for higher energy neutrons which is the major reason for using high pressure filling.

Solid-State Detectors

Fast neutrons interacting with polystyrene in organic solids will produce recoil protons which can cause detectable events if the target material con-

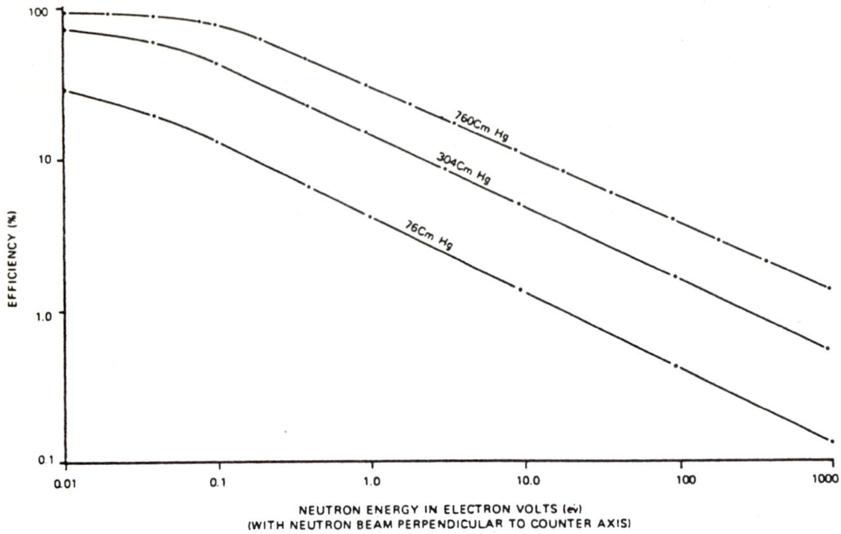

Figure 9.5. Efficiency curves for helium-3-filled proportional counters for neutron energies from 0.01 to 1,000 eV.

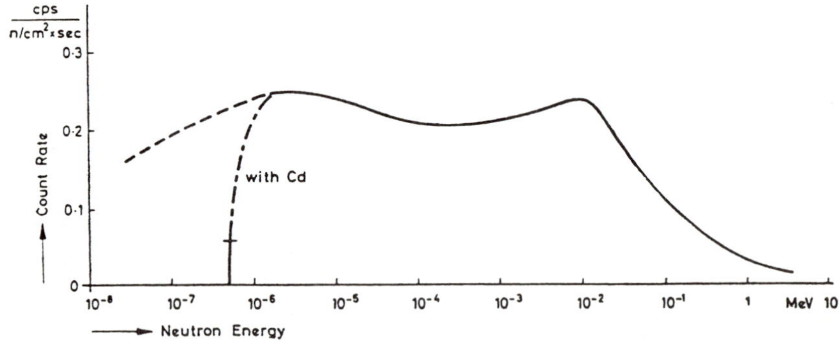

Figure 9.6. Energy response curve of boron-ZnS-loaded plastic scintillator in a 6.5-cm-diameter paraffin sphere, with and without Cd shielding (after Basson).

stitutes a plastic scintillator or a track-etch detection foil. A simple proton scintillator that is quite insensitive to gamma radiation can be produced by incorporating zinc sulfide in transparent Lucite or Perspex (Hornyak Button); other scintillating plastics have been prepared with good fast-neutron sensitivity. Figure 9.6 illustrates the fairly flat energy response of a scintillator plastic loaded with ^{10}B and ZnS and surrounded by a 6.5 cm diameter paraffin sphere.

Other scintillator systems using either large plastic phosphors or large tanks filled with scintillating solutions have been employed for the detection of high-energy neutrons or neutrinos by employing time and amplitude discrimination of recoil-induced pulse events.

Another approach that is useful for portable neutron dosimeters is based on the radiation damage induced by fast neutrons in semiconductor diodes. Figure 9.7a shows the change in forward voltage sensitivity with neutron dose; it is seen that the calibration is somewhat sensitive to ambient temperature. The neutron dose is computed and displayed in a digital system outlined in Figure 9.7b. The radiation damage is permanent and cumulative and the system is intended for high dose ranges; accuracy is of the order ± 20%.

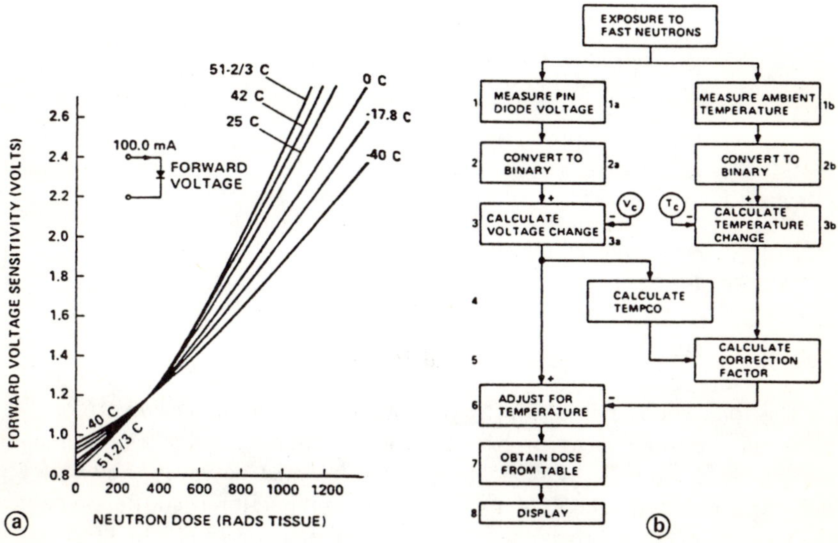

Figure 9.7. Direct-indicating digital neutron dosimeter using p-i-n semiconductor diodes (Ramondetta and Groeber, 1978): (a) dose calibration curves for 50-mil n-type PIN diodes; (b) flowchart for built-in computer calculation.

In summary, Table 9.3 compares some of the more important methods for the determination of neutron energy spectra, while Figure 9.8 compares their detection sensitivities.

Table 9.3 Survey of Methods for the Determination of a Neutron Energy Distribution (after Kiefer and Maushart)

Reaction	Measuring Detector	Reading	Energy Range	
H (n, p)	Proportional counter	Pulse-Height Distribution	Directional	0.1-2 MeV
	Plastic scintillator			0.7-20 MeV
	Nuclear emulsion			0.3-20 MeV
$^{6}Li\ (n, \alpha)^{3}H$	^{6}LiI(Eu) crystal			1-14 MeV
	^{6}LiF between solid-state detectors			1-14 MeV
	^{6}Li-loaded nuclear emulsion			0.4-0.9MeV
$^{3}He\ (n, p)^{3}H$	^{3}He proportional counter			
	^{3}He between solid-state detectors			0.1-1 MeV
Threshold Detector	Activation detectors Au + Cd, Au, Pu, U, Np(nf), S(n, p), ^{12}C(n, 2n), ^{209}Bi(n, f)	Activity Measurement	Nondirectional	Subdivision into energy ranges
	Variable ^{10}B-absorber layer around fission chambers			
	^{235}U-fission chamber			
	^{239}Pu-fission chamber	Pulse Rate		
Moderator-type Detector	Moderator combination 5, 7.5 and 20-cm diam. paraffin wax spheres around ^{6}LiI crystal			n_{th}-8 MeV

The "Long Counter" of Hanson and McKibben

The long counter is a directional counter system for fast neutrons. It relies on the moderation of fast neutrons to thermal energies at which they are absorbed strongly by ^{10}B atoms. The original design is shown in Figure 9.9. It consists of a BF_3 detector inside a moderator (paraffin or polyethylene) which itself is surrounded by a thermal-neutron shield to keep out any initially low-energy neutrons. Such detectors have been adapted to the calibration of neutron sources, with an internally mounted neutron source. The count rate, when exposed to an external neutron source, is given by

$$C = a + \frac{b}{(x + c)^2} \tag{9-7}$$

where a is a measure of response to room-scattered neutrons, b is a measure of the flux density sensitivity, x is the distance of the center of the source to the front of the counter system, and c depends on the location of the

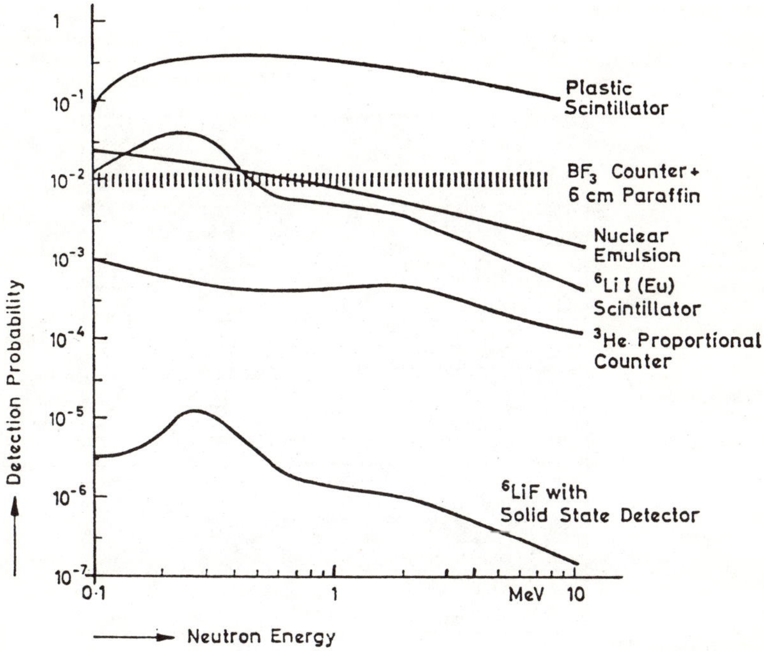

Figure 9.8. Comparison of optimum detector probability of various fast neutron detectors (after Wallace, 1963).

effective counter volume in the assembly and is to some extent a function of the mean neutron energy. The efficiency of the detector is essentially independent of energy over the range from 10 keV to about 3 MeV, aided by the cylindrical holes drilled into the moderator which increase epithermal neutron penetration. The absolute sensitivity is of the order of 200-300 counts/min per unit neutron flux (n/cm^2-sec).

Neutron Rem-Meters

When a thermal-neutron detector is surrounded by a paraffin sphere of at least 6-in. (15-cm) radius, it is found to give a roughly dose-equivalent response over a wide neutron energy range (thermal to 15 MeV) even if calibrated at only one energy. This property has given rise to a whole family of related instruments that are collectively called Rem-meters. They have the advantages of approximately dose-equivalent response, the ability to measure low dose rates (about 0.5 mrem/hr), and little or no geometrical dependence. They are available in various forms as portable instruments or fixed-area monitors.

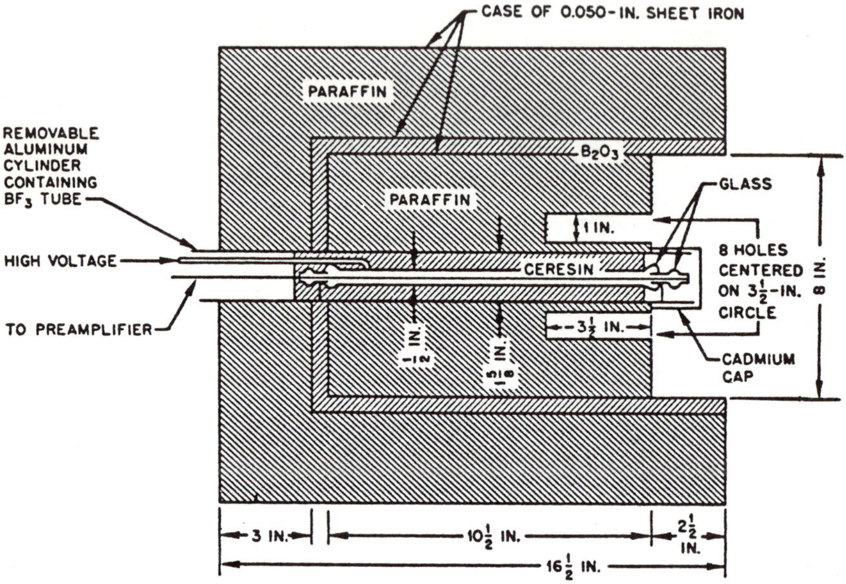

Figure 9.9. The Long Counter of Hanson and McKibben.

Figure 9.10 shows the principal features of such an instrument. The detector is usually a LiI(Eu) scintillation crystal, utilizing the ^{6}Li(n,α) reaction, though occasionally BF_3 proportional counters are used. The detector is surrounded by a moderator material, usually polyethylene or paraffin wax. They are fairly heavy, movable and must be handled with care. The energy response can be flattened by introducing an internal absorber, in a geometry similar to the long counter, consisting of a perforated boron plastic cylinder or cadmium foil surrounding the detector. These methods have been discussed by Patterson and Thomas. The use of a small BF_3 counter instead of LiI(Eu) tends to reduce the detector's sensitivity to gamma-rays.

Multisphere Detectors (Bonner Spheres)

Investigations by Bonner and his collaborators have shown that a similar system, with a small ^{6}LiI(Eu) crystal placed in the center of a series of polyethylene spheres, ranging in diameter from 5 to 30 cm, can be used as a method of neutron spectrometry. Figure 9.11 shows the moderating effects

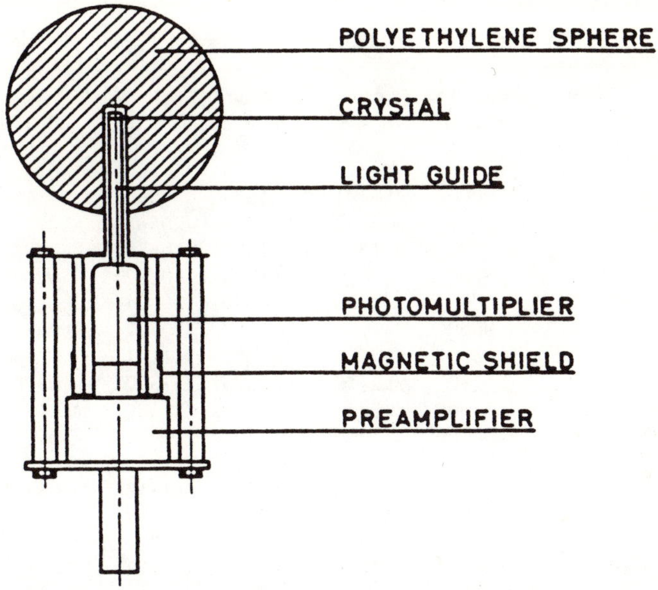

Figure 9.10. Diagram of single-sphere rem-meter without an internal absorber (after Nachtigall).

of five different sphere diameters from 2 in. (5 cm) to 12 in. (30 cm) over a wide neutron energy range. The neutron dose equivalents, in rem/unit fluence, superimposed as a shaded curve indicates that a 25-cm sphere detector approximates the dose response curve for energies up to 5 MeV.

For the purpose of using the detector as a spectrometer, one has to know the response of each sphere over the range of energies. Table 9.4 lists the response factors calculated by O'Brien, Sanna and McLaughlin for a series of spheres with a 4-mm x 4-mm crystal. The count rate for each detector has to be related to the energy spectrum by a mathematical unfolding technique, several of which have been reported in the literature. The Bonner ball system is less convenient and bulkier than are threshold detectors; however, it permits measurements on the differential neutron spectrum over a wider range of energies.

In-Core Detectors

A special situation for neutron detection exists for measurements inside nuclear reactors. In this case the detector has to operate in a particularly unfavorable environment owing to the elevated temperatures prevailing, the

Table 9.4 Response Matrix for Bonner Sphere Counting System[a]

Neutron Energy (eV)	Counting Efficiency (counts per neutron per square centimeter)						
	Bare Detector	Cd Covered Detector	2-in. Sphere	3-in. Sphere	5-in. Sphere	8-in. Sphere	12-in. Sphere
1.0 (-2)[b]	0.1220	0.0000	0.1372	0.1059	0.0540	0.0156	0.0024
1.6 (-2)	0.1220	0.0000	0.1401	0.1087	0.0560	0.0162	0.0025
2.5 (-2)	0.1220	0.0000	0.1457	0.1140	0.0588	0.0171	0.0026
4.0 (-2)	0.1180	0.0000	0.1513	0.1225	0.0639	0.0185	0.0029
6.3 (-2)	0.1160	0.0000	0.1620	0.1446	0.0712	0.0207	0.0032
1.0 (-1)	0.1140	0.0000	0.1760	0.1530	0.0824	0.0240	0.0037
1.6 (-1)	0.0110	0.0000	0.1977	0.1787	0.0975	0.0285	0.0044
2.5 (-1)	0.1020	0.0000	0.2207	0.2050	0.1141	0.0333	0.0051
4.0 (-1)	0.1160	0.1160	0.2410	0.2326	0.1327	0.0386	0.0059
6.3 (-1)	0.1100	0.1100	0.2553	0.2560	0.1480	0.0433	0.0065
1.0 (0)	0.0840	0.0840	0.2560	0.2701	0.1605	0.0479	0.0071
1.6 (0)	0.0760	0.0760	0.2532	0.2809	0.1710	0.0517	0.0080
2.5 (0)	0.0680	0.0680	0.2480	0.2853	0.1818	0.0541	0.0082
4.0 (0)	0.0600	0.0600	0.2382	0.2872	0.1897	0.0572	0.0088
6.3 (0)	0.0520	0.0520	0.2257	0.2880	0.1971	0.0598	0.0091
1.0 (1)	0.0420	0.0420	0.2121	0.2877	0.2033	0.0617	0.0096
1.6 (1)	0.0360	0.0360	0.1991	0.2847	0.2904	0.0647	0.0100
2.5 (1)	0.0280	0.0280	0.1890	0.2800	0.2150	0.0675	0.0105
4.0 (1)	0.0200	0.0200	0.1767	0.2743	0.2203	0.0707	0.0111
6.3 (1)	0.0100	0.0100	0.1630	0.2672	0.2252	0.0732	0.0113
1.0 (2)	0.0020	0.0020	0.1528	0.2608	0.2292	0.0763	0.0117
1.6 (2)	0.0000	0.0000	0.1427	0.2535	0.2236	0.0789	0.0122
2.5 (2)	0.0000	0.0000	0.1318	0.2451	0.2349	0.0816	0.0127
4.0 (2)	0.0000	0.0000	0.1201	0.2362	0.2357	0.0829	0.0129
6.3 (2)	0.0000	0.0000	0.1106	0.2227	0.2363	0.0842	0.0130
1.0 (3)	0.0000	0.0000	0.1013	0.2187	0.2375	0.0865	0.0132
1.6 (3)	0.0000	0.0000	0.0935	0.2107	0.2392	0.0901	0.0142
2.5 (3)	0.0000	0.0000	0.0860	0.2050	0.2409	0.0935	0.0150
4.0 (3)	0.0000	0.0000	0.0785	0.1915	0.2412	0.0956	0.0153
6.3 (3)	0.0000	0.0000	0.0703	0.1850	0.2418	0.0983	0.0158
1.0 (4)	0.0000	0.0000	0.0655	0.1780	0.2423	0.1023	0.0167
1.6 (4)	0.0000	0.0000	0.0594	0.1707	0.2445	0.1069	0.0171
2.5 (4)	0.0000	0.0000	0.0535	0.1625	0.2453	0.1106	0.0181
4.0 (4)	0.0000	0.0000	0.0470	0.1532	0.2474	0.1178	0.0197
6.3 (4)	0.0000	0.0000	0.0421	0.1457	0.2499	0.1266	0.0220
1.0 (5)	0.0000	0.0000	0.0380	0.1372	0.2536	0.1402	0.0256
1.6 (5)	0.0000	0.0000	0.0322	0.1258	0.2591	0.1582	0.0312
2.5 (5)	0.0000	0.0000	0.0247	0.1120	0.2644	0.1792	0.0396
4.0 (5)	0.0000	0.0000	0.0178	0.0950	0.2641	0.2063	0.0533
6.3 (5)	0.0000	0.0000	0.0127	0.0788	0.2520	0.2356	0.0745
1.0 (6)	0.0000	0.0000	0.0085	0.0600	0.2310	0.2705	0.1040
1.6 (6)	0.0000	0.0000	0.0060	0.0390	0.2050	0.2720	0.1500
2.5 (6)	0.0000	0.0000	0.0038	0.0287	0.1550	0.2640	0.1856

Table 9.4, continued

Neutron Energy (eV)	Counting Efficiency (counts per neutron per square centimeter)						
	Bare Detector	Cd Covered Detector	2-in. Sphere	3-in. Sphere	5-in. Sphere	8-in. Sphere	12-in. Sphere
4.0 (6)	0.0000	0.0000	0.0024	0.0191	0.1153	0.2380	0.2067
6.3 (6)	0.0000	0.0000	0.0013	0.0130	0.0685	0.1950	0.1995
1.0 (7)	0.0000	0.0000	0.0010	0.0074	0.0563	0.1415	0.1742
1.6 (7)	0.0000	0.0000	0.0003	0.0041	0.0337	0.0992	0.1420
2.5 (7)	0.0000	0.0000	0.0002	0.0020	0.0205	0.0737	0.1141
4.0 (7)	0.0000	0.0000	0.0001	0.0010	0.0130	0.0476	0.0853
6.3 (7)	0.0000	0.0000	0.0000	0.0004	0.0056	0.0265	0.0543
1.0 (8)	0.0000	0.0000	0.0000	0.0003	0.0087	0.0152	0.0266
1.6 (8)	0.0000	0.0000	0.0000	0.0001	0.0001	0.0035	0.0150

[a]O'Brien, K., Sanna, R., and McLaughlin, J.,Inference of Accelerator Stray Neutron Spectra from Various Measurements, U.S.A.E.C. First Symposium on Accelerator Radiation Dosimetry and Experience, USAEC-CONF-651109, p. 346 (1965).

[b]Digit in parentheses denotes power-of-ten multiplier.

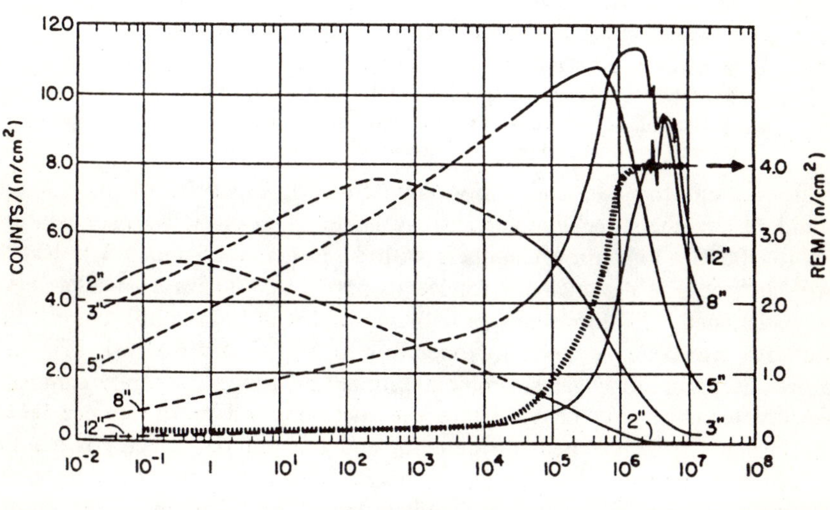

Figure 9.11. Response curves for Bonner spheres for different-diameter polyethylene spheres (Bramblett *et al.,* 1960). Shaded curve shows corresponding neutron dose equivalent (NCRP, 1957).

intense neutron flux levels, requiring relatively low sensitivity, the steady burnup of converter material resulting in gradual loss of efficiency, the need to discriminate against high levels of gamma radiation, the possibility of radiation damage to the detector and its associated components, such as cables and preamplifiers, and the need for an unusually high degree of reliability for operation over very long periods, up to 30 years.

No one detection system satisfies ideal requirements of this type; one must either choose locations outside the reactor core proper to retain sensitivity and minimize radiation damage or one must select as simple a system as possible that can function inside the core. Gas-filled detectors in general will not function in very high radiation fields as their filling gas would remain permanently ionized. Conventional cable insulation fails at high temperatures and most organic insulators break down in high radiation fields.

Among gas-filled detectors, compensated ionization chambers and fission-counter proportional detectors have been developed for use inside the reactor vessel. The principles of operation are the same as for normal ionization chambers and proportional counters; however, the major design changes needed to make conventional ionization chambers and fission counters suitable for in-vessel use are:

1. replacement of organic insulation with inorganic insulation;
2. replacement of aluminum parts with stronger, higher melting point materials when temperature requirements or corrosion conditions make aluminum unacceptable;
3. increasing clearances to accommodate expansion of various parts;
4. replacement of standard coaxial or triaxial cable and connectors with high-temperature connectors and cables or with integral cables; and
5. application of neutron-sensitive electrode coatings that will survive thermal cycling over a wide temperature range.

Ionization chambers and fission counters with sensitivities comparable to standard chambers are commercially available for operation at temperatures up to 300°C. Ionization chambers with a sensitivity of about 5×10^{-14} A/n/cm^2-sec and designed for service to 550°C at 600 psi(4.14 MPa) have been designed. Fission counters with sensitivity of about 0.01 counts/n-cm^2-sec have also been designed for operation at 550°C and 600 psi. To improve discrimination against ambient gamma radiation, a separate gamma-sensitive chamber can be built into the enclosure and the difference between the currents from the neutron detector and the gamma detector is measured to obtain a neutron flux signal. Figure 9.12 illustrates the principle of a gamma-compensated ionization chamber. Some problems may arise from neutron activation of chamber materials which may contribute to the detected signal. Care must obviously be taken in selecting chamber materials

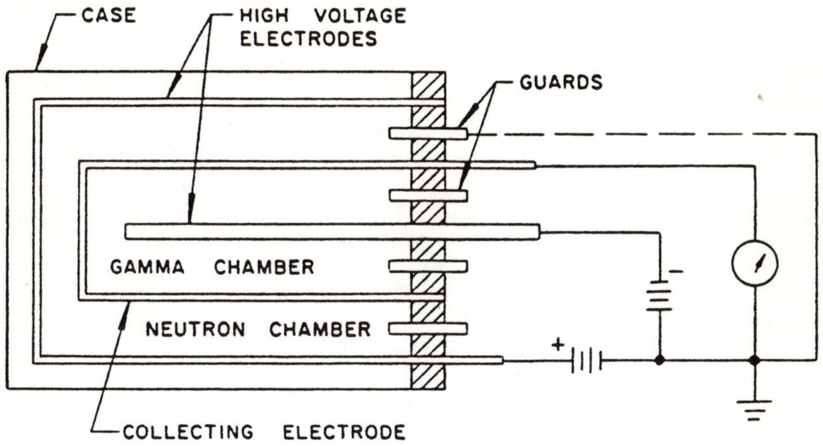

Figure 9.12. Schematic diagram of gamma-compensated ionization chamber and associated circuit (after Boland).

of low cross section and short activated half-lives to minimize this effect. Ionization chambers are the fastest instruments available for reactor measurements and are capable of measuring a wide range of flux levels.

In fission chambers one of the electrodes is coated with a fissionable material which converts the incident neutrons into heavily ionizing radiation. The chambers are usually operated in a pulsed mode and the ac signal produced can be readily amplified. The external circuit must be able to discriminate between signals produced by fission events and other ionizing radiations or noise.

For in-core detection where high gamma-ray levels, burnup and radiation damage limit the usefulness of the gas-filled chambers, other methods have been developed. These include:

- thermocouple measurements
- self-powered (Hilborn) detectors, and
- gamma and neutron thermometers.

All of these are dc measurements with essentially passive solid-state devices. The thermocouple systems employ absorber materials, usually a fissile material, for the measurement of transient neutron flux variations. If the sensitivity of the absorber to gamma radiation is negligible compared with that for neutrons, and if the heat losses between the fissile element and surrounding materials are small during the period of irradiation, then the time-dependent neutron flux $\phi_{(t)}$ is related to the thermocouple output voltage ΔV as

$$\phi_{(t)} = \frac{c}{K\sigma_f NQ} \Delta V \qquad (9\text{-}8)$$

where σ_f = microscopic fission cross section
N = number of fissile atoms in absorber
c = heat capacity of absorber element
Q = heat absorbed per fission
K = thermocouple constant (temp-voltage)

In practice, heat losses occur along the thermocouple wires and there is a finite response time of the order of 0.2-2 msec in responding to step changes in reactor power or neutron flux, due to the finite heat capacity of the detector assembly.

Another widely employed neutron detector for monitoring in-core radiation levels is the Hilborn detector. In this class of detector a positive charge is produced on an electrode by the emission of energetic electrons during exposure to radiation. This emission of electrons is equivalent to a weak current that can be detected by means of a charge-sensitive electrometer or high-impedance femtoammeter. Consequently, no external voltage source is required, giving rise to the name *self-powered detectors.* Electron emission may be due to neutron capture followed by beta decay, neutron capture followed by electron emission from excited nuclei, or Compton and photoelectron emission after absorption of energetic gamma radiation. Most practical devices of this type are based on internal conversion (n,β) reactions; ideally this process should be prompt with minimum delay times. If burnup of the emitter material is minimal and if radioactive decay occurs with a single decay constant, the device current I, in amperes, is given by

$$I(t) = K\,N\,\sigma\,\phi\,q\,(1 - e^{-\lambda t})$$

where N = the number of emitter atoms
σ = neutron activation cross section
ϕ = neutron flux
q = electric charge per neutron capture
λ = decay constant for beta emission
K = detector constant

The detector constant K depends on geometry, neutron self-shielding and beta self-absorption. In practice, corrections must be applied for burnup of emitter atoms, flux nonuniformity, radiation damage to the insulator and parasitic neutron absorption. Figure 9.13 schematically shows the construction of such devices. In its usual form, the detector consists of a coaxial cable with a central emitter electrode, oxide insulation and an outer collector electrode. The detector may have active lengths of 3-20 cm and outside diameters of 1.2-3 mm. The detector is usually fabricated as an integral assembly with its connecting cable. Table 9.5 presents the characteristics of some practical emitter materials. Rhodium, vanadium and cobalt are the most popular materials; aluminum, cadmium and scandium have been proposed as well. Platinum emitters have been used when combined neutron and gamma-ray sensitivity was desired. Allowing for burnup

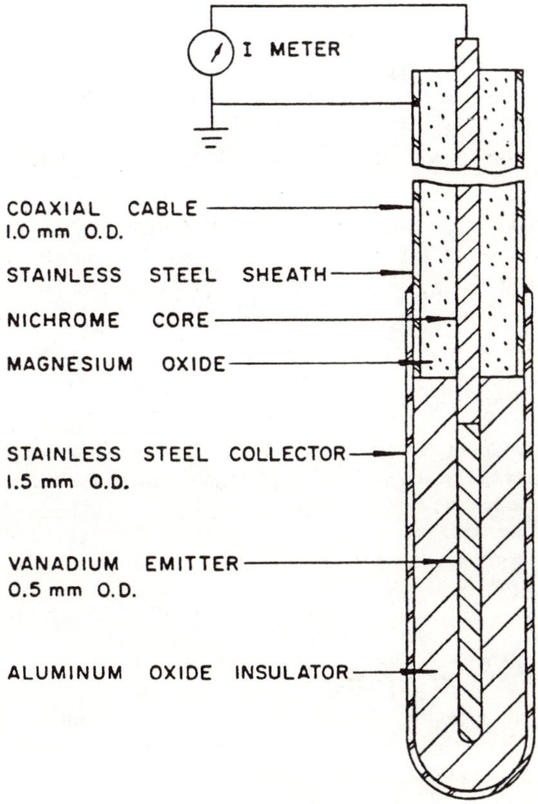

Figure 9.13. Schematic diagram of self-powered detector.

Table 9.5
a. Characteristics of Emitter Materials

Element	Thermal Neutron Cross Section (barns)	Half-Life (sec)	Maximum β Energy (MeV)	β-Current Generated (A/nv-g)	Burnout Rate in 10^{13} nv (%/ month)	Neutron Sensitivity[a] (A/nv-cm)
^{7}Li	0.033	0.85	13.0	4.5×10^{-22}		
^{27}Al	0.23	158	2.9	8.2×10^{-22}		
^{51}V	4.5	226	2.6	8.5×10^{-21}	0.013	7.7×10^{-23}
^{103}Rh	150	42[b]	2.4	1.5×10^{-19}	0.23	1.2×10^{-21}
^{59}Co	37[c]	63			0.094	1.7×10^{-23}
Cd	2500[c]	3,000			1.3	1.6×10^{-22}
^{45}Sc	23[c]	20				

b. Typical Detector Specifications

Detector:			
Emitter material	Rh	V	Co
Emitter diameter (mm)	0.7	2.0	2.0
Insulator	Al_2O_3 (or MgO)	Al_2O_3 (or MgO)	Al_2O_3 (or MgO)
Collector material	Inconel	Inconel	Inconel
Sensitive length, mm (typical)	50	100	210
Outer detector diameter (mm)	2.5	3.5	3.5
Characteristics:			
Thermal neutron sensitivity A/nv	7.7×10^{-21}	4.8×10^{-21}	5.4×10^{-21}
Gamma sensitivity A/R/h	10^{-17}	4.0×10^{-17}	5.6×10^{-17}
Max. operating thermal neutron flux, n/cm^2-sec	10^{15}	10^{15}	10^{15}
Build-up of Co^{60} current			<0.5%/month
Insulation resistance (detector + 15 m cable			
out-of-core 20°C	$>10^{12}\Omega$	$>10^{10}\Omega$	$>10^{12}\Omega$
in-core 250°C	$>10^{8}\Omega$	$>10^{8}\Omega$	$>10^{8}\Omega$
Temperature, max (°C)	500	500	500
Response time (0-63%)	1.1 min	5.5 min	prompt
Burn-up rate at 10^{14} n/cm^2-sec	3.9% per month	0.12% per month	1.0% per month

[a]Values quoted are for 0.020 in. diameter emitter in 1/16 in. diameter sheath.
[b]^{103}Rh also has an 8% yield with a half life of 264 sec.
[c]Cross-section given is for total neutron interactions, not for production of beta rays; electron emission predominantly by prompt conversion processes.

of emitter material and assuming that output signals are read over times long compared with decay constant, the output characteristics can be expressed as

$$I_{(t)} = K \sigma \phi N_o e^{-k\sigma\phi t}$$

where N_o = the initial number of emitter atoms
k = self-shielding correction constant

Depending on the size and thermal capacity of the detector the response is relatively slow, of the order of seconds, in some cases. Their small size makes them easy to install.

Two devices for in-core measurements relying directly on heat conversion are the gamma thermometer and the neutron thermometer. Figure 9.14 illustrates a gamma thermometer. It consists of a container, with highly polished internal walls. filled with a gas of low thermal conductivity. When exposed to a major gamma-ray flux the gas is heated, mainly by Compton interactions, and the temperature rise is measured by the thermocouple, usually in reference to the coolant temperature. Such thermometers have the advantage of being simple and inexpensive to operate; they have no burnup limitations, and can provide highly local data. Disadvantages include relatively large size, slow response time, and dependence of the sensitivity on the heat sink temperature.

For neutron measurements a neutron thermometer has been developed. This consists of a small cylinder or disk of a nitride ceramic, such as

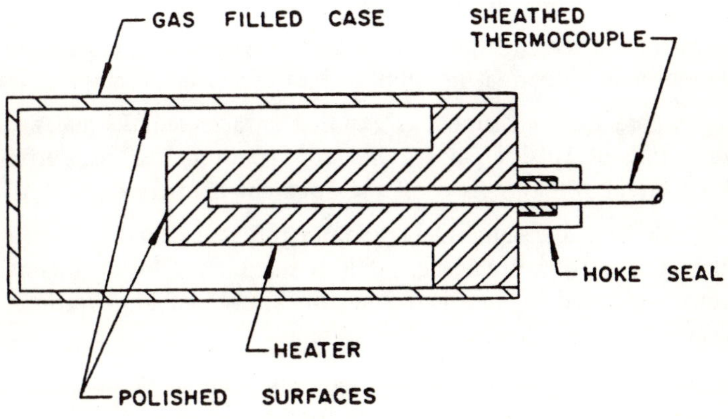

Figure 9.14 Diagram of gamma thermometer.

SiN, AlN or TiN cemented to a thermocouple. Neutrons are absorbed by the $^{14}N(n,p)^{14}C$ reaction, which has a thermal cross section of 1.8. barn and the protons are absorbed inside the cylinder, raising its temperature. Such a detector is relatively insensitive to gamma-rays, capable of operation at high temperatures and able to respond more rapidly than the self-powered detector, depending on cylinder size. In other respects, its sensitivity also depends on heat sink temperature; expected radiation damage effects are negligible.

REFERENCES FOR FURTHER READING

Attix, F. H., and E. Tochilin, Eds. *Radiation Dosimetry,* Vol. 3 (New York: Academic Press, Inc., 1969).

Beckurts, K. H., and K. Wirtz. *Neutron Physics* (New York: Springer-Verlag New York, Inc., 1964).

Boland, J. F. *Nuclear Reactor Instrumentation* (New York: Gordon & Breach, Science Publishers, Inc., 1970).

Griffith, R. V., D. E. Hankins, R. B. Gammage and L. Tommasino. *Recent Developments in Personnel Neutron Dosimeters—A Review.* Health Phys. 36:235-260 (1979).

International Atomic Energy Agency. *Neutron Fluence Measurements,* Technical Reports Series No. 107 (Vienna: IAEA, 1970).

Patterson, H. W. and R. H. Thomas. *Accelerator Health Physics* (New York: Academic Press, Inc., 1970).

PROBLEMS

9.1 Calculate the induced neutron activity in a 100-mg indium foil exposed to a thermal neutron fluence (nvt) of 10^{16}.

9.2 Estimate the useful lifetime of a BF_3 detector if it is exposed continuously in a thermal neutron flux of 5×10^{12} n/cm^{2}sec.

9.3 An in-core fission chamber is exposed to a mixed thermal neutron-γ-ray flux of 10^{13} n/cm^{2}sec and 10^{12} photons/sec, respectively. What effect will this have on the apparent neutron detection rate?

9.4 A 5 cm long, 0.7 mm thick rhodium rod is exposed to a neutron flux of 5×10^{13} n/cm^{2}sec. If it is surrounded by an Inconel collector separated by two half-value thicknesses of MgO, estimate the electron current flowing in the system.

9.5 A neutron thermometer employs a 1-cm right cylinder of SiN (50:50%) to determine neutron fluence. If heat losses along the thermocouple wires are negligible, estimate the heating of the cylinder in a flux of 10^{13} n/cm^{2}sec ($\sigma_{(n,p)}$ = 1.81 barn for N-14) .

9.6 Calculate the time of flight of 50-eV neutrons between detectors 50 m apart.

Assuming reasonable detector response times, how rapidly should such a beam be pulsed for detection with acceptable precision?

9.7 Construct a threshold-foil detection scheme to cover the fast-neutron range. Indicate the foils used and what corrections are needed to obtain comparable sensitivities over the full energy range.

9.8 A spherical neutron area-monitor employing an 8-in. (20-cm) paraffin sphere is calibrated with a Pu-Be source. If this sphere is replaced by a 5-in. (12.5-cm) sphere, what effect will this have on the efficiency and count rate in a reactor neutron field? How is the sensitivity to 0.1-eV neutrons affected by this change and why?

9.9 A plastic track-etch detector foil is exposed to a fast-neutron flux of 10^{13} n/cm^{2}sec. If the probability of producing a "bulls-eye" collision is 10^{-4} and it requires 10^5 knock-on protons to produce a developable track, estimate the track density after 30-days exposure.

9.10 A 10-mg californium-252 source is immersed in a 1-m^3 manganese sulfate bath. Estimate the optimum time of exposure of this source to obtain adequate activity in a 100-ml liquid sample in an appropriate detector. State assumptions made.

CHAPTER 10
CALIBRATION AND STANDARDS

INTRODUCTION

The reliability of any radiation measurement depends both on the inherent consistency of the measurement procedure as well as on the accuracy of the instruments used. In practice, all instruments have to be calibrated since the results typically appear as numerical multiples of some appropriate unit. Very few radiation measurements are conducted in absolute terms based on "*first principles*." Instead, almost invariably one relies on a meter reading or a comparison standard that can be used to *calibrate* the system, *i.e.*, establish a firm scale of measurement.

The quantities determined in most radiation measurements are intensity ("number flux"), energy, absorbed dose or source strength ("output" or "exposure"). The measurements employed to determine these quantities depend on the particular detector used, the type of radiation, the energy spectrum and the surrounding medium, and are usually sensitive to variations in ambient temperature, cosmic radiation background, atmospheric pressure, electrical supply variations, as well as to human factors.

Instruments measuring intensity or dose are usually calibrated by means of sources of known activity; secondary standards can be derived by intercomparing carefully handled instruments with others that have been calibrated under "standard conditions" or under conditions permitting recourse to first principles, usually at one of the national standards laboratories, such as the U.S. National Bureau of Standards (NBS) or the British National Physical Laboratory (NPL). These institutions will certify the calibration and accuracy of "secondary standard" instruments that, in turn, can be used to intercompare other instruments.

Most radiation equipment depends on "counting" pulses or integrating charge accumulated across a capacitor. One of the major sources of error in counting is the timing system, particularly if it depends on a mechanical clutch timer. For this reason electronic clock circuits are used increasingly. Timers can be checked by the use of a pulse generator which provides signal pulses of known frequency that can be fed into the counting system. Various precision pulse generators are available commercially which provide pulses of precise repetition frequency of accurate and adjustable voltage amplitude. In many cases pulse length can be adjusted as well to permit checking of dead time and the time resolution of scaling circuits and coincidence gating circuits. If provided with a delay circuit, such a system can help set up elaborate gating and time-of-flight calibrations. By varying the amplitude in fixed increments, the pulser can check and calibrate multichannel analysers for channel width, channel uniformity and discriminator stability.

SOURCE CALIBRATION

In X-ray therapy systems and in radioisotope technology it is usually necessary to check the radiation output of machines and sources at regular intervals. Although variations in output in theory can be monitored, for instance, by means of fixed or portable dosimeters, in many cases shifts in energy spectrum may occur that may make the meter indication unreliable because its own calibration basis may no longer apply.

Historically, most early measurements of source strength were ultimately referred to five carefully prepared radium sources, prepared by O. Hönigschmid in 1916, that have been distributed to five national standard laboratories and have been intercompared at long intervals. Those sources, being essentially gamma-ray emitters, contributed little to other types of radiation measurement.

For standardizing other sources, reference must be made to "absolute" counting methods, employing detectors of accurately known efficiency to count samples of accurately known activity. For alpha and beta emitters, this can be done in a 4π-geometry internal, gas-flow proportional counter of the type illustrated in Figure 10.1. The source, typically a carefully weighed carrier-free source, such as polonium-210 or phosphorus-32, is deposited on a thin foil of mylar in the middle of the detector volume. Corrections must be applied for self-absorption, backscatter from the support, secondary wall effects and residual impurities in the counting gas. Once standardized, the source can then be used to calibrate other detectors or other sources of comparable characteristics.

For beta emitters that have one or more γ-rays associated with each disintegration, β-γ coincidence measurements can greatly improve detector sensitivity and accuracy. In such a system, the 4π beta detector is placed above a NaI(Tl) scintillator detector. Connecting the outputs

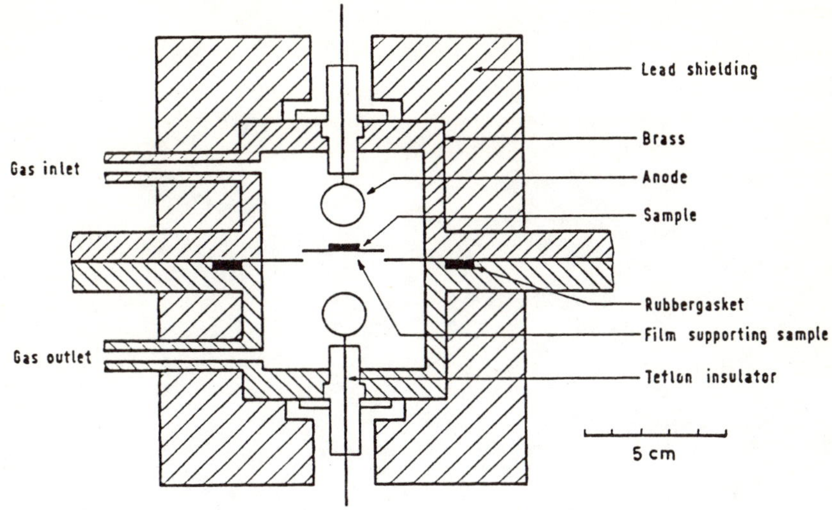

Figure 1
4 π β Counter

Figure 10.1 4π Absolute beta counter.

from both detectors, after amplification, in coincidence eliminates most background events that affect only one of the detectors. Corrections have to be made for accidental coincidences and dead-time losses, and the decay scheme for the radionuclide being measured must be known.

For short-lived gamma-ray sources, relative standardization can be done conveniently with a well-type ionization chamber, like the one shown in Figure 10.2. This detector can calibrate vials containing radioactive solutions, as well as solid sources. The source must be well centered, and its vertical position also should be established reproducibly.

If a certified secondary standard source is available, in the form of a planchet or *ampule,* detectors can be calibrated against them provided care is taken to provide reproducible positioning, to take into account backscatter from source containers and shield walls, and to stabilize all operating conditions, background, voltages, temperature and humidity to the greatest extent possible.

NEUTRON SOURCES

Neutron sources can be calibrated by means of a "manganese bath." In this method a source is immersed in a tank, 90-100 cm in diameter, that is filled with manganese sulfate solution that is kept agitated. The

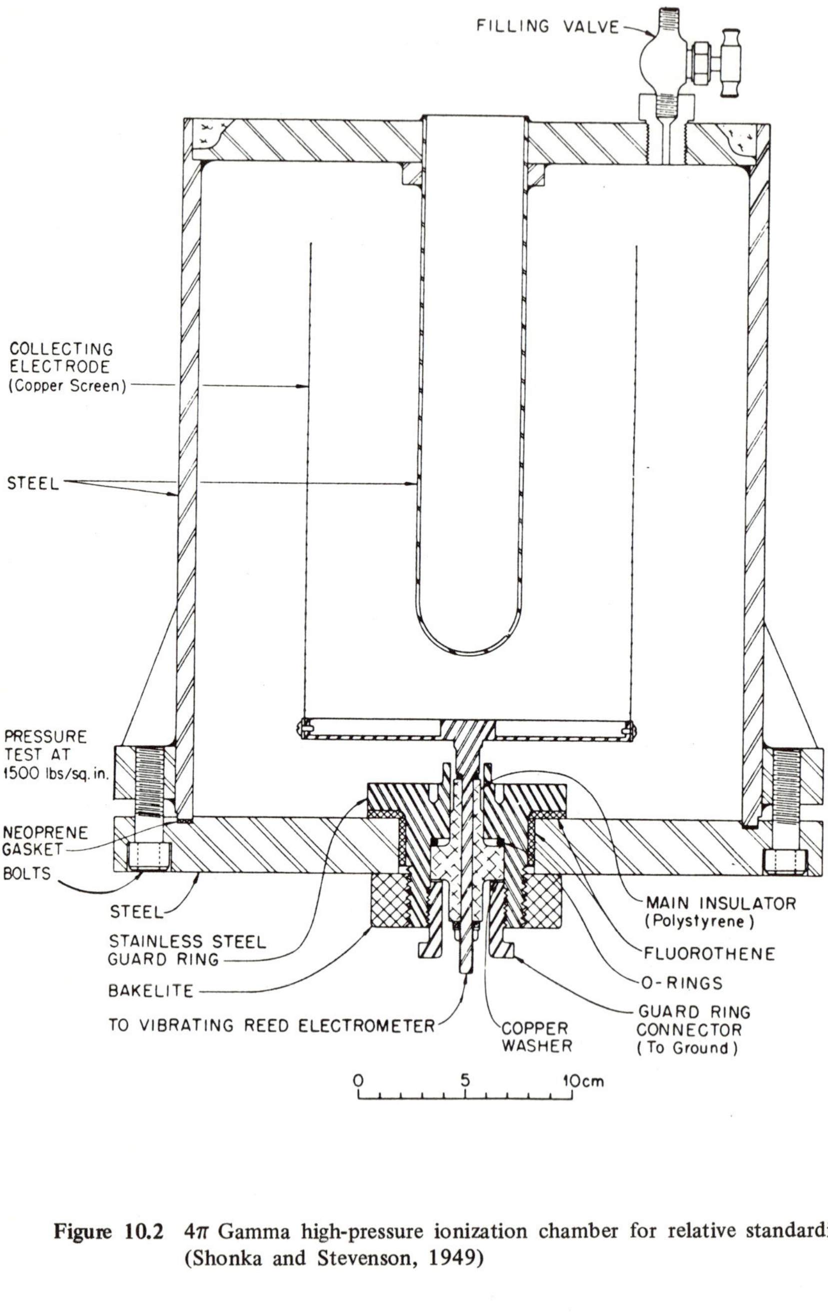

Figure 10.2 4π Gamma high-pressure ionization chamber for relative standardization. (Shonka and Stevenson, 1949)

neutrons are moderated in the solution and activate it, producing radioactive Mn-56. When this activity has built up to equilibrium, a portion of the solution is removed and the active manganese concentrated and counted in a calibrated detector. At equilibrium the source emission rate is

$$A = A_{Mn} \frac{N_{Mn}\,\sigma_{Mn} + N_S\sigma_S + N_H\sigma_H}{N_{Mn}\,\sigma_{Mn}} \tag{10-1}$$

where σ = thermal capture cross sections for Mn, S and H in the solution

N = concentrations.

With concentrations near saturation, half the neutrons are captured by hydrogen. Figure 10.3 illustrates the process.

For machine sources of neutrons using exhaustible tritium sources, it is important to monitor the neutron output constantly. This is done most conveniently by means of an "associated particle" detector, an example of which is shown in Figure 10.4. The principle of the method relies on the fact that in the $^2H(^3H,n)^4He$ [D(T,n)α] reaction both a neutron and an alpha particle are emitted simultaneously. Although the number of α particles reaching the detector is appreciably less than the number of neutrons emitted, because of self-absorption and geometry factors, there will still be a fixed relation between the two, so that the detector can give a fair indication of relative neutron output once the system has been calibrated, usually by an activation foil.

X-RAY MACHINES

Because of the importance of ensuring a known and well-defined radiation output from X-ray machines in clinical service, regular measurements must be conducted to determine the spectrum and the dose, usually expressed in Roentgens per minute at one meter. Johns and Cunningham have outlined one such procedure as follows.

Exposure Calibration of an X-Ray Generator

1. Select a treatment shielding cone and set it up horizontally to avoid scattering from walls and equipment. A 10 x 10 cm cone is a useful reference cone for a 250-kV machine; all other cones used should be measured and intercompared.

2. Set up a dosimeter chamber to be aligned at the center of the treatment field. The surface of the chamber should just touch the end of the cone.

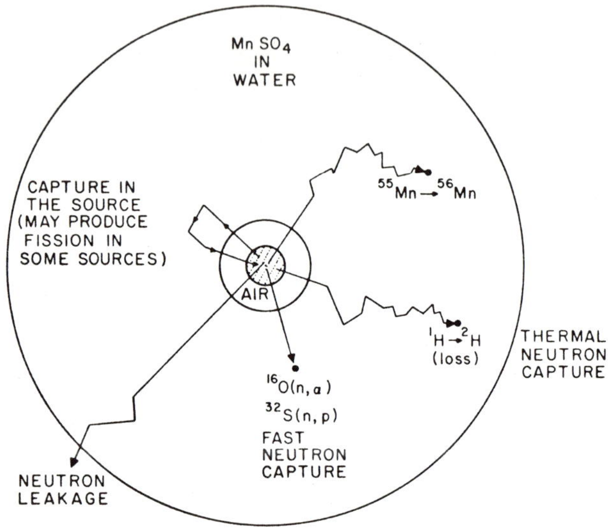

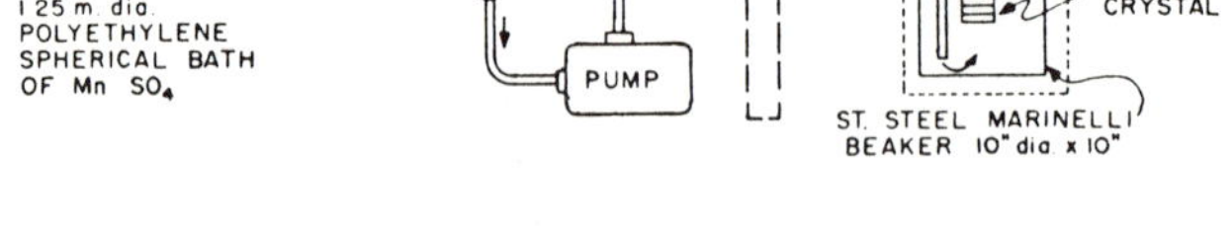

Figure 10.3 Manganous sulfate bath source calibration: (a) schematic of detection process; (b) circulating system and shielded γ-ray detector.

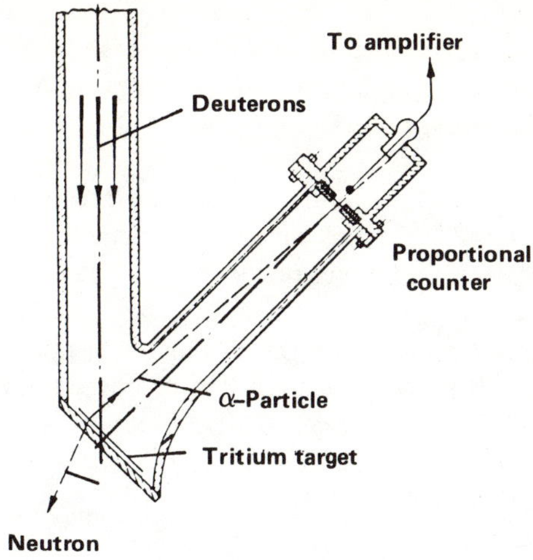

Figure 10.4 Associated particle detector system.

3. Choose the tube voltage, current and filter and determine the half-value layer for these conditions. If possible, adjust the filter to obtain an even value for the half-value layer to facilitate the use of standard depth dose tables.

4. Select the appropriate dosimeter chamber to match the quality of radiation indicated by the half-value layer. The chamber should be supplied with a calibration chart.

5. If the dosimeter is of the charger type, insert it into the charger, set the scale to zero and record the reading obtained after removal from the charger. Recheck this reading after a few minutes to detect any charge leakage.

6. Place the dosimeter in position, turn on the machine to the selected condition, set the timer to open the shutter for a specific exposure. Maintain the current manually at the predetermined value. The dose reading should be of the order of 60% of full scale; vary exposure times on repeated runs as necessary. Record the tube voltage.

7. Read the dosimeter in the reader/charger unit and record.

8. Repeat this procedure several times for both long and very short exposures, alternately, and tabulate all readings.

9. Correct all readings to the standard current and voltage conditions by assuming that for small variations dose is proportional to current and to the square of the voltage.

10. Check from the short exposure runs if any errors were introduced by the finite time it takes for the shutter to open. Correct the calibration exposure values for this if necessary.

11. Since the sensitive volume of the dosimeter is not exactly at the end of the cone, use the inverse square law to correct for this distance difference, measured from the target. This correction is important only for short source-to-skin distances. (For a long collimator, the inverse square law does not apply strictly.)

12. Other corrections: The chambers usually are calibrated at 22°C and 760 torr pressure. Any other conditions must be corrected for by the factor

$$\frac{760}{P}\left(\frac{273 + T}{295}\right)$$

In small chambers whose air volume is small, ionization in the stem may generate an appreciable portion of the ion current. This can be checked by shielding the stem and cap, unless the calibration chart shows that this effect was included at the time.

Similar procedures apply to the calibration of the exposures from a gamma-ray therapy machine though, of course, no current and voltage conditions exist. It must be remembered that at the higher energies involved, cone leakage and backscatter may contribute a substantially high stray field unless precautions are taken. The ion chambers employed should be sufficiently thick-wall to give equilibrium conditions and the calibration charts supplied with them must be obtained in a cobalt-60 field.

CALIBRATION OF DETECTION EQUIPMENT

Since many measurements of exposure and dose are done by means of direct-reading instruments, the nature and reliability of the indication is important particularly for personnel protection applications. Undue reliance on factory calibration can be disastrous, particularly if the detector is calibrated in one energy range but used in another.

Furthermore, both detector and amplifier system will deteriorate in time and frequent checks of the calibration against standard instruments or recalibration with standard sources are essential.

The following outline has been prepared by R. L. Kathren to provide general guidance for the handling, service and calibration of radiation monitoring devices:

An effective program can take many forms depending on the number of instruments to be handled over a fixed period of time. A major factor is the availability of funding for the establishment of permanent calibration facilities. These factors notwithstanding the objectives of such a program are relatively constant. The main objectives of a calibration program are:

a. to ensure sound functioning of the instrument;
b. to ensure that the measurements of the instrument are reliable and accurate in all ranges and for all energies in which the instrument is expected to provide a response;
c. to ensure further that the instrument functions reliably in the dose-rate range specified and that its energy-dependent and directional-dependent characteristics are fully understood; and
d. to confirm the performance of the instrument under extremes of environmental conditions in which it is certified to perform satisfactorily.

The latter two objectives are not easily satisfied in very simple calibration arrangements. These checks require rather specialized techniques and equipment which fit better in a centralized facility.

All radiation monitoring instruments (both stationary and portable) should be carefully calibrated as soon as they are received for use. The instrument should also be inspected for mechanical and electrical problems which might lead to a malfunction in the future. Another important factor in the use of radiation monitoring instruments is the need for regular calibration of the instruments at carefully predetermined intervals. Thus, in order to satisfy the objectives outlined above, a calibration program should have established procedures for the following:

a. scheduling of routine calibration for all instruments;
b. receipt and handling of the instruments;
c. calibration of test irradiation systems;
d. calibration of radiation monitoring instruments;
e. return of calibrated instruments to service or removal from service for repair;
f. reporting of results, calibration factors, etc., to the users; and
g. keeping of records of calibrations, repairs, manufacturers technical information, etc.

The keeping of records can be an extremely important aspect of such a program. For example, detailed records can identify those instruments (either by type or manufacturer) which perform above specifications and also those instruments in which chronic problems exist. Such information is valuable when considering the purchase of new or replacement instrumentation.

AN EVALUATION PROGRAM FOR PORTABLE INSTRUMENTS*

Proper selection of portable radiation monitoring instruments is as important as proper use. The ultimate objective, of course, is to procure instruments that will meet prescribed performance specifications with the lowest overall cost per unit.

Selection implies prior evaluation, and a meaningful evaluation must be based on knowledge of the performance characteristics, limitation and economics of the instrument under consideration. Unfortunately, adequate and accurate data on which to base a meaningful evaluation are unavailable in most instances. Perhaps the best available are the published specifications of the manufacturer, but these are often incomplete or couched in language that is not wholly comprehensible, and may even be misleading. And, since each manufacturer may report or measure the same performance characteristics differently, the published specifications of various manufacturers may not be comparable. Many instrument purchases, however, are based solely on the claims of the manufacturer or his representative, published or otherwise.

A second body of relatively accessible data is word-of-mouth from those who have had prior experience with a given instrument. In this case, the potential for misinformation is quite high; the spoken word gets notoriously garbled in transmission through several persons.

What is needed is a well-oriented and direct evaluation program to determine appropriate physical, electronic and radiological characteristics on which a performance and cost evaluation can be based. A typical evaluation program for portable radiation survey meters is outlined in Table 10.1. Obviously, no single program would fit all portable instruments now in use, or even those commercially available. However, the skeletal program in Table 10.1 should provide a sufficient body of information to enable the health physicist to make a sound selection,

---

*This section is based on a paper by Zuerner and Kathren in *Health Physics Operation Monitoring*, C. A. Willis and J. S. Handloser, Eds. (New York: Gordon & Breach, Science Publishers, Inc., 1978).

Table 10.1 Typical Evaluation Factors for Portable Radiation Survey Meters (after Zuerner and Kathren)

I. MECHANICAL
   A. Physical Construction
   B. Shock and Moisture Resistance
   C. Human Factors Engineering

II. ELECTRONIC
   A. Power Supply
      1. Stability
      2. Temperature Dependence
      3. Battery Life
   B. Input Sensitivities
   C. Linearity
   D. Electromagnetic Interference
      1. Magnetic Fields
      2. AC-Induced Fields and Transients
      3. Radio Frequency
      4. Electrostatic Fields
   E. Switching Transients
   F. Capacitance Effects
   G. Geotropic Effects
   H. Temperature Dependence
   I. Extracameral Effects
   J. Sound and Vibration Effects

III. RADIOLOGICAL
   A. Range
   B. Sensitivity and Detection Limit
   C. Accuracy
   D. Reproducibility
   E. Saturation
   F. Energy Dependence
   G. Temperature and Pressure Dependence
   H. Angular Dependence
   I. Response to Unwanted Radiations

consistent with the needs of the operation, and the dictates of budget. The items listed in Table 10.1 on the whole are self-explanatory. Human factor engineering includes absence of safety hazards, such as sharp edges or electrical shock, reasonable weight and balance for portable monitors, and ease of servicing.

Many of the electronic factors can only be evaluated through experience in use. However, such factors as sensitivity to ambient temperature, limited battery life, vibration sensitivity, *e.g.*, in some ionization chambers and multiwire counters, may be crucial to the suitability and proper functioning of detectors in specific applications. In high-energy physics research, insensitivity to magnetic fields, other types of radiation and electrical interference may be an important selection consideration.

Accuracy

Sensitivity and accuracy are most important in selecting and calibrating equipment. In meter instruments the sensitivity is assumed to be that corresponding to half the scale units on the dial. However, the fact that a meter, whether digital or dial-type, can give indication to a given significant number is no guarantee that the instrument is, in fact, that accurate; the true accuracy may have to be established in a separate counting test.

Although accuracy is a statistical phenomenon, calculable from the known variance of the numerous factors that affect the reading of the instrument, such a calculation (and indeed even securing the data on which to base the calculation) is often impractical. Hence, the term accuracy in practice refers to the total uncertainty (± 3 standard deviations) in the instrument reading, with the measurement made under nearly ideal conditions, and with appropriate corrections made for such factors as temperature, pressure, energy and geometry.

Accuracy is best determined with a calibrated source. Instrument readings over the entire range are compared with the "true" value as determined from the source strength. The following are some actual data obtained with an ion chamber survey meter:

True mR/hr	Instrument mR/hr	Instrument Range
15	14	1X
30	30	1X
45	47	1X
50	45	10X
100	95	10X
300	310	10X
450	460	10X
1,500	1,600	100X
3,000	3,000	100X
4,500	4,800	100X

This instrument was rated as having an accuracy of ± 10%.

Reproducibility

Reproducibility or precision is a means of describing the consistency of readings. It is determined by making at least five readings in a given field, removing the field between each reading. This should be done over at least three points on the range of the instrument. The following data were obtained with a GM survey meter on the 10X range:

"True" CPM	Observed CPM
1,000	1,050; 1,050; 1,100; 1,050; 1,100
3,000	2,900; 3,000; 2,950; 2,900; 3,300
5,000	5,100; 4,900; 4,950; 5,200; 5,000

Reproducibility of better than ± 5% is indicated by the data.

Note that reproducibility is not the same as accuracy; an instrument that always reads exactly 1.50 the true value would be 100% reproducible, but would have an accuracy of + 50%.

Saturation

When placed in radiation fields above their intended range, some instruments will saturate. The effect manifests itself in one of two ways: either the instrument reading rises to a fraction – usually about 80-90% – of the maximum full-scale reading and remains there regardless of how intense the radiation level gets, or, the instrument is driven off scale initially, then the reading drops back down to zero as the field intensity increases. Each of these conditions can result in a serious situation with possible overexposures to personnel. The former effect occurs most commonly in ion chamber instruments; the latter in Geiger-type survey meters.

Saturation, or overloading, is easily checked by placing the instrument in an appropriate radiation flux, and increasing the intensity until the effect is noted or a predetermined level – usually 100 times the maximum range of the instrument – is reached.

Energy Dependence

Energy dependence – also known as spectral sensitivity – refers primarily to the response of an instrument to X- or gamma radiation of different energies. Energy dependence is caused by many factors, the two most important ones being photoelectrons (and, to a lesser degree,

Compton electrons) from the detector wall, and self-absorption within the detector wall. These effects are, of course, competitive.

To evaluate energy dependence, the detector is placed in a field of known strength and energy and the response compared with the true exposure rate. Ordinarily, the instrument is calibrated with a source having an energy of approximately 1 MeV, and the instrument response is normalized to this energy.

Several effective energies in the range of 10 keV to greater than 1 MeV are necessary to achieve an accurate indication of energy dependence. An X-ray machine – preferably with 300 kVp capability – can be used to obtain effective energies in the 50-250 keV_{eff} region by means of heavily filtered spectra; a wide-beam K fluorescent source can be used to conveniently obtain essentially monoenergetic photons in the region 8-100 keV. The higher energies are provided with nuclides, commonly ^{137}Cs (662 keV), ^{226}Ra + daughters (∿ 800 keV), and ^{60}Co (1.25 MeV). A typical series of energies for a Cutie-pie-type instrument might be 8, 17, 23, 40, 60, 80, 100, 125, 175, 200, 662, 1,250 keV.

Energy dependence should be determined with both open and closed window and often for other conditions, such as with the beam end-on or through a thicker side wall. Energy dependence should be reported as a ± percentage over a specific energy range or a ± percentage normalized to a specific energy. Graphical presentation, in the form of a plot of response per unit exposure as a function of energy, is a superior method of showing energy dependence.

Angular Dependence

Because most detectors are not spherical, angular dependence may cause serious discrepancies in readings. In particular, angular dependence becomes significant when a fixed geometry or windowed detector is used in a ambient field or in such a manner that the window cannot be pointed towards the direction of the field.

Angular dependence is best checked through a full 360 degrees in perpendicular planes, one parallel to the horizontal and the other to the vertical axis of the detector. Measurements should be made at 15° increments; if necessary the determination can be made over 180° or 90°, as indicated by the geometry. The statement of angular dependence should be expressed as ± a percentage from a fixed point, usually the so-called "normal" of the detector. Again, a graphical representation is superior as is true for energy dependence.

Response to Unwanted Radiations

Portable survey meters are usually intended to monitor one type of radiation. Photon monitors such as ionization chambers, should be relatively insensitive to other penetrating radiations, for example, neutrons. Similarly, alpha or neutron monitors should be insensitive to photon radiations.

Checking the rejection of unwanted radiations is often more art than science. For example, energy dependence must be considered, particularly when evaluating the rejection of photons by neutron or alpha scintillation instruments. Neutron interference may show a similar energy dependence, with a response being elicited, say, only by thermal neutrons; also, the undesired response may not appear except in a mixed field. As a general rule of thumb, it is usually necessary to check response to unwanted penetrating radiation up to levels of about 10 R/hr in the case of photons, 10^7 n/cm^2/sec of fast neutrons, and 10^8 n/cm^2/sec of slow or thermal neutrons.

Dose Calibrations

Because of the close numerical resemblance between exposure values in air, expressed in roentgens, and dose values in tissue expressed in rads, many instruments are in use that are calibrated directly in rads, or even in rems, to serve as direct-reading dosimeters. Chapter 4 contained some examples. Many of these instruments are calibrated in 250-kVp X-ray fields where such relationships are superficially valid; however, obvious misuse occurs if such dose readings are then assumed to be valid for γ-rays or X-rays of different energy. Another complication will arise if doses are expressed in grays, or sieverts, where no simple relationship is preserved with X-ray exposure readings. For neutron monitors, also, the conditions of calibration must be clearly understood to indicate whether a dial reading can be accepted as a direct indication of dose.

In general all direct-reading dosimeters or area monitors, whether X-ray-, γ-ray- or neutron-sensitive, must be calibrated in a reproducible configuration with a source of known output and energy. Such calibrations should be done in a large room, with both source and detectors well removed from floor and walls to minimize backscatter contributions. If the source can be considered to be a point source, known exposure conditions can be obtained at fixed distances from the source. The apparent reading can then be plotted against the true exposure value to obtain a calibration curve or to adjust the instrument. Conversion to dose readings may require intercomparison with an absolute dosimeter, such as the Fricke dosimeter (Chapter 8).

REFERENCES FOR FURTHER READING

American National Standard Institute. *Radiation Protection Instrumentation Test and Calibration*, ANSI N 323.

International Atomic Energy Agency. *Handbook on Calibration of Radiation Protection Monitoring Instruments*, Technical Report Series 133 (Vienna: IAEA, 1971).

International Atomic Energy Agency. *Metrology of Radionuclides* (Vienna: IAEA, 1960).

Johns, H. E., and J. R. Cunningham. *The Physics of Radiology*, 3rd ed. (Springfield, IL: C. C. Thomas, 1971).

Kathren, R. L., and H. V. Larson. "Radiological Calibration and Standardization for Health Physics," *Health Physics* 16:777 (1969).

Kiefer, H., and R. Maushart. *Radiation Protection Measurement* (Oxford: Pergamon Press, Inc., 1972).

National Bureau of Standards. *Measurement of Neutron Flux and Spectra for Physical and Biological Applications*, NBS Handbook 72 (Washington, D.C.: NBS, 1960).

National Council on Radiation Protection and Measurements. *A Handbook of Radioactivity Measurements Procedures*, NCRP Report No. 58 (Washington, D.C.: NCRP, 1978).

CHAPTER 11

ELECTRONIC SYSTEMS

No description of any detector system is complete without a discussion of the electronic system associated with the detector proper. This interdependence arises from two factors: (1) the fact that in the majority of detectors the product of radiation absorption in the detector volume is the accumulation of electric charge or the momentary appearance of a charge pulse, and (2) the requirement in most cases for treating and evaluating the detection event in terms of its occurrence frequency, its magnitude or any other experimental parameter, all of which usually is handled most conveniently by means of electronic circuits. The evolution and revolution in electronics that has accompanied the increasing speed and complexity in functions that have become possible through developments in semiconductor electronics have had their impact in the field of radiation detection too, leading to the elimination of many time-consuming operations and to impressive systems capable of automatic sample selection, programming of experimental sequences and computation, video presentation, and storage and recording of a wide variety of detection operations. Though a detailed description of many such systems goes beyond the scope of this book and can be found in the commercial literature, there are certain basic functions of the electronic systems that are inherent in the use of many detectors and must be fully understood for intelligent use of the detectors. It is dangerous and generally unsatisfactory to consider the electronic portion of a detector system solely as a "black box," with certain connections to be made and buttons to be pushed for operation.

At its most basic, the electrical system associated with most detectors has to fulfill three functions:

1. supply the voltage necessary for the detector to function;
2. amplify and modify the detection signal from the detector to facilitate extraction of its information content; and
3. display or store the information in a convenient format.

These functions may be combined in a single unit or they may be supplied by separate component units. In portable equipment the electronic portions are often integrally coordinated with the detector itself, whereas in laboratory equipment separate units are often employed for greater flexibility.

To give a concrete example, in a NaI(Tl) scintillation detector system a high voltage of the order of 900-1,200 volts will be needed for operation of the photomultiplier. The output pulse appearing at the anode of the photomultiplier is too small to be counted directly and, hence, an amplifier must be provided. If the detector and counting system are physically separated, a small compact amplifier stage, called a "preamplifier," may be built into the detector head, to anticipate and overcome any attenuation or loss in signal amplitude resulting from a long coaxial cable. (The preamplifier may also serve as an impedance transformer between the photomultiplier anode circuit and the cable impedance.) The main amplifier then further enhances the signal pulses to a level where they can drive a pulse counting circuit or can be sorted according to amplitude in a pulse-height discriminator. If the pulses are to be counted, they pass to a "scaler," which counts them and selects for further handling every tenth, hundredth or thousandth event. A timing system must be provided to determine the number of counts within a given time interval or the time it takes to reach a certain total count. This total count or time interval must then be displayed to be read by the experimenter, or it may be printed out.

The quality of the system depends on the reliability of all the components and its stability under a wide range of disturbing parameters. Electronic systems can be designed to minimize the influence of supply voltage variations, ambient temperature, humidity, vibration and shock to a certain degree. However, for consistent operation it is usually advisable to run detection equipment with an external voltage stabilizer and in an air-conditioned room, if conditions require this. A stable ground-potential connection should be provided if at all possible; it is not usually satisfactory to rely on cable braiding for this purpose alone. Each unit should be directly connected to ground or a grounded bus; circulating currents must be avoided. Frayed cable braiding and improperly soldered connections are the source of most common intermittent faults and cable connections must be treated with care and should be inspected from time to time.

To provide flexibility in designing electronic systems and to permit easy change of units and functions, U.S. manufacturers have adopted a modular approach in designing units for the amplification, shaping, discrimination and correlation of detector signals. These modules have been designed to fit interchangeably into a base unit, usually referred to as a *NIM bin* (Figure 11.1) which provides standard voltages and interconnections through a series of multipole connectors. The NIM bin itself fits a standard 19-in. (47.5-cm), between uprights, instrument rack, so that a number of bin units can be mounted in vertical arrays. By combining different modules inside any bin a wide range of circuit functions can be accommodated.

Not all detectors require very elaborate systems. For instance, a Geiger counter merely needs a moderately stable voltage supply, a scaler or ratemeter, and a display consisting of a count register or a milliameter. For this reason it is simplest to review each detector according to the nature of the supply needed and the kind of output signal provided; this has been done in Table 11.1. One can then specify the major electronic components needed, which can be categorized as:

a. voltage supplies
b. amplifiers
c. ratemeter circuits
d. counting circuits
e. pulse-shaping circuits
f. discriminators
g. coincidence and anticoincidence gates
h. clocks and timers
i. memories
j. analytical and computational treatment
k. output units

POWER SUPPLIES

As Table 11.1 shows, most detectors need a supply voltage of the order of several hundred volts or more. Depending on the type of detector this voltage should be stabilized against fluctuations in the supply voltage, against variations in load current, and against electrical drift and temperature variations. Most power supplies operate by full-wave rectification of the mains voltage, directly or transformed up to a high voltage. The rectified voltage is then smoothed by a series of capacitors and stabilized by a feedback amplifier, which itself compares a portion of the output voltage against a reference voltage. This reference voltage may be derived from the ignition potential of a glow tube or the breakdown voltage of a Zener diode (Figure 6.2). For Geiger tubes a simpler stabilization system is often used that employs just a gas-filled reference tube. Figure 11.2 shows two representative power circuits.

Figure 11.1 Picture of NIM bin with power supply.

The feedback system shown is adequate for stabilization against sudden changes in detector current or interferences in the supply system. Slow voltage drifts are more difficult to compensate and may require close control of operating temperatures and careful grounding of the system: vacuum-tube systems show greater stability if they are kept running to minimize warm-up effects. This is less significant with transistor circuits; however, they are more sensitive to high voltage transients. For this reason one should avoid, for example, disconnecting input cables to semiconductor amplifiers with the supply voltage turned on.

Table 11.1 Electronic Needs of Detector Types

Detector	Function	Power Supply: Voltage	Power Supply: Stability	Output Signal	Amplifier	Count Mode R (ratemeter) C (counter)	Discrim PHA	Remarks
Ion Chamber	portable	V charger	battery	Visual				needs charger
	pressure	300-1,200 V	high	∿ 1 mV	vibr. reed + amplifier	R		
Prop. Counter	sealed	300-1,500 V	high	∿ 10 mV	yes	R/C	yes	
	gas-flow	600-1,000 V	high	∿ 10 mV	yes	C	yes	
GM Tube	monitor	800-1,200 V	medium	0.5-5 V	no	C/R	no	
NaI(Tl) Scintillation	counter	700-1,200 V	high	∿ 100 mV	yes	C	no	
	spectrometer	700-1,200 V	high	∿ 100 mV	yes	C	yes PHA	
Liquid Scintillation	counter	700-1,600 V	high	∿ 100 mV	yes	C	coinc.	
Ge(Li), Si(Li)	spectrometer	1,500-3,000 V	medium		yes	C	PHA	cooler (liq. N_2)
Surface Barrier	α detector	100- 800 V	medium		yes	C/R	no	
TLD	monitor	-		light		integrates		needs reader
Photographic Film	monitor	-		visible tracks		integrates		needs developer
Track-etch Film	monitor	-		visible track		integrates		needs developer
Self-Powered	neutron flux	-		∿ 10^{-13} amp	yes	R		
Glass Dosimeter	dose monitor	-		light		integrates		needs reader
γ,n Thermometer	reactor flux	-		mV(thermocouple)	yes	integrates		

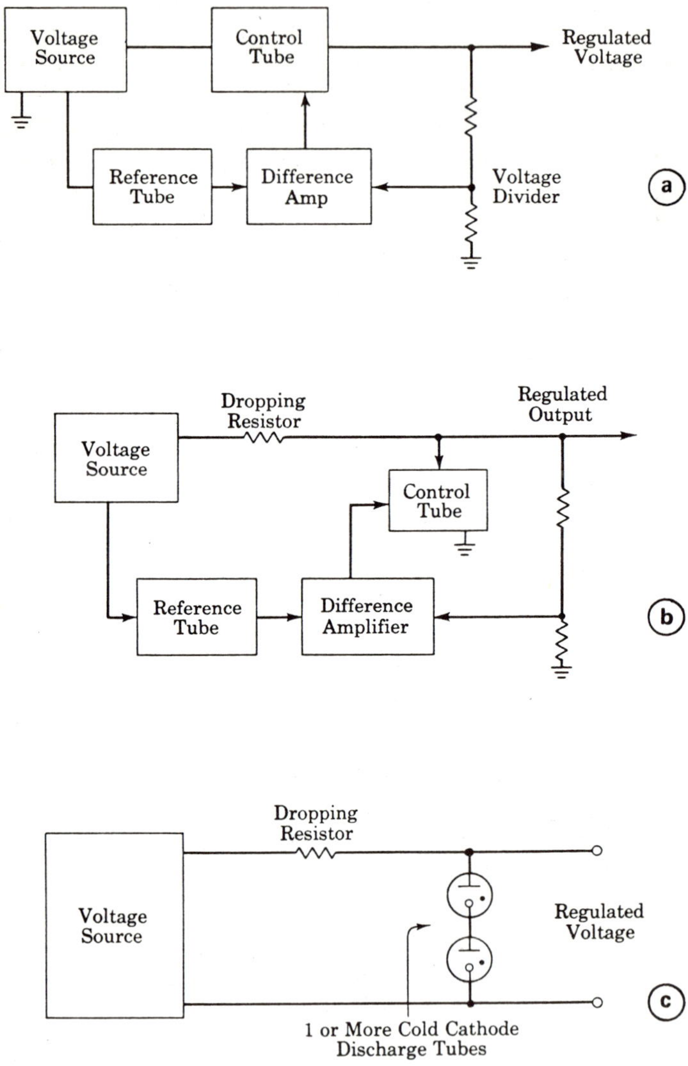

Figure 11.2 Voltage regulator circuits: (a) series control; (b) shunt control; (c) gas tube regulation.

THE DETECTOR SIGNAL

The *signal* from most detector systems consists of a burst of electric charge collected at a suitable electrode to which a biasing voltage has been applied. In the case of gas-filled detectors there actually are two such pulses, one at the positive electrode, the anode, due to the arrival of the electrons, the other, at the negatively charged cathode, due to the slower positive ions, producing essentially a pulse of equal charge content, though one of different shape and length owing to the slower rate of arrival. The pulse heights, rise times and amplitudes thus contain information on the nature of the particles collected, their mobilities and total ionization, *i.e.*, number of ion pairs produced by primary and secondary processes that survived early recombination. Figure 11.3 shows how such an ion pair might look: the area under the curves would be the same since they are of equal though opposite charge; by using polarity or amplitude discrimination one could count the electron peak only, thus counting a "faster" signal. Since the two pulses would oppose each other the signal would have to be fed to an amplifier, a vacuum tube or a suitably biased transistor, which would amplify pulses of one polarity only while suppressing the other. If the amplifier is *linear* with respect to the accepted pulse, then input and output signals will be fully proportional with respect to amplitude and pulse shape and any technical

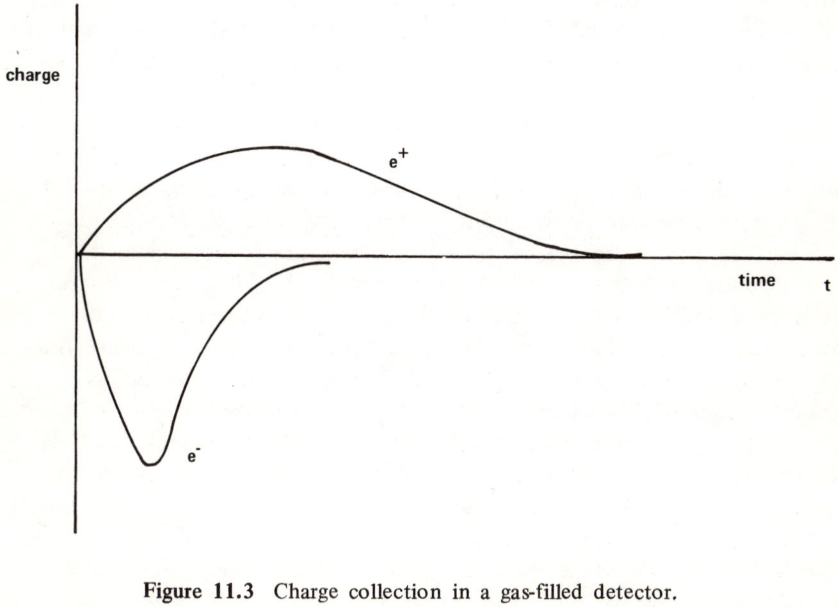

Figure 11.3 Charge collection in a gas-filled detector.

information content will be preserved. The same would apply if particles of different energy or LET were detected in the same detector volume; each event would give rise to signals of different shape, *i.e.*, different rise time and amplitude, and a circuit would have to be selected to either amplify linearly and preserve these differences, or to differentiate and select them so that only signals of a desired rise time, polarity or amplitude would be accepted.

Similar considerations apply to other detectors. In the case of scintillation detectors, the output signal from the photomultiplier is a large electron burst reaching the collector, one pulse per primary event. Depending on the circuit configuration (Figure 5.4), the signal may be positive- or negative-going with respect to ground. Its amplitude is usually of the order of tens of millivolts and it must be amplified in a preamplifier before being conducted to a further amplifier and discriminator. As the signal appears across a large anode resistor (50-500 kilohms), the preamplifier also serves as an impedance transformer to match that impedance to the rather low impedance of a coaxial cable. Otherwise significant distortion and loss in amplitude could result from a mismatch. In the case of semiconductor detectors the device impedance is usually fairly low; nevertheless, considerable amplification is necessary since the signal amplitude itself tends to be small.

For beta- and gamma-ray spectrometry, the total charge collected (the "area under the curve") is assumed to be proportional to the energy of the incident particle. The shape of the pulse reaching any discriminator or analyzer is determined by the successive, statistically fluctuating contributions of the process of ion-pair (or hole-electron pair) formation, the mobility and collection time of each, the amplifier characteristics and its noise contribution, and finally voltage and environmental fluctuations. If the resultant pulses show a near-Gaussian distribution around their mean value, the pulse amplitude may serve as a direct indicator of the particle energy; otherwise the whole signal has to be evaluated.

Collection time and, hence, rise time can be useful to discriminate between different particles, such as alpha particles and electrons, or protons and neutrons, that have different ion-pair densities along their track or tend to produce ions in well-defined regions of the detectors as against volume-wide ionization. This difference may be used in ionization chambers, liquid scintillators or solid detectors. In these cases rise time discrimination is best performed by pulse differentiation, a process of passing the signal through a short-time constant RC circuit (Figure 11.4a). If the pulse rises steeply, a sharp pulse results whose amplitude reflects the input pulse. At the end of the pulse a negative-going short pulse provides a measure of the pulse length. If the input rises slowly, the

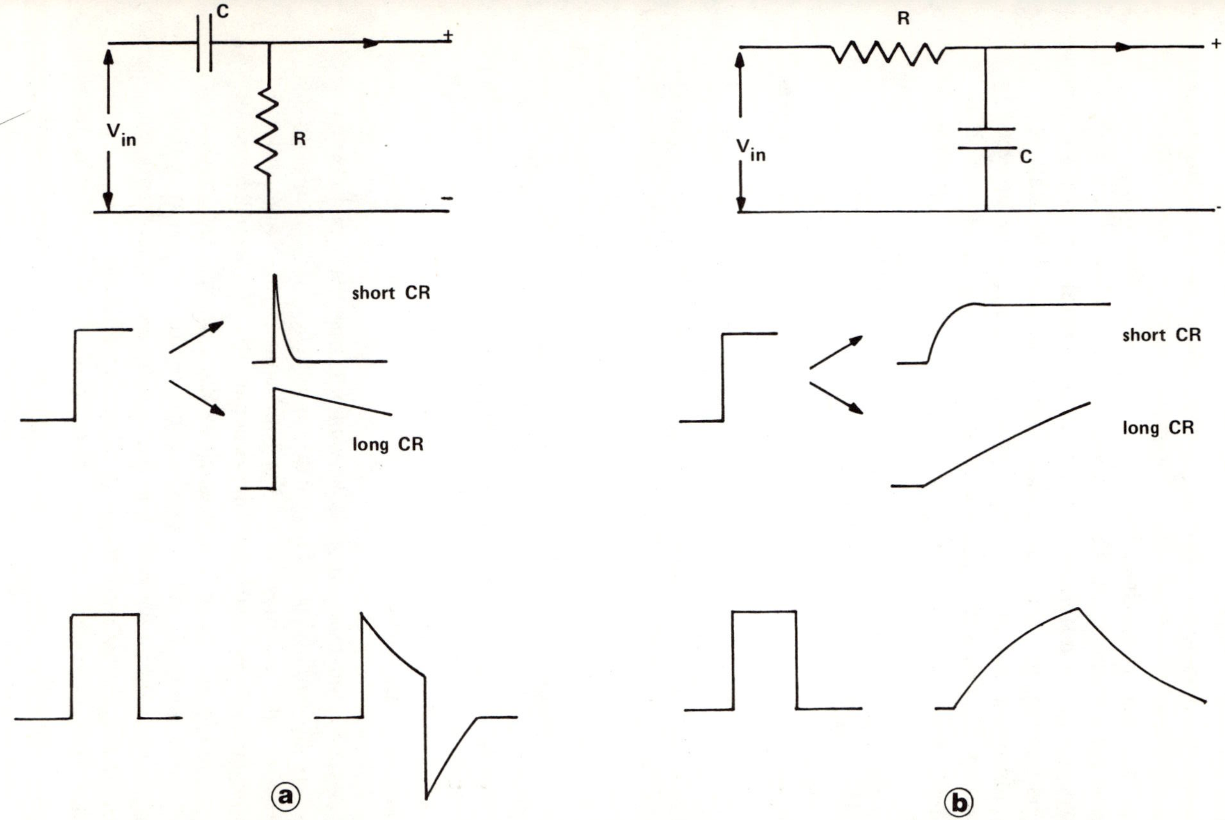

Figure 11.4 Pulse-shaping circuits: (a) differentiation; (b) integration.

charge across the capacitor C starts leaking off before full passage of the signal has occurred. Hence the differentiated pulse will be smaller and can be discriminated. Simple RC differentiation does not easily produce pulses of acceptable form. Frequently a square wave form is preferred and that is obtained by the use of an inductive delay line. Delay-line pulse shapers, like that shown in Figure 11.5, yield highly symmetrical, doubly differentiated pulses. Such pulses are excellent for timing and gating, but for noise reduction and pulse-height analysis a slightly more rounded pulse shape is preferred. This can be obtained by next passing the pulses through several RC integrating sections with carefully chosen component values.

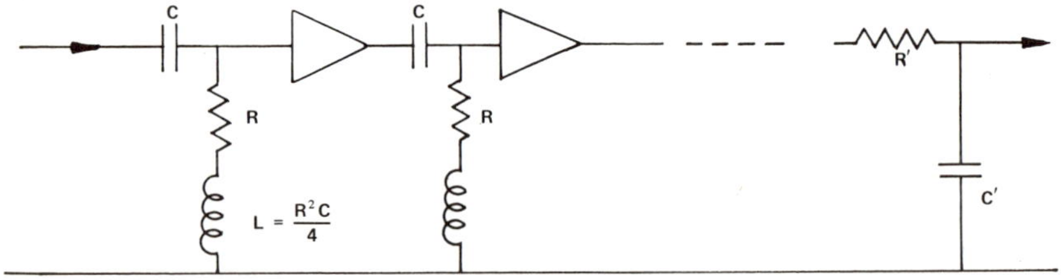

Figure 11.5 Differentiation with inductive compensation.

RATEMETER CIRCUITS

Figure 11.4b illustrates another way of utilizing the signal content—integration. By employing a fairly large capacitance, the voltage appearing across C will be a measure of the total amount of charge deposited in a time interval less than the time constant RC. If the pulses are of equal amplitude and pulse length, such as those from a Geiger counter, the voltage V across C will be proportional to the number of pulses charging it, *i.e.*, the *pulse rate*. For this reason such instruments are referred to as pulse-rate meters or briefly ratemeters.

Figure 11.6 illustrates the basic circuit involved. A series of n pulses of equal amplitude V_0 is assumed to enter the circuit from the left. After clipping any unwanted pulse overshoot and removing any d.c. component with the rectifiers, each pulse will deposit a charge C_0V_0 on C, provided $C >> C_0$. The charge on C, of course, fluctuates as it leaks

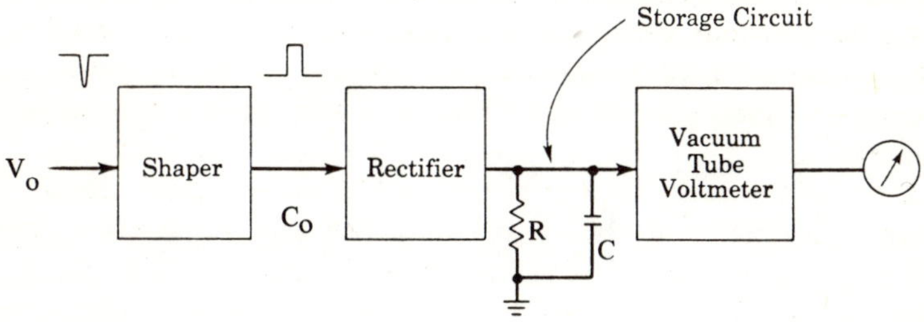

a. Basic linear ratemeter.

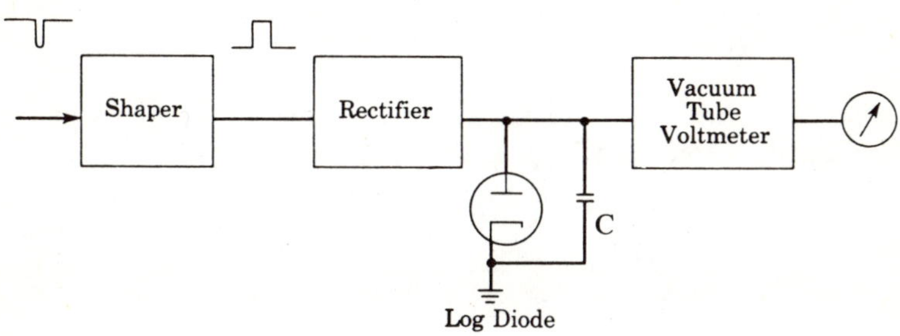

b. Log diode ratemeter.

Figure 11.6 Ratemeter circuits.

off through resistor R and is replenished by successive pulses. If n is sufficiently high to keep C_o fully charged and $1/n >> 5$ CR, then there will be an average charge $\bar{q}$ on C giving rise to an average potential $\bar{V}$ that can be measured. At equilibrium the charging rate equals the leak rate, or

$$\bar{V} = nqR = nC_oV_oR \tag{11-1}$$

For fixed values of C_o, R, and V_o, $\bar{V}$ is then measure of the pulse rate. Each pulse counted at the instant t_o excites a transient across R

$$v(t) = \frac{C_o}{C} e^{-(t-t_o)\ RC} \qquad t > t_o \tag{11-2}$$

$$v(t) = 0 \text{ for } t_o < t$$

The probability of an event occurring between u and u + du ($u = t - t_o$) is ndu and its contribution to the total voltage is

$$dV = ndu$$

or

$$V = \int_{-\infty}^{+\infty} nv_{(u)}du = nV_oC_oR \qquad (11\text{-}3)$$

as before.

In practice, the arriving pulses do so at a random rate and the output voltage *will* fluctuate. The mean fluctuation voltage $(\overline{\Delta V})^2$ is given by

$$(\overline{\Delta V})^2 = (\overline{V' - V})^2 = \int_{-\infty}^{+\infty} v^2_{(u)} \; ndu = \frac{nC_o^2V_o^2R}{2C} \qquad (11\text{-}4)$$

since ndu is the mean square deviation of the number of counts in the interval du and, therefore, $v^2_{(u)}n$ is the contribution of this mean square deviation. From error theory the fractional probability error

$$\sigma = 0.67 \frac{\Delta V}{V} = \frac{0.67}{\sqrt{2nRC}} \qquad (11\text{-}5)$$

A practical consequence of this is that a large value of the time constant RC gives a steadier reading for a given pulse rate, but also takes a longer time to reach equilibrium or to respond to changes in pulse rate. In practical instruments the meter range can be changed by switching R; the value of C should be kept moderately low to avoid excessively long time constants.

In dosimetry and reactor experiments the pulse rate and dose ranges may vary over several orders of magnitude. To avoid excessive changes in response time, logarithmic ratemeters have been developed whose output voltage is proportional to log n. This can be achieved by using semiconductor components with logarithmic characteristics or by the use of selective feedback (Figure 11.6b).

AMPLIFIERS

Since the signal from most detectors is too small to be easily counted or sorted, the use of an amplifier is usually required. An amplifier is a passive electronic system that must increase the pulse amplitude by a constant factor, the "gain," regardless of how large or small the pulses are, without changing the pulse shape and regardless of their rate of arrival. This constancy in gain is referred to as *linearity*; a well-designed amplifier should retain its linearity over a wide range of inputs. While these objectives were fairly easy to accomplish with low pulse rates and relatively slow detectors, with rise times of the order of several microseconds at best, the advent of semiconductor systems and faster detectors has required nanosecond or better response times in recent years; it is in this area that substantial progress has been made, stimulated by parallel development in microelectric circuits. The problem involved will be appreciated when one analyzes a fast-rising pulse by Fourier analysis. Any square pulse can be analyzed into a spectrum of sine waves whose envelope it forms; the steeper the rise the higher the corresponding frequency of the sine waves involved. Consequently, to handle faster, more rapidly rising pulses faithfully, the amplifiers must have a much broader frequency response. This is more easily done in transistor circuits than with vacuum tubes, as the higher "gain bandwidth" product of transistors permits more feedback to be applied with a corresponding improvement in gain stability.

Most linear amplifiers intended for quantitative radiation detection are made up of one or more negative feedback loops, which increase amplifier gain stability and linearity by reducing the dependence of the gain on transistor, or tube, transconductance. Figure 11.7 shows a block diagram of such a feedback loop. If the amplifier without feedback has a voltage gain K and if a fraction b of the output signal V_o is fed back

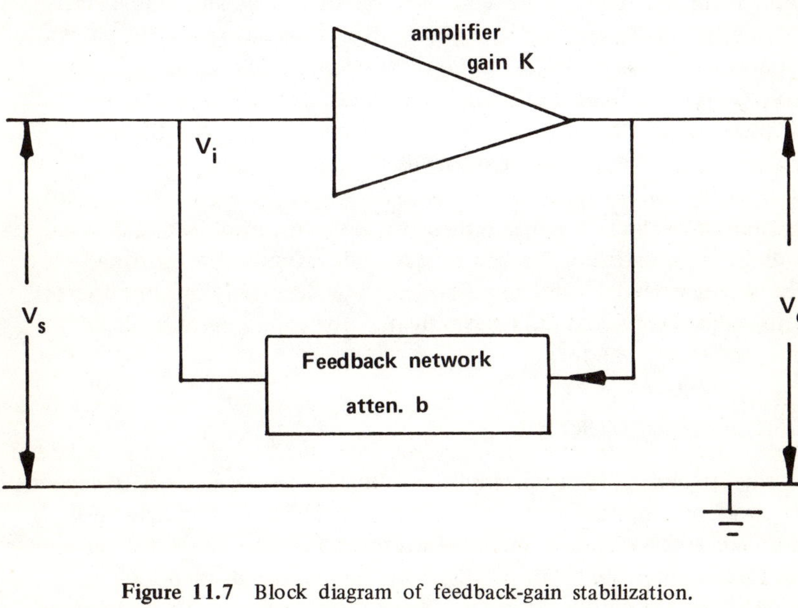

Figure 11.7 Block diagram of feedback-gain stabilization.

to the input, opposite in phase to the input signal V_s, then the net signal voltage at the input may be written

$$V_i = V_s - bV_o \tag{11-6}$$

Since $V_o = KV_i$, Eqn. (11-6) can be written

$$V_o = K(V_s - bV_o)$$

or

$$KV_s = V_o(1 + bK) \tag{11-7}$$

The gain with feedback, G, is then

$$G = \frac{V_o}{V_s} = \frac{K}{1 + bK} \tag{11-8}$$

If the gain without feedback K is large enough for $bK >> 1$, then

$$G \approx \frac{1}{b} \tag{11-9}$$

If b is determined by passive linear components and if bK is large, the gain with feedback tends to depend only on the linear and stable feedback factor b and is relatively independent of fluctuating and often nonlinear transconductance values. Besides improving amplifier gain stability, negative feedback increases the amplifier bandwidth because G remains almost constant at all frequencies for which $bK >> 1$, even though K may fall off at low and high frequencies.

The negative feedback loop may be purely resistive or it may consist of one or more stages of amplification, as long as the out-of-phase feedback condition is satisfied. High feedback reduces amplifier gain and must be compensated for by more stages. A wide variety of amplifiers is on the market which should be evaluated for gain stability, linearity, dynamic range and bandwidth.

PULSE-HEIGHT DISTORTION

The fidelity with which an amplifier handles a series of signals depends on the range of amplitudes, their constancy in shape and their arrival intervals. Since the pulses have finite width and height and arrive at a random rate, their height can be distorted by chance overlap and by limitations in the amplifier linearity. Noise, gain drift due to temperature and shifts in d.c. baselines also add to the distortion. The amplitude distribution of small signals should not be disturbed by the presence of large signals; however, if the gain is adjusted to accommodate small signals, any large signal may drive the amplifier to saturation, giving rise to "overload" conditions. If the amplifier does not recover quickly from overload paralysis, caused by driving the stage into the limiting current mode, smaller pulses occurring during the recovery interval will not be measured correctly or may be lost altogether. Any pulse overloading the amplifier can be observed, on an oscilloscope, as developing a pronounced flat top and usually a longer decay time than other pulses. Overload conditions can be minimized by use of a rectifier diode to limit the voltage swing or by limited feedback though at some loss of linearity.

Another cause of distortion is *pile-up*, the lack of resolution between successive pulses if any two are too closely spaced. To some extent such events are unavoidable at high count rates, but obviously the detector systems must be matched to expected count rates if accurate results are to be obtained. In a preamplifier, pile-up can drive the output stage over a wide dynamic range, thereby causing an apparent broadening of spectral peaks. In a main amplifier, if pulses are shaped and differentiated, pulses following too closely may be found to be superimposed on the undershoot or decay tail of the previous pulses thus changing their effective amplitude. If they are not resolved such a pair may give rise to a "sum peak."

NOISE

In many detectors, such as proportional counters and self-powered detectors, very high gain amplifiers must be employed to obtain manageable signal amplitudes. The ultimate limit to the gain that can be employed or the smallest signal that can usefully be detected is set by the electronic *noise* of the system. Noise is a description of all random events, usually of thermal origin, that appear as voltages across the input terminals and get amplified. Any signal voltage then is effectively superimposed on this random noise voltage. Its importance depends on the relative size of the signal with respect to the noise (signal-to-noise ratio) and the uncertainties in the signal amplitude due to other causes, such as the uncertainty in the number of ion pairs produced in a detector or photomultiplier dark current.

The principal source of noise is the thermal motion of electrons in the resistor components of the input circuit and any other components in parallel. This noise is usually called "Johnson noise"; it gives rise to an effective noise voltage

$$V_N = \sqrt{4k\ TR\ \Delta f} \tag{11-10}$$

where k = Boltzmann's constant, 1.381×10^{-23} joule per °K
T = absolute temperature, °K
R = resistance in ohms
Δf = frequency bandwidth of amplifier, in which the noise is measured

In vacuum tubes, statistical fluctuations in the cathode current of the amplifier input tube give rise to a noise component called "shot noise" or Schottky effect. This can be minimized by using a high-transconductance tube at the amplifier input. In semiconductors, a Schottky effect noise can also be observed at elevated temperature due to random electron transitions into the conduction band.

An unrelated, though similar-appearing effect is due to *microphonics*, the effect of mechanical vibration of electrodes or other detector components. This is pronounced in certain ionization chambers, multiwire detectors and streamer chambers and can be cured only by a damped enclosure or redesign of the detector to avoid mechanical resonances.

Each source of noise can be characterized by one circuit parameter called the "noise resistance," which for a uniform "white" noise spectrum represents the equivalent series or parallel resistor generating an equivalent amount of thermal noise. For a charge-sensitive detector, with a simple RC integrating and differentiating filter of time constant τ the equivalent RMS noise charge $\bar{q}$ is given by

$$\bar{q}_p = \frac{\sqrt{4kT\tau}}{R_p} \tag{11-11}$$

and

$$\bar{q}_s = \frac{\sqrt{4kTR_s}}{\tau} C_i \tag{11-12}$$

where R_p = resistance in parallel to input
R_s = resistance equivalent to series noise source
C_i = combined input capacitance of detector and preamplifier

The time constant τ is a measure of bandwidth and midfrequency. If the noise contributions can be split between series and parallel components an optimal time constant is obtained

$$\tau_{opt} = \sqrt{R_p R_s \times C_i} \tag{11-13}$$

and a minimum total noise

$$\bar{q}_{min} = \sqrt{4kTC_i}\left(\frac{R_s}{R_p}\right)^{1/4} \qquad (11\text{-}14)$$

Low-noise detectors are encountered in charge-sensitive circuits, low count-rate systems such as carbon dating systems and liquid scintillation systems. Noise can be minimized, though not eliminated, by low-temperature environments, selection of low-noise preamplifier units, especially field-effect transistors, use of pulsed feedback circuits and anticoincidence circuits with fully independent amplifier circuits (to eliminate shot noise).

PULSE SHAPING

As we have seen, the signal pulses from various detectors may vary considerably in shape and size and for some applications more reliable counting may occur if the pulse shapes were reformed to meet a standard shape or to define more clearly, for instance, the time of first appearance of the initiating particle. Such reshaping of the signal pulses is referred to as "pulse shaping" and it typically occurs, when needed, after initial amplification of the signal.

The generation of a uniform pulse shape is desirable if the pulses are intended to drive a scaler, *i.e.*, if the pulses are just to be counted, or a ratemeter. In both cases counting is easiest if the pulses are squared off to form a flat-topped pulse of fixed length. This is usually done by means of a "flip-flop" circuit or Schmitt trigger circuit; Figure 11.8 shows two of many variations, one a vacuum tube version, one a transistor form. This is one example of a bistable trigger circuit. In the quiescent state, the left half, T_1, is conducting, while T_2 is biased to cut off. A negative input signal will cut off T_1, its current drops, raising the anode potential which in turn initiates conduction in T_2. After a time, determined by the time constant CR_{L_2}, the anode voltage of T_2 drops to a level no longer adequate to attract current which is diverted to the grid. As T_2 stops conducting, the cathode coupling lowers the cathode potential of T_1 reverting the system to its initial condition. Figure 11.7b shows a transistor form of the same circuit. The pulse length and amplitude are set by the choice of circuit components and bias voltages. With appropriate components a nearly square pulse can be obtained (Figure 11.7c), though the top is often slightly sloped unless rectifiers (diode clamps) are added to clip overshoots and top slopes of the pulses. There are many variations on the basic flip-flop circuit to lengthen the output pulse, minimize the standing current through the "on" side, or to stabilize conduction of each side to permit use of the

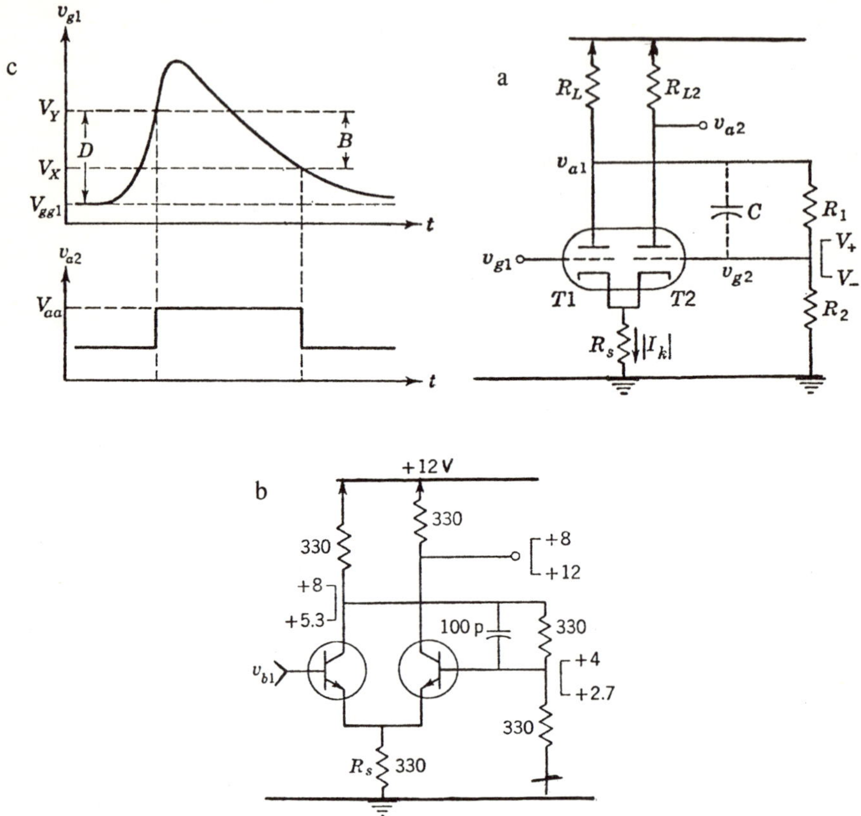

Figure 11.8 Schmitt trigger circuits: (a) tube circuit representative components; (b) transistor circuit; (c) pulse shapes.

circuit as a binary switch. Details on such circuits are beyond the scope of this book but can be found in any text on pulse electronics.

For driving a scaler circuit, which typically is a bistable pulse circuit itself, input pulses of uniform rise time and amplitude are desirable though not necessary. For ratemeter circuits, where each pulse ideally should contain the same charge, pulse shaping is almost essential.

If the signal itself is longer than desirable for rapid counting, sorting or analysis, it can undergo further pulse shaping after the first conformation step. For instance, it may be differentiated, resulting in a sharp spike associated with the rise of the square pulse, and a second reverse spike from the switch-off. By clipping or selecting one or the other polarity and using that pulse to undergo further pulse shaping, a shorter pulse can be produced that would be suitable to start a clock, to initiate a time-of-flight measurement, to start the sweep analysis of a pulse height analyzer or any other function. Commercial units are available to provide gate pulses, delayed pulses, stretched pulses and many kinds of control and logic circuits.

By changing the polarity of signals from background counts or discriminated Compton events and storing them in a memory, automatic subtraction of background counts and the Compton shelf can be obtained for routine spectroscopy involving repetitive samples of comparable composition.

By introducing reference pulses of fixed amplitude from a pulse generator that would normally be placed in the same channel of a pulse-height analyzer (PHA) one can stabilize the PHA calibration and eliminate drift effects due to small fluctuations in discriminator bias. In some analyzers, the discriminators themselves, one for each channel, are modified Schmitt triggers with a highly stabilized bias voltage and a fixed delay circuit increasing delay time from channel to channel. For PHAs with 1,000 channels or more such circuits must be extremely fast to provide the capacity and pulse resolution required.

GATING AND TIMING PULSES

For coincidence systems, one of the signals serves as a "gate" for the other. In that case the gating pulse must be shaped to define exactly its starting and ending time. The second signal, which may contain information on the initiating particle, then passes the gate if it arrives during the proper interval without undergoing any shaping itself.

Most modern PHAs are based on the amplitude-time conversion principle. The input is stretched to provide a constant-amplitude signal whose amplitude provides a gating pulse which permits clock pulses from a pulse generator to pass successive channels until the end of the gate pulse locks the time signal in the highest channel reached.

Gating pulses, in conjunction with a precise pulse generator serving as a clock, can be particularly useful in coincidence measurements and flight-time determinations. Coincidence systems serve in the determination of nuclear decay schemes and counter telescope arrangements, which distinguish particle beams from scattered radiation, as well as in arrangements for the elimination of dark current and unwanted background events.

The characteristic of principal interest in coincidence circuits is the simultaneity of events recorded by different detectors or traveling along separate electronic pathways. Two events are said to be simultaneous if the interval between them is too short to be discernible. Conversely, they are rejected as unrelated if their time separation exceeds the circuit resolving time τ. To distinguish truly correlated events from others it is desirable to keep the resolving time short; however, it must not be made infinitely short because the time of occurrence of nuclear events cannot be measured with absolute precision, since every detection process introduces an uncertainty into the time of interaction. If this uncertainty is much greater than τ, some truly correlated events will be rejected.

To choose the optimum resolving time one must be able to predict, at least approximately, the number of chance coincidences that would be recorded. Let two detectors, D_1 and D_2, receive uncorrelated signals at count rates n_1 and n_2, respectively. All signals from D_2 that are separated from a D_1 signal by an interval less than τ are recorded as chance coincidences. Each of the D_2 signals must fall within a set of time intervals of length 2τ which symmetrically bracket the D_1 signals. These intervals represent a fraction $2\tau n_1$ of the total data collection time, provided n_1 is small enough that the intervals overlap only rarely. By multiplying this fraction by the D_2 count rate, n_2, one obtains the chance coincidence count rate R

$$R = 2\tau n_1 n_2 \qquad (11\text{-}15)$$

Both n_1 and n_2 are proportional to the source strength so that R is proportional to the square of the source strength, I^2. On the other hand, the true coincidence count rate must be proportional to the source strength, so that the ratio of true-to-chance coincidences is proportional to 1/I. Reducing the source strength will reduce the proportion of chance coincidence counts, but will also reduce the statistical accuracy of the true coincidence counts unless the counting time is increased significantly.

By a similar argument it can be shown that for a system with three inputs the threefold chance coincidence rate

$$R = 3\tau^2 n_1 n_2 n_3 \qquad (11\text{-}16)$$

The general expression for an m-fold coincidence circuit is

$$R = m\tau^{m-1}\ n_1 n_2\ \ldots\ldots\ n_3 \qquad (11\text{-}17)$$

Practical coincidence circuits may rely on superposition of signals to overcome a bias clamp voltage, or on the input of separate signals into a pair of coupled tubes or transistors, with a common load resistor, whose combined output signal then may or may not exceed a succeeding discriminator bar. Some simple coincidence circuits were shown in Figure

5.14 and Figure 11.9 shows diagrammatically two other systems. In the differential coincidence circuit the input pulses are stretched and shaped to make the resolving time independent of input pulse shape. Coincident pulses arrive at the midpoint simultaneously and cancel each other, while uncorrelated pulses provide voltages of different polarities across the difference amplifiers.

In the fast-slow system (Figure 5.14), both input pulses are fed to two coincidence circuits which have widely different resolving times. The

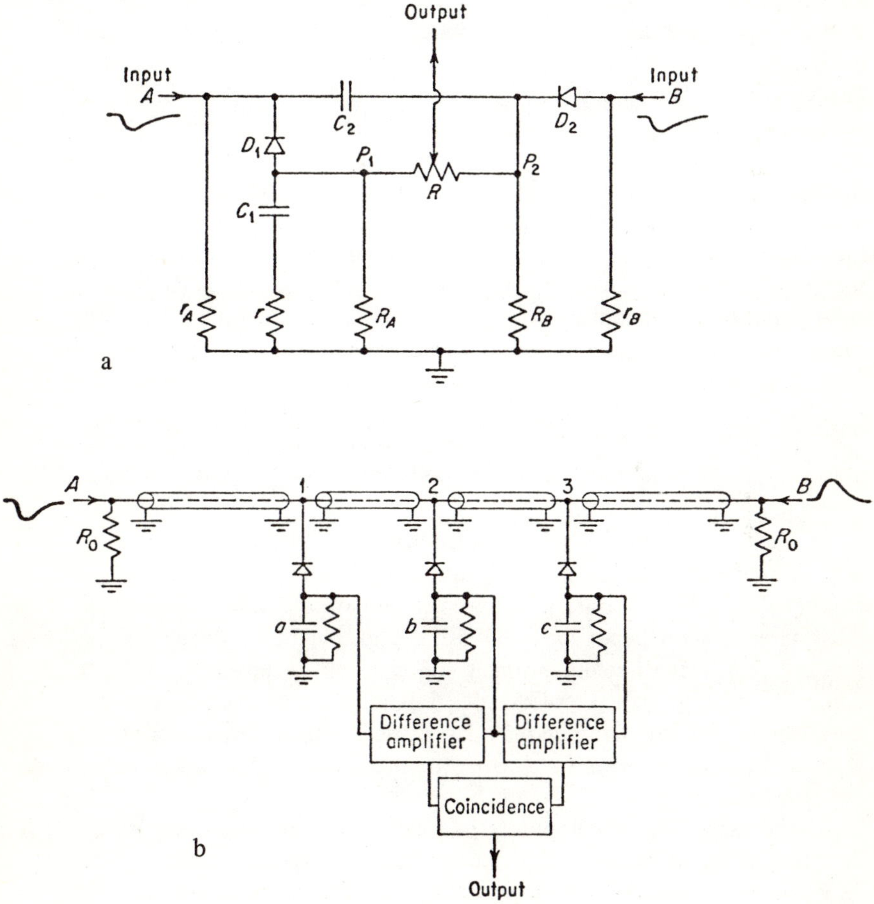

Figure 11.9 Two simple coincidence circuits: (a) Bay balanced circuit; (b) Meunier differential coincidence circuit.

combined outputs result in a triple coincidence only if time and amplitude conditions are satisfied at the same time.

Component improvements in recent years have made it possible to design circuits with faster switching times, resulting in sharper time discrimination in timing circuits and coincidence systems. This has made it possible to reverse the gating process. Instead of using one signal from the detector to initiate the timing system or provide the acceptor gate, one can use a fixed timer to pulse-on the detector at a known instant at a certain repetition rate and look for those detected events that appear in two detectors identifiably from the same initiating event. This is the procedure that can be used with a neutron chopper or a pulsed accelerator; it results in a considerable reduction in background and chance coincidences, especially if the pulse rate can be varied. By combining gating circuits with adjustable delay circuits, it becomes possible to trace secondary and later particle chains in high-energy physics. Similar use of timing pulses and pulse stretching is employed in many analog-to-digital conversions. Such circuits can be used to drive the increasingly popular digital readouts of ratemeters, TLD readers, voltmeters and thermal indicators. In wire chambers and spark counters time-to-pulse conversion and the reverse operation are used to provide input identifiers to a computer to allow reconstruction of particle paths and their characteristics. Many commercial instruments will store reference spectra on tape for recall and comparison with sample spectra, and, if suitably encoded, they will identify and print out any peaks above a certain height that coincide with major peaks in the reference spectrum.

In nuclear medicine computer-aided data acquisition and processing is providing methods for storing count data of successive scans and displaying the whole contrast image at the end of the scan either on a CRO screen or on paper tape. By using peak height to trigger beam intensity controls three-dimensional contrast can be displayed, of if a color TV tube is used, a color-coded intensity distribution image. The data can be further manipulated to provide contrast enhancement to sharpen isodose contours to throw into relief detected areas of maximum radiation dose.

The growing use of computers in treating, storing and comparing radiation detector pulses and to adapt them for specific modes of display has revolutionized the detector industry. It has also removed the working of the system somewhat from the user, confronting him instead with a dazzling array of keyboards and display controls whose function must be learned from a manual. In the process there is a certain danger that the detection process by which the input was obtained gets forgotten, and its inherent limitations and uncertainties overshadowed by the apparently unlimited precision offered by the computer or the digital readout.

In the end, in any measurement of a physical or chemical property there can be no substitute for insight and understanding of the method employed, the quantity being measured and the constraints imposed by Murphy's law and the basic laws of statistics.

REFERENCES FOR FURTHER READING

Chase, R. L. *Nuclear Pulse Spectrometry* (New York: McGraw-Hill Book Company, 1961).

Fairstein, E., and J. Hahn. "Nuclear Pulse Amplifiers–Fundamentals and Design Practice," *Nucleonics* 23(7):56 (1965); 23(9):81 (1965); 23(11):50 (1965); 24(1):54 (1966); 24(3):68 (1966).

Littauer, R. *Pulse Electronics* (New York: McGraw-Hill Book Company, 1965).

Watt, D. E., and D. Ramsden. *High Sensitivity Counting Techniques* (New York: Macmillan Publishing Co., Inc., 1964).

APPENDIX A

NOTES ON DETECTOR GEOMETRY

In many detector applications it is important to estimate the fraction of the radiation emitted from a source that strikes or passes through the detector volume. If the source is a point source facing a distant flat detector, this calculation involves computation of the solid angle Ω subtended by the source at the detector. In practice, though, the detector has finite thickness and volume and the source, too, may be extended, as a disk, cylinder or sphere.

Radiation from a point-like radioisotope source is emitted equally in all directions, into a solid angle 4π. The solid angle then will be represented by the fraction of 4π that the area subtended by the detector represents in terms of the total area of the sphere of unit radius surrounding the source. For an "isotropic" source, for any given sphere size, the average radiation flux in photons or particles per unit time per unit sphere surface area will be the same over the entire surface.

The fractional solid angle subtended by the detector, averaged over the detector volume, is referred to as the *geometry* G of the source-detector arrangement, $G = \Omega/4\pi$. This factor can be calculated easily only for a few simple configurations; for others a point-by-point integration is required. Such computations are described in most books on Shielding calculations.

For the case of a point source on the axis of a cylindrical detector, as shown in Figure A.1, the geometry factor G, in terms of the angle ϕ between the detector axis and a line from the source to the detector extremity, can be shown to be

$$G = \frac{1}{2}(1 - \cos\phi) \qquad \text{(A-1)}$$

For source-to-detector distances, d, that are large compared to the detector area A, Equation A-1 is well approximated by the ratio of the area of the detector to the area of a sphere with radius d.

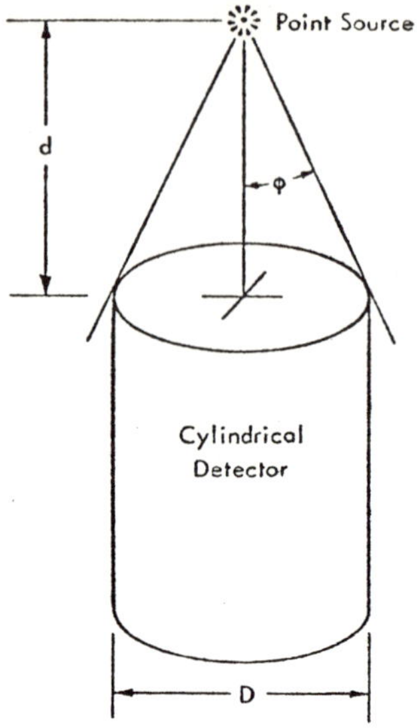

$$f_G = \frac{1}{2}(1 - \cos \varphi) \cong \frac{D^2}{16d^2}$$

Figure A.1. Schematic representation of a point source and cylindrical detector arrangement.

$$G \cong \frac{A}{4\pi d^2} \tag{A-2}$$

This approximation is also used when the source is not a point source but is small compared to the dimension d. The general problem can be seen by reference to Figure A.2.

In the notation of Figure A.2

$$\Omega = \int_{r=0}^{R} \int_{\phi=0}^{2\pi} (\cos \theta/d^2) r dr d\phi. \tag{A-3}$$

Except for the single case in which the point is located on the axis of the disc, no closed-form expression for this double integral has been obtained. Even in cases where infinite-series representations have been developed, none of them converge rapidly over the practical range of values of h/D and R/D, the two parameters which uniquely determine Ω. Evaluation of the integral either by term-by-term integration of a series expansion or by numerical quadrature methods result in length computations and are not suitable for incorporation in mathematical models which often require many calculations of Ω. An alternative approach is to prepare a

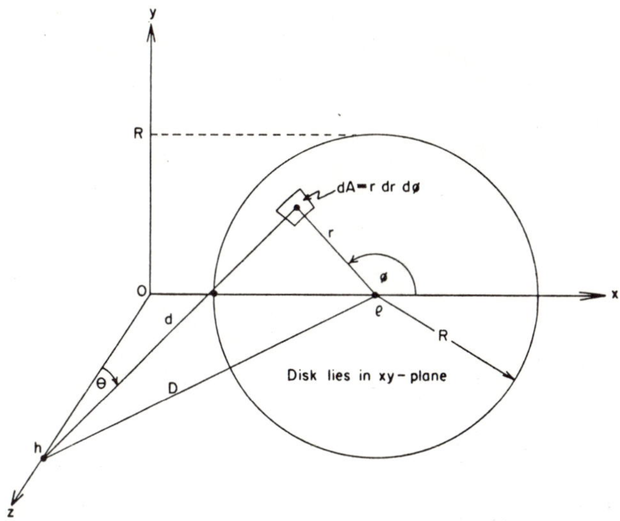

Figure A.2. Coordinate system and notation for circle.

table of values of Ω over a range of values of h/D and R/D and to obtain the inbetween cases by interpolation. The important case of a circular disk course, shown in Figure A.3, has been treated by Gardner and Verghese [*Nucl. Inst. Methods*, 93:163 (1971)]. They approximated the circle by a polygon to generate a table of solid angle values for the case of a circular disk facing a circular disk. The two disks are parallel to each other and their centers are intersected by a single line normal to their surfaces. The arrangement of the two disks and the notation used is given in Figure A.3. The pertinent solid angle is given by:

$$\Omega_D(S/h,\ R/h) \ = \ \int_0^S \Omega(h,R,\rho)\rho d\rho / \int_0^S \rho d\rho \qquad \text{(A-4)}$$

or

$$\Omega_D(S/h,\ R/h) \ = \ (2/S^2) \int_0^S \Omega(h,R,\rho)\rho d\rho \qquad \text{(A-5)}$$

The integration indicated by Equation A-5 was performed using a standard Simpson's rule numerical method on the computer. The resulting values for a range of S/h from 0 to 3.0 and a range of R/h from 0.1 to 3.0 are presented in Table A.1. The interval size for the integrations for ρ was

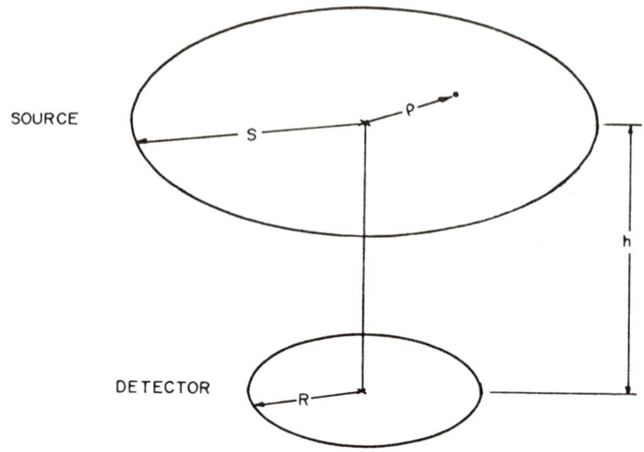

Figure A.3. Notation for circular disk facing a circular disk.

always 0.01 so that the error due to the numerical integration is of the order of $-(10^{-10}\ h^5/144)\ (d^4y/d\rho^4)$ where y is $2\Omega(h,R,\rho)\rho/S^2$ and is in the range from y at $\rho = 0$ to y at $\rho = S/h$. Integrations with different interval sizes indicated that the controlling error was that due to evaluating $\Omega(h,R,\rho)$, so the estimated accuracy of the values in Table A.1 is $\leqslant 0.1\%$.

Table A.1. Circular Disk to Circular Disk Solid Angle Values for Various Values of S/h and R/h

	S/h										
R/h	0.1	0.3	0.6	0.9	1.2	1.5	1.8	2.1	2.4	2.7	3.0
0.1	0.03095	0.02925	0.02476	0.01986	0.01567	0.01242	0.00997	0.00818	0.00671	0.00562	0.00477
0.2	0.12116	0.11481	0.09776	0.07880	0.06239	0.04954	0.03980	0.03243	0.02682	0.02248	0.01908
0.3	0.26328	0.25045	0.21526	0.17492	0.13924	0.11091	0.08925	0.07282	0.06025	0.05053	0.04290
0.4	0.44680	0.42708	0.37143	0.30515	0.24470	0.19577	0.15795	0.12906	0.10689	0.08970	0.07618
0.5	0.65997	0.63413	0.55894	0.46532	0.37664	0.30304	0.24532	0.20085	0.16656	0.13988	0.11887
0.6	0.89147	0.86103	0.76970	0.65039	0.53234	0.43134	0.35064	0.29780	0.23903	0.20094	0.17087
0.7	1.13154	1.09827	0.99552	0.85472	0.70853	0.57890	0.47299	0.38940	0.32402	0.27271	0.23208
0.8	1.37246	1.33801	1.22890	1.07237	0.90146	0.74364	0.61123	0.50503	0.42117	0.35498	0.30236
0.9	1.60860	1.57431	1.46338	1.39749	1.10699	0.92312	0.76401	0.63394	0.53006	0.44749	0.38158
1.0	1.83613	1.80299	1.69389	1.52472	1.32082	1.11458	0.92977	0.77527	0.65020	0.54996	0.46954
1.2	2.25712	2.22791	2.12937	1.96798	1.75641	1.52114	1.29270	1.09086	0.92173	0.78331	0.67081
1.4	2.62808	2.60346	2.51930	2.37672	2.17806	1.93761	1.68279	1.44153	1.22982	1.01554	0.90407
1.6	2.95056	2.93026	2.86038	2.73981	2.56524	2.33980	2.08017	1.81417	1.56660	1.34999	1.16650
1.8	3.22972	3.21311	3.15579	3.05603	2.90856	2.70986	2.46499	2.19324	1.92198	1.67254	1.45443
2.0	3.47156	3.45798	3.41108	3.32925	3.20716	3.03871	2.82114	2.56226	2.28368	2.01134	1.76304
2.2	3.68179	3.67064	3.63219	3.56511	3.46478	3.32491	3.13924	2.90661	2.63803	2.35664	2.08615
2.4	3.86541	3.85622	3.82453	3.76934	3.68692	3.57171	3.41685	3.21650	2.97185	2.69712	2.41599
2.6	4.02668	4.01905	3.99275	3.94708	3.87909	3.78425	3.65634	3.48823	3.27513	3.02110	2.74323
2.8	4.16911	4.16271	4.14074	4.10268	4.04625	3.96786	3.86237	3.72305	3.54286	3.31863	3.05769
3.0	4.29558	4.29018	4.27168	4.23974	4.19257	4.12738	4.04007	3.92495	3.77498	3.58357	3.34978

APPENDIX B

LIST OF ELEMENTS

Atomic Number	Symbol	Name	Atomic Weight	Atomic Number	Symbol	Name	Atomic Weight
0	n	neutron	–	52	Te	tellurium	127.60
1	H	hydrogen	1.0079	53	I	iodine	126.9045
2	He	helium	4.00260	54	Xe	xenon	131.30
3	Li	lithium	6.941	55	Cs	cesium	132.9054
4	Be	beryllium	9.01218	56	Ba	barium	137.34
5	B	boron	10.81	57	La	lanthanum	138.9055
6	C	carbon	12.011	58	Ce	cerium	140.12
7	N	nitrogen	14.0067	59	Pr	praseodymium	140.9077
8	O	oxygen	15.9994	60	Nd	neodymium	144.24
9	F	fluorine	18.99840	61	Pm	promethium	–
10	Ne	neon	20.179	62	Sm	samarium	150.4
11	Na	sodium	22.98977	63	Eu	europium	151.96
12	Mg	magnesium	24.305	64	Gd	gadolinium	157.25
13	Al	aluminum	26.98154	65	Tb	terbium	158.9254
14	Si	silicon	28.086	66	Dy	dysprosium	162.50
15	P	phosphorus	30.97376	67	Ho	holmium	164.9304
16	S	sulfur	32.06	68	Er	erbium	167.26
17	Cl	chlorine	35.453	69	Tm	thulium	168.9342
18	Ar	argon	39.948	70	Yb	ytterbium	173.04
19	K	potassium	39.098	71	Lu	lutetium	174.97
20	Ca	calcium	40.08	72	Hf	hafnium	178.49
21	Sc	scandium	44.9559	73	Ta	tantalum	180.9479
22	Ti	titanium	47.90	74	W	tungsten	183.85
23	V	vanadium	50.9414	75	Re	rhenium	186.2
24	Cr	chromium	51.996	76	Os	osmium	190.2
25	Mn	manganese	54.9380	77	Ir	iridium	192.22
26	Fe	iron	55.847	78	Pt	platinum	195.09
27	Co	cobalt	59.9332	79	Au	gold	196.9665
28	Ni	nickel	58.71	80	Hg	mercury	200.59
29	Cu	copper	63.546	81	Tl	thallium	204.37
30	Zn	zinc	65.38	82	Pb	lead	207.2
31	Ga	gallium	69.72	83	Bi	bismuth	208.9804
32	Ge	germanium	72.59	84	Po	polonium	–
33	As	arsenic	74.9216	85	At	astatine	–
34	Se	selenium	78.96	86	Rn	radon	–
35	Br	bromine	79.904	87	Fr	francium	–
36	Kr	krypton	83.80	88	Ra	radium	–
37	Rb	rubidium	85.4678	89	Ac	actinium	–
38	Sr	strontium	87.62	90	Th	thorium	232.0381
39	Y	yttrium	88.9059	91	Pa	protactinium	–
40	Zr	zirconium	91.22	92	U	uranium	238.029
41	Nb	niobium	92.9064	93	Np	neptunium	–
42	Mo	molybdenum	95.94	94	Pu	plutonium	–
43	Tc	technetium	–	95	Am	americium	–
44	Ru	ruthenium	101.07	96	Cm	curium	–
45	Rh	rhodium	102.9055	97	Bk	berkelium	–
46	Pd	palladium	106.4	98	Cf	californium	–
47	Ag	silver	107.868	99	Es	einsteinium	–
48	Cd	cadmium	112.40	100	Fm	fermium	–
49	In	indium	114.82	101	Md	mendelevium	–
50	Sn	tin	118.69	102	No	nobelium	–
51	Sb	antimony	121.75	104	Ku	kutusovium	–
52	Te	tellurium	127.60	105	Ha	hahnium	–

INTERNATIONAL NUMERICAL MULTIPLE AND SUBMULTIPLE PREFIXES

Multiples and Submultiples	Prefixes	Symbols
10^{18}	exa	E
10^{15}	peta	P
10^{12}	tera	T
10^{9}	giga	G
10^{6}	mega	M
10^{3}	kilo	k
10^{-1}	deci	d
10^{-2}	centi	c
10^{-3}	milli	m
10^{-6}	micro	μ
10^{-9}	nano	n
10^{-12}	pico	p
10^{-15}	femto	f
10^{-18}	atto	a

SYMBOLS, UNITS AND EQUIVALENTS

Symbol	Unit	Equivalent
Å	angstrom	10^{-10} meter
A	ampere(s)	1 coul/sec
a	annum, year	
Bq	becquerel	1 disintegration per sec
Ci	curie	3.7×10^{10} dps–2.22×10^{12} dpm $= 3.7 \times 10^{10}$ becquerel
cpm	counts per minute	
dpm	disintegrations per minute	
dps	disintegrations per second	
eV	electron volt	1.6×10^{-12} ergs = 0.16 aJ
g	gram(s)	3.527×10^{-2} ounces = 2.205×10^{-3} pounds
Gy	gray	100 rad = 1 J/kg
Hz	hertz	cycle per second
kVp	kilovolt peak	
m	meter(s)	39.4 inches = 3.28 feet
m^3	cubic meter(s)	1000 liters
mCi/mi^2	millicuries per square mile	0.386 nCi/m^2 (mCi/km^2)
MeV	million electron volt	0.16 picojoule
mi	mile(s)	1.609 m
ml	milliliter(s)	
nCi/m^2	nanocuries per square meter	2.59 mCi/mi^2
R	Roentgen	2.58×10^{-4} coul/kg
rad	unit of absorbed radiation dose	100 ergs/g = 0.01 gray
s	second	0.01666 min
yr	year	3.15×10^{7} sec

CONVERSION FACTORS

Multiply	by	to obtain
Atmospheres	33.90	ft of water
	10,332	kg/m^2(Pa)
	1,013.2	millibar
	760	Torr
Barn	10^{-24}	cm^2
Becquerel (Bq)	2.70×10^{-11}	Curie
Btu	1.055×10^{10}	erg
	252	g-cal
	1,055	joule
	2.93×10^{-4}	kWh
	6.59×10^{15}	MeV
cm	0.394	inch
cubic ft.	0.0283	m^3
cubic meter (m^3)	1,000	liter
	264.2	gallon (U.S.)
	35.31	ft^3
curie (Ci)	3.70×10^{10}	disint/sec (Bq)
	2.22×10^{12}	disint/min
erg	9.49×10^{-11}	Btu
	2.78×10^{-14}	kWh
foot	1.894×10^{-4}	mile (statute)
	0.3048	meter
gallon (U.S.)	3,785	cm^3
	0.1337	ft^3
	3.785	liter
	0.8327	gallon (Imperial)
gram	2.205×10^{-3}	lb
gray	100	rad
g-calorie	3.969×10^{-3}	Btu
	2.61×10^{13}	MeV
hour	4.167×10^{-2}	day
	3,600	sec
inch	2.540	cm
	1.578×10^{-5}	mile
inch of water	2.458×10^{-3}	atm
	1.866	Torr
joule	9.48×10^{-4}	Btu
	2.778×10^{-4}	watt-hr
kilogram	2.250	lb
	9.842×10^{-4}	ton (long)
	1.102×10^{-3}	ton (short)

CONVERSION FACTORS, continued

Multiply	by	to obtain
kilometer	0.6214	mile
	1,094	yard
liter	0.0353	ft^2
	0.2642	gallon (U.S.)
meter	3.281	ft
	39.37	inch
	6.214×10^{-4}	mile
millimeter	0.03937	inch
meter/second	3.281	ft/sec
	2.237	mile/hr
micron (μm)	10^{-3}	mm
MeV	1.517×10^{-16}	Btu
	1.602×10^{-6}	erg
	1.602×10^{-13}	joule
	3.83×10^{-14}	g-cal
Pascal (Pa)	10^{-5}	bar
pounds/in^2	0.068	atm
	6,894.7	N/m^2(Pa)
	144	lb/ft^2
rad	100	erg/g tissue
rem	0.01	joule/kg
roentgen (R)	2.58×10^{-4}	coul/kg of air
sievert	100	rem
ft^2	929	cm^2
	0.111	yd^2
ton (long)	1,016	kg
	2,240	lb
	1.120	ton (short)
ton (short)	907.2	kg
	2,000	lb
Tonne (Mg)	1,000	kg
	2,205	lb
watt	3.413	Btu/hr
	6.24×10^{12}	MeV/sec
year	3.154×10^{7}	sec

INDEX